Smart Grids 2016

MATEC Web of Conferences Volume 91 (2017)

Tomsk, Russia
10 - 14 October 2016

Editors:

Pavel A. Strizhak
Geniy V. Kuznetsov
Alena O. Zhdanova

ISBN: 978-1-5108-3508-5

Printed from e-media with permission by:

Curran Associates, Inc.
57 Morehouse Lane
Red Hook, NY 12571

Some format issues inherent in the e-media version may also appear in this print version.

Printed by Curran Associates, Inc. (2017)

For additional information, please contact EDP Sciences – Web of Conferences at the address below.

EDP Sciences – Web of Conferences
17, Avenue du Hoggar
Parc d'Activité de Courtabœuf
BP 112
F-91944 Les Ulis Cedex A
France

Phone: +33 (0) 1 69 18 75 75
Fax: +33 (0) 1 69 28 84 91

contact-edps@webofconferences.org

Additional copies of this publication are available from:

Curran Associates, Inc.
57 Morehouse Lane
Red Hook, NY 12571 USA
Phone: 845-758-0400
Fax: 845-758-2633
Email: curran@proceedings.com
Web: www.proceedings.com

TABLE OF CONTENTS

STATISTICAL ANALYSIS OF CONSEQUENCES CAUSED BY THE COLLISIONS OF SOARING DROPS OF ORGANIC COAL-WATER FUEL

Timur R. Valiullin[1,*], *Jean Claude* Legros[1,2], and *Pavel* P.Tkachenko[1]

[1] Tomsk Polytechnic University, 634050 Tomsk, Russia
[2] Université libre de Bruxelles, CP 165/62, Brussels, Belgium

Abstract. The paper examines the processes of collision of the soaring organic coal – water fuel (OCWF) drops in the specialized combustion chamber in case of direct injection of suspension with the subsequent crushing, decay and coagulation. High speed video registration is used. The fuel composition is prepared with the use of a coal conversion waste (filter-cake "G"), turbine oil waste, water and a plasticizer. The statistical analysis of consequences of impingement of OCWF drops at their movement in an oxidizer flotation (temperature is about 800-840 K) with their following deformation is carried out. The conditions when the processes of coagulation, disintegration or crushing of drops are dominated are established.

1 Introduction

More and more growing interest in creating and applying organic coal –water fuel (OCWF) [1] on the basis of numerous waste oil products, waste oil, low-grade coal, the filter-cakes and others. Effective utilization of waste by incineration in thermal power plants is an urgent task at present. The basis for the creation of OCWF was the use of coal-water fuel [2] .Now accumulated a large amount of experimental research data [3, 4, 5].

To determine the characteristics of ignition particles OCWF (or groups of particles) without the use of holders developed an experimental technique that provides its wool-oxidant environment using specialized combustion chambers [6]. For the aggregate, group, set or stream of droplets soaring to play a decisive role droplet collision processes (crushing, coagulation, decomposition). It is therefore of interest in experimental research of factors the effects of collisions drops OCWF flying at them in the oxidant flow. Consequences of collisions of drops of fuel were studied by the statistical analysis of experimental data. Assessments of results allow predicting to what share of probability these events can happen.

The purpose – establishment and registration of characteristic consequences of collision of drops of OCWF with each other soaring in a flow of an oxidizer and their subsequent ignition in the combustion chamber as a result of statistic analysis of the obtained experimental data.

2 Experimental setup and study technique

The technique is similar to the experimental studies described in the paper [6]. Organic coal – water fuel were prepared using the following components (with the indicated concentrations of mass) filter-cake grade "D" (89 %), the exhaust turbine oil (10 %) and a plasticizer (1 %). To a suspension of the fuel injection flow (7) unused T-mixer (5), one of which is fed the compressed air channels (4), and in the second – OVUT (6). The air pressure was 200 kPa. The amount of fuel used was 4 ml. This system made it possible to generate fuel droplets of different sizes. The average droplet size of the injected OCWF R_d varied in the range of 0.1-1 mm.

The specialized combustion chamber from optically transparent quartz glass (Fig. 1) was used to a research of processes of crushing, coagulation and decay of drops of OCWF with characteristic formation of agglomerates. The mechanism (3) by rotating speed about 400-600 RPM, causing oxidizer flow turbulence, set in the lower part of the combustion chamber was used to support of optimum conditions of a flying, speed of movement of particles and agglomerates of organic coal –water fuel rotational propeller. Processes of crushing, coagulation and decay of the soaring OCWF particles, and also their sizes registered using system the high speed Phantom video camera with a frequency of 500 frames per second.

The initial rate of motion of the drops and the speed of the oncoming stream of the oxidant varies between 2.5-3.3 m / s and 3.2 m / s, respectively. The average temperature T_g oxidizer into the combustion chamber was about 820 K. The

* Corresponding author: radisovich@yandex.ru

oxidizer temperature measurement is carried out by two chromel-alumel thermocouples (range 273-1373 K, accuracy ± 3 K, the inertia 3 sec.), through the technological hole (10).

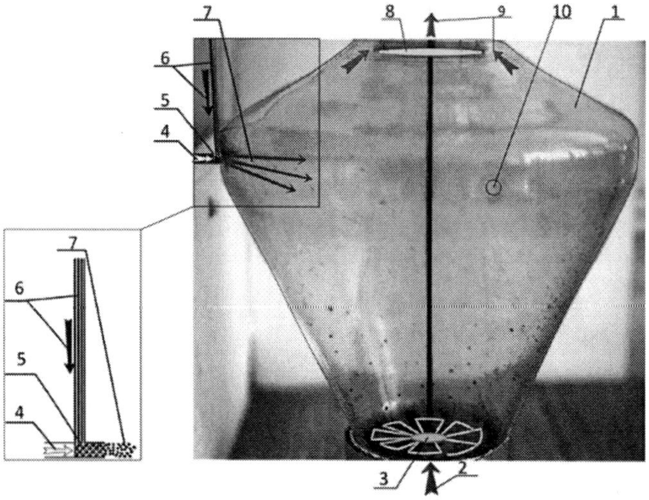

Fig. 1. Specialized combustion chamber: 1 – chamber wall heat-resistant optically transparent quartz glass; 2 – initial oxidant stream; 3 – rotary vane mechanism; 4 – Compressed air supply channel; 5 – T-mixer; 6 – channel for supplying OCWF; 7 – OCWF stream of particles; 8 – oxidant flow restrictor; 9 – oxidant stream at the outlet from the chamber; 10 – technological hole for temperature measurement.

3 Experimental results

As a result of processing and statistical analysis of the experimental data set three possible consequences of a collision soaring drops OCWF: crushing drops into smaller drops, coagulation (clumping OCWF particles agglomerate formation) and the decay of particles (collision of coagulated particles without substantially changing the size) for a certain interval time. The processes of fragmentation and coagulation were reported from the beginning of the injection of fuel into the combustion chamber suspension until its completion, the duration of which was about 8 sec. During the collapse of the agglomerates formed OCWF observed before the start of steady particle ignition. Typical effects of the collision and deformation of two drops of fuel are shown in fig.2, fig. 3 and fig. 4. All figures are depicted sequence of video frames (a, b, c) with the large-scale increase in the area of the collision in the inset on the left.

The next stages of phase transformations are the ignition and combustion of the fuel. From fig. 5 (a) shows that the two particles are of irregular shape with the local region are moving towards the ignition. Further, these particles collide (fig.5, b), there is an intensive volatile products with the formation of a common flame front. After the collision, the particles scatter OCWF (fig. 5, c) in the part and the process can be repeated. These effects largely depend on the size and configuration of the particles, as well as associated aerodynamic forces. To intensify the ignition and combustion processes it is important to consider the consequences of colliding drops incendiary. The main mass of the particles at the same time are lit separately.

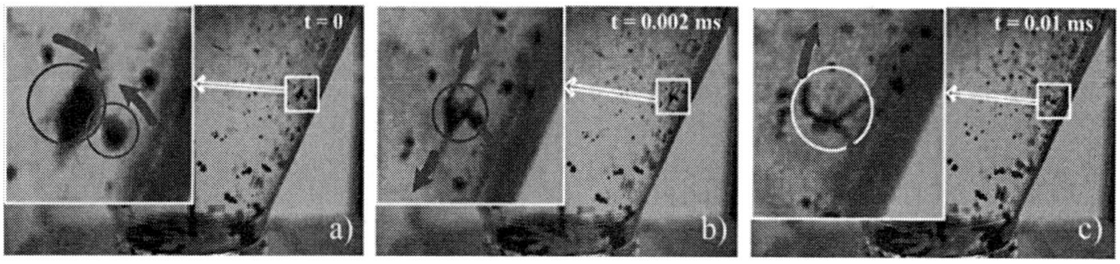

Fig. 2. Chipping OCWF drops.

 DOI: 10.1051/matecconf/20179101001

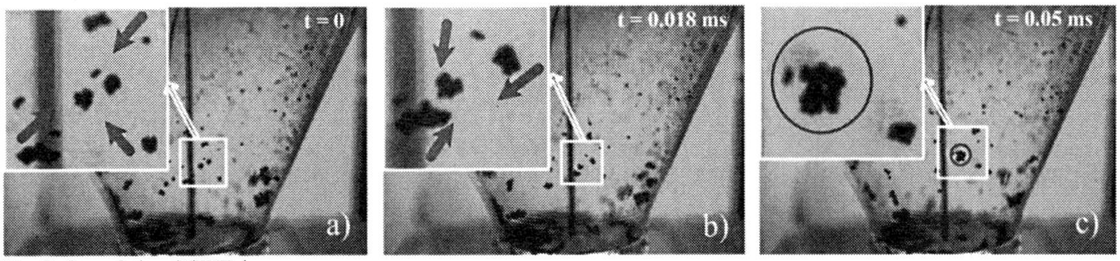

Fig. 3. Coagulation OCWF drops.

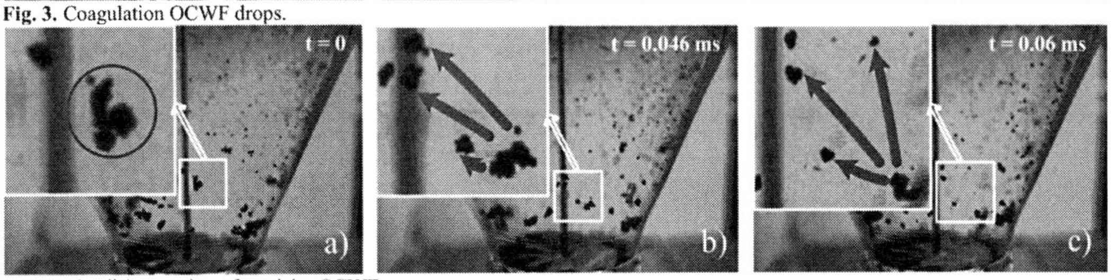

Fig. 4. The disintegration of particles OCWF.

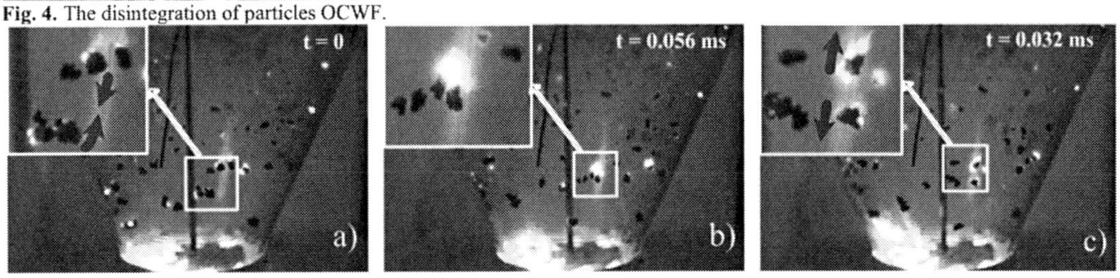

Fig. 5. Collision incendiary particles OCWF.

On the basis of experimental data is a bar graph fig. 5 probability P_1 (crushing), P_2 (decaying), P_3 (coagulation) of occurrence of the collision of two or more OCWF drops in the wool-oxidant stream. It was found that all these processes take place at different vibrations R_{d1} and R_{d2} particle size, defined their ratio, respectively, for the crushing of 0.2-1.7 mm, the decaying of 0.2-1.9 mm, the coagulation of 0.1 - 2.3 mm. Probability values of P_1, P_2, P_3 was evaluated by treating the amount of 50 collisions.

At calculation of values formulas have been applied:

$$P_1=N_1/(N_1+N_2+N_3),\ P_2=N_2/(N_1+N_2+N_3),\ P_3=N_3/(N_1+N_2+N_3),$$

where N_1, N_2, N_3 – number of collisions as a result of which three options of events are implemented: crushing, decaying and coagulation.

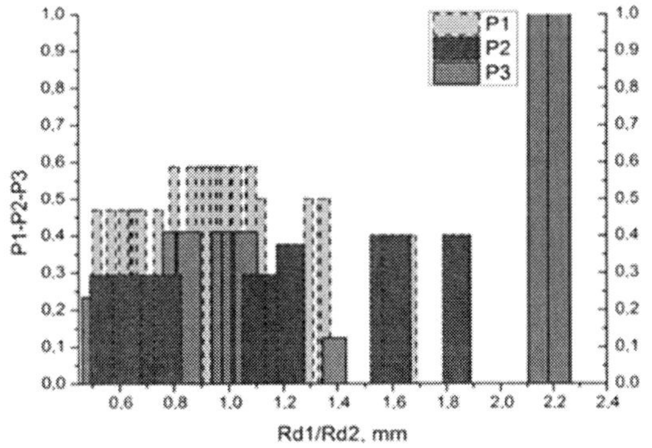

Fig. 6. Statistical analysis of the consequences of the collision of two drops OCWF in their wool-oxidant stream, depending on their size.

The analysis of the histogram shows that probabilities of emergence of crushing (P_1) and coagulation (P_3) of drops of OCWF at a ratio of the R_{d1}/R_{d2} sizes from 0.2 and 2.2 are made by about 50 % and 30 %, respectively. About 20 %

fall to the share of probability of disintegration of agglomerates. While a ratio of the R_{d1}/R_{d2} sizes of two pushing together drops fuel suspensions it is close to 1, probabilities of crushing and coagulation are almost comparable, 60 and 40 %. At the same time at the same time at a stream of an oxidizer there are damp and dry particles of fuel. As viscosity of drops of OCWF at a stage of injection of fuel suspension rather big, is observed domination of processes of crushing and coagulation. With parallel oxidant effect of temperature decreases the viscosity OCWF drops and an increasing number of dehydrated particles. Therefore there is a high probability of the formation of agglomerates. But it is not ruled out the probability of disintegration of agglomerates formed during their combustion chamber flying. In the intermediate case (R_{d1}/R_{d2} between 1.5 and 2) the probability of coagulation is practically not observed, and the likelihood of fragmentation and disintegration of the same. Once the ratio R_{d1}/R_{d2} size increases to 1/1.5, the coagulation of the droplets become quite rare. When the ratio of fuel particle sizes larger $R_{d1}/R_{d2} = 2$ the probability of coagulation rapidly increased to $P_3 \sim 1$ and the process begins to dominate.

4 Conclusion

The analysis of singularities collision OCWF hovering particles can be concluded that for efficient ignition and combustion nozzle unit OCWF structure should provide optimum ratio of air flow rate and uniformity of droplet size. This will lead to a decrease in coagulation factors that respectively increase fullness and reduce fuel burn incomplete combustion. In the experiments carried out, the ratio R_{d1}/R_{d2} size was not more than 1/1.5. It is important to note that the probability of droplet collisions OCWF processes not only depends on the droplet size and the velocity of their movements, the amount of fuel particles per unit volume of the chamber, and defined oxidant parameters.

The study was performed by a grant from the Russian Science Foundation (project 15-19-10003).

References

1. A. E. Kontorovich, M. I. Epov, L. V. Eder, Russian Geology and Geophysics **55**, 534 (2014)
2. I. M. Zasypkin, V. I. Murko, V. I. Fedyaev, M. P. Baranova, Therm. Sci. **16**, 1229 (2012)
3. G. S. Khodakov, Thermal Engineering **54**, 36 (2007)
4. A. Kijo-Kleczkowska, Fuel **90** 865 (2011)
5. D. O. Glushkov, D. P. Shabardin, P. A. Strizhak, K. Yu. Vershinina, Fuel Process. Technol. **143** 60 (2016)
6. T. R. Valiullin, P. A. Strizhak, S. A. Shevyrev, Therm. Sci. **2015** 1 (2015)

MODERNISATION OF BOILER UNIT AUTOMATION SYSTEM AT ACTING POWER STATION

Aleksei A. Abramovskikh[1,*], *Ulyana* A. Ulyanova[1], *Sergey* V. Lavrinenko[1], *Anastasiya* A. Matveeva[1], and *Dmitry* Malishev[1]

[1]Tomsk Polytechnic University, 634050, Tomsk, Russia

Abstract. The article deals with the modernization process of automated control system of operating process of boiler unit at South Kuzbas hydro electric power station. The boiler unit has been modernized through the installation of automated control system and control on hierarchical arrangement consisting of the levels where the tasks of input and ouput control signals are distributed. Controllers produced by Ram Company and sensors produced by Elemer company were used for modernization of boiler unit. Automatic and process control was made on the base of MicroSCADA software. It has been currently estimated that the use of automation system will result in decrease in fuel consumption by 1 % and reduce nitrogen oxide and carbon monoxide emission by 3 %. System payback time would be no more than two years.

1 Introduction

Constant energy resource price rise for energy resources in particular for fossil fuel is one of the main problems in modern power industry [1]. Severe climatic environment in Russia has become one of the reasons of extensive development of boiler systems with different fuel types including alternative fuel [2, 3]. They provide essential conditions for accommodation facilities and business. Therefore, technical problem solving connected with effective boiler operation control and its development is one of the most important ways of science and technology development [4].

High wear of public infrastructure is a current problem for many regions in Russia [5]. The work on heat networks and boiler systems modernization does not only allow to prevent areal communal fail but also leads to the increase of energy efficiency – old equipment is replaced with the new one with improved performance [6-7]. According to experts estimates there will be collapse in the Communal Service Market without modernization of local boiler plants. First of all many of them will break down, and no one operation organization will take on running worn-out equipment. Secondly a lot of people will have to use furnace heat, that is go back in time several decades. Modernization of boiler plants also implies not only the change over of equipment but also the parts of heat

[*] Corresponding author: aleksey82@tpu.ru

networks. All this leads to reduction of heat losses, and as a consequence keeps the amount charges down [8]. This question is also relevant in the context of atomic power stations where special attention is paid to the control systems parameters [9]. For nuclear energy modeling of different processes is widely used for security purposes [10] with further analysis of more optimal adjustments [11].

2 Objectives of the research

One of the major challenge associated with operation of boiler unit is to ensure balance between generated and consumed energy. In its turn processes of vaporization and energy transmission in boiler unit are definitely connected with the amount of flow of working medium and heat storage medium. Burn rate, efficiency and fuel burning stability depends on the air supply and air distribution between the fuel particles. It is conventionally accepted to divide the process of fuel consumption into three stages: breakdown, burning and reburning [12, 13]. These stages happen consistently, sometimes with overlap. Burning calculation means the determination of air amount in cubic meters which is necessary for burning of unit of mass or the volume of fuel amount and the composition of heat balance and burning temperature test [14]. Nowadays the equipment is out dated and is not maintainable. The majority of control circuits are switched to manual mode.

The aim of modernization of boiler unit SB-10 (steam boiler) at joint stock company South Kuzbas hydro electric power station is to remove defects of the existing system and to provide the increase of fuel consumption rate on steam generation and hot water for consumers. The aim of this boiler unit is gathering information about process value and working conditions, operation of working conditions and also archiving of measured parameters values and graphing its changes.

Power regulation of boiler units and pressure control in boiler drum is in maintenance of balance between vapor extraction and water supply. The level of water in boiler drum is the key parameter that characterizes the balance. The reliability of boiler unit work depends on the quality of level control. When the pressure gets high, the decrease below the acceptable limits can lead to circulation loss in water tubes. It can result in temperature increase of pipe walls and their over burn.

The increase can also result in emergencies, because water flow in steam heater can happen and it breaks down. The highest demands are made on accuracy of maintenance of required level. Feed water control is defined by the equality of water supply. It is necessary to provide boiler with equal water supply because frequent and deep change in feed water flow can result in serious thermal tensions in the metal of economizer.

Boiler drums with natural circulation are intrinsic to some considerable storage capacitance which manifests itself in transient states. If in transient mode the level of water in boiler drum is determined by the condition of burden material balance then in transient states big amount of insurges influences level position.

The main insurges are the change of feed water flow, change of boiler output if consumer load is changed and change of firebox load, the change of feed water flow temperature.

The regulation of ratio fuel-air is required both from physical and economical point of view. Fuel burning is known to be the most important processes that takes place in boiler plant. The chemical part of fuel burning is oxidation reaction of fuel elements by oxygen molecules. The oxygen is used for burning. The air to firebox is supplied in a particular proportion with fuel by means of forced flow-fan. The ratio fuel – air is approximately 1:1.3. In case of lack of air incomplete fuel burning takes place. Unburned fuel is released into the atmosphere that is neither economically nor ecologically acceptable. The cooling of firebox will be in furnace chamber in case of extra fuel, but the fuel will burn out. In this

 DOI: 10.1051/matecconf/20179101002

case the rests of air will form nitrogen dioxide which is impossible from ecological point of view, because this combination is dangerous for people and environment. Parameters alarm and protection working on boiler shut down are essential because operator is unable to keep track of all parameters of working boiler. An emergency situation can result from this. On running the water from out of boiler drum the level of water is decreased in it, as a consequence circulation is lost and pipe over burn of bottom screen happens. Active protection prevents generator breakdown. On decreasing generator load, burn rate in firebox decreases. Burning becomes non-steady and can stop. There is protection of flame distinguishing.

In boiler unit SB-10 (Fig. 1) the burning process takes place in furnace chamber where the feedwater goes down the pipes, saturated steam travels through the waterwall pipes. Combustion products, air – fuel come on furnace chamber through the burners. The boiler unit contains multiloop superheater of a mixed type with cross flow of coil with sweep of gases. Superheater is used for transformation of saturated steam into hot vapour [15]. Economizer is a continuous loop type device of a boiling type, a smooth pipe battery with horizontal position of coils with cross flow of sweep of gases. Economizer is aimed at preheating of feed water. It consists of two blocks. The second block of air heater is located between the blocks of economizer which also consists of two blocks. Air heater is a pipe four-way type, with longitudinal sweep of gases. Air heater is used to heat the air which is meant for combustion.

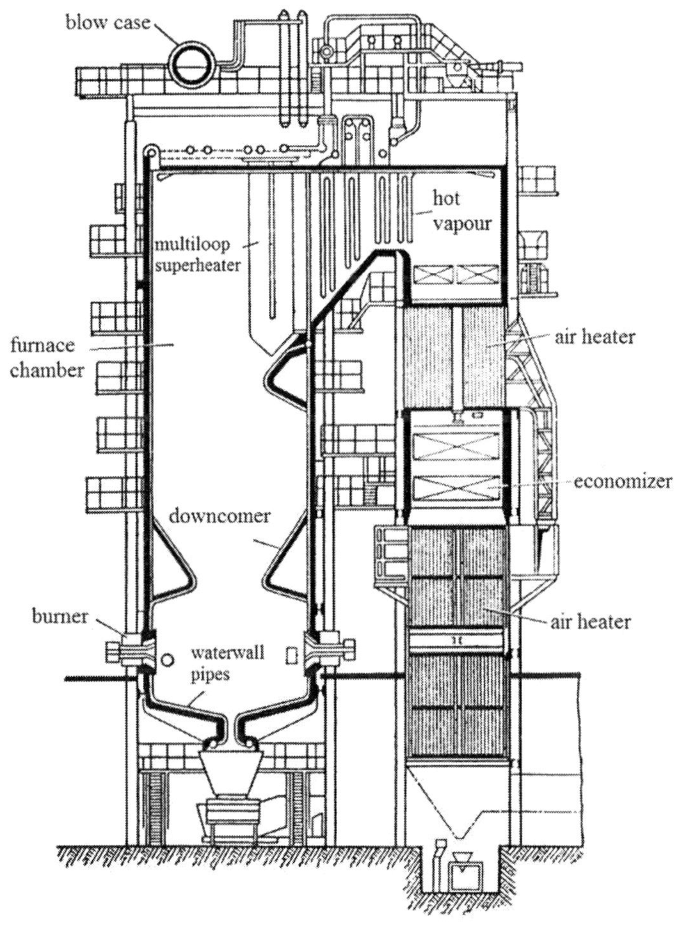

Fig. 1. Boiler PK-10 in elevation.

 DOI: 10.1051/matecconf/20179101002

3 Methods of research and results

Modernization of SCADA boiler unit was made on hierarchical arrangement consisting of three levels of automatic performance:

1 level - "Field level of the system" (basic automatic devices – fuel gages, pressure and temperature transmitters, power units). They provide measurement of input and output signals of boiler unit, control of regulating parts and information transmission to the next system level.

2 level – "Lower level of the system" (programmable controller with input/output devices) is connected with the field level of the system with the use cables.

3 level – "High level of the system" - computer-based working stations of operators with software tools including SCADA programs and standard interface.

SCADA-system means specialized software program implementing man-machine interface [16].

Controllers produced by Ram company were used for modernization of boiler unit. These controllers have proved to be the best on the Russian market as quality price indicator.

Fuel gages, pressure and temperature transmitters, power units that have been utilized in research are produced by Elemer company. These devices are used in parameter control system of boiler unit and they are widely recognized as reliable devices. Micro-SCADA is software program SCADA, that implements automatic operation and control of technological process and specializes in automatic performance in certain field [8]. The work of boiler units, the key process conditions, alarm messages can be tracked on the monitor schematically (fig. 2, 3).Visualization system besides displaying the main parameters such as pressure, temperature, level, fuel consumption, regulatory body statue and current load of electric devices makes measurement of boiler capacity in units of evaporation, fuel rate and relative hour fuel consumption rate. Mnemonic scheme in the form of pictograms shows whether actuating units are in manual or automatic mode.

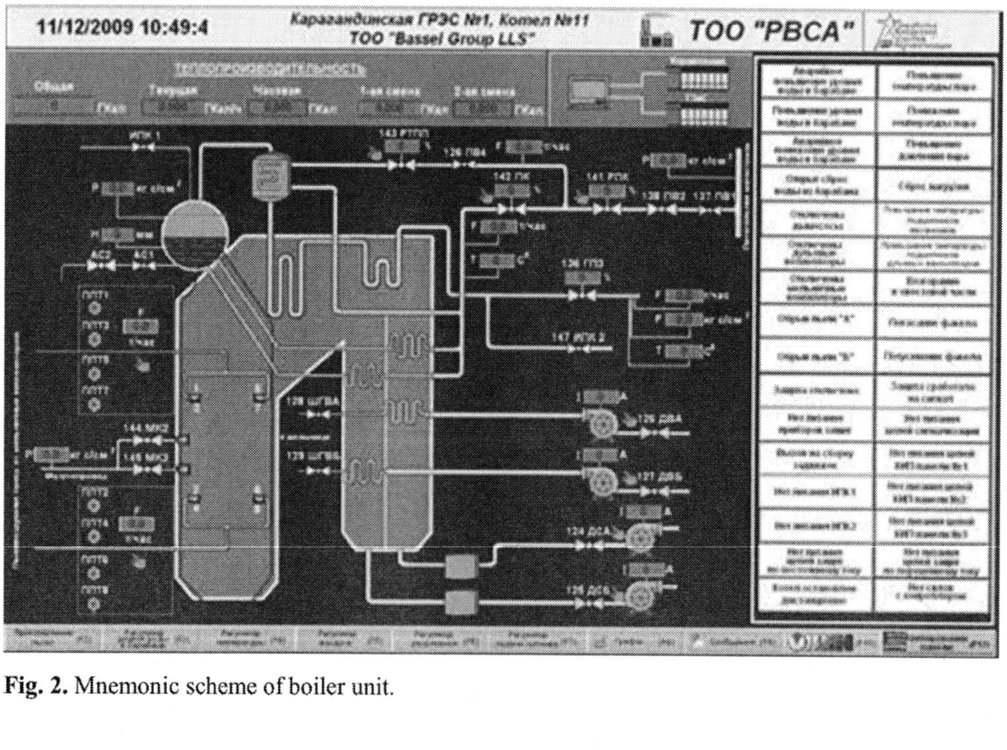

Fig. 2. Mnemonic scheme of boiler unit.

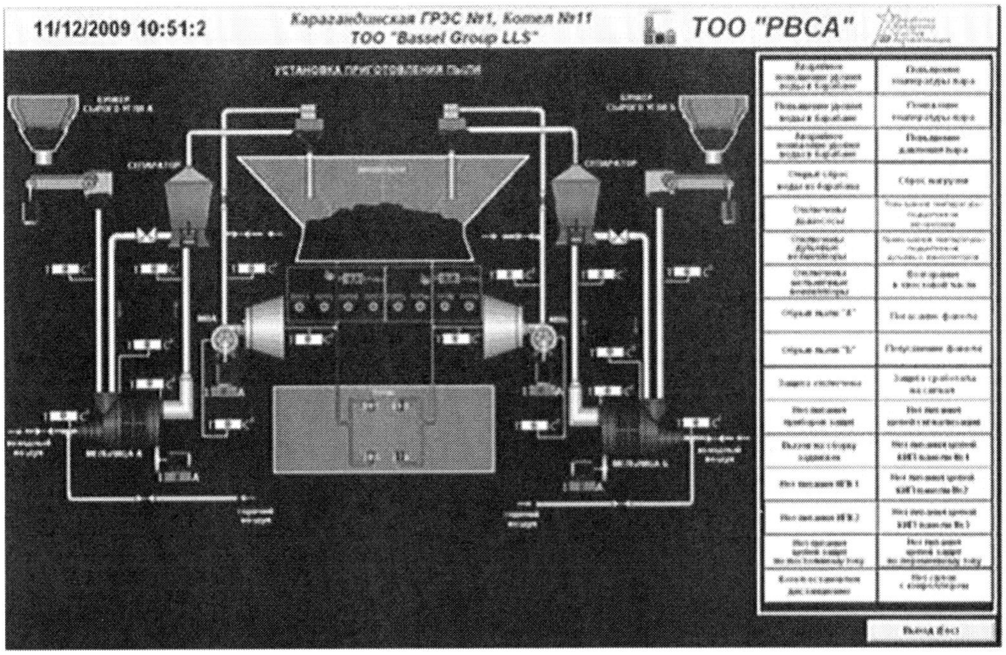

Fig. 3. Mnemonic scheme of Pulverized coal system.

Emergency control automatics provides automatic stop of boiler in critical operations when some parameter exceeds the value by conditions of process safety or possible equipment breakdown. The operation of alarm system is accompanied by audio and light alarm, display of messages on to the primary and secondary stations of operator. The system shows the reasons of accidental situation. All events are recorded on the message window in chronometrical sequence with recording the time of occurrence, acknowledgement and signal loss.

4 Conclusion

Functional structure of improved system of boiler unit control has been developed. All technical facilities have been chosen out of optimal ratio price-quality. Improvement of process automation system of the process of boiler unit operation PK-10 a joint stock company South Kuzbas hydro electric power station will result in decrease in fuel consumption by 1 % and reduce nitrogen oxide and carbon monoxide emission by 3 %. System payback time would be no more than two years.

The study was realized in "National research Tomsk polytechnic university" in framework of federal target program "Research and development in prior directions of scientific-technological complex development in Russia in 2014-2020 year", unique identifier of R&D project RFMEFI58114X0001.

References

1. D. A. Pavlov, E. N. Semikova, Modern high-end technologies, **8-2**, 309 (2013)
2. A. N. Golovanov, A. S. Yakimov, A. A. Abramovskikh, Y. R. Sukov, Thermophysics and Aeromechanics, **15**, 4, 651 (2008).

3. A. M. Grishin, A. N. Golovanov, Ya. V. Sukov, A. A. Abramovskikh, Journal of Engineering Physics and Thermo-physics, **80**, 6, 1154 (2007)
4. A. A. Silaev, S. N. Zadvorskiy, Bulleting of Volgograd state technical university, **13 (177)**, 141 (2015)
5. A. Zenkov, S. Yankovsky, A. Matveeva, S. Lavrinenko, A. Gromov, MATEC Web of Conferences, **72**, 01131 (2016)
6. E. Pavelyev, S. Lavrinenko, MATEC Web of Conferences, **72**, 01085 (2016)
7. K. Savina, A. Abramovskih, V. Haymin, S. Lavrinenko, MATEC Web of Conferences, **72**, 01042 (2016)
8. V. Martov, Sanitary engineering, heating and conditioning, **7 (151)**, 94 (2014)
9. A. V. Vorobiev, A. M. Antonova, M. P. Vinogradov, EPJ Web of Conferences, **110**, 01053 (2016)
10. A. M. Antonova, A. V. Vorobiev, A. S. Orlov, MATEC Web of Conferences, **23**, 01069 (2015)
11. A. Uvarov, A. Antonova, A. Vorobjev, MATEC Web of Conferences, **37**, 01062 (2015)
12. N. Galashov, S. Tsibulskii, A. Matveev, V. Masjuk, EPJ Web of Conferences, **110**, 01019 (2016)
13. A. A. Somov, D. A. Mel'Nikov, D. V. Gvozdjakov, V. E. Gubin, Yu. S. Borovikov, A. S. Matveev, MATEC Web of Conferences, **37**, 01054 (2015)
14. M. I. Reznikov, Yu. M. Lipov, *Boiler units of power stations* (Mir Publishers Moscow) (1989)
15. V. G. Aleksandrov, *Steam boilers of low and average power* (1966)
16. *Operating Manual TRACE MODE 6 u T-FACTORY*, (Москва 2011)

MATEC Web of Conferences **91**, 01003 (2017) DOI: 10.1051/matecconf/20179101003
Smart Grids 2016

SIMULATION OF SURFACE-TYPE CONDENSING UNITS FOR HEAT RECOVERY FROM THE FLUE GAS WITH AIR HEATING

Victor V. Bespalov [1,*], *Leonid* A. *Beljaev* [1], and *Leonid* S. Kuchman [1]

[1]Tomsk Polytechnic University, 634050, Tomsk, Russia

Abstract. The technology of flue gas heat recovery has been reviewed. The mathematical model requirements were analyzed and determined. The mathematical model of the installations, that implement this technology, was developed and described in paper. As a result, the software application that implements a mathematical model has allowed to design the different installations and can be useful to design organizations in the energy industry.

1 Introduction

The increased interest of industrial and energy enterprises in energy saving defines the development and introduction of new technologies.

The flue gas heat recovery from the natural gas combustion can provide an additional 15÷20% of the boiler thermal capacity [1] or, respectively, reduce the gas flared amount. Scientists develop the recovery technologies and improve the heat transfer calculation methodologies with water vapor condensation contained in the flue gases [1-3]. The technology developed allows to increase the flue gas heat recovery depth through the use air as a heated medium [4, 5]. For designing the installations, that implement this technology, are required the mathematical model that allows to carry out the thermal and constructional calculations of the entire system, including the component parts. The mathematical model must include the heat balance calculations of elements and the heat transfer calculations through the heat exchange surfaces. The balance equations in joint solution allow to find the thermal power of the elements and the flows temperatures. The heat transfer equations associate the thermal power of each element with its heat exchange area.

The proposed installation scheme (Fig. 1) [4] consists of a recuperative heat exchanger and condensing heat recovery unit. Gas-gas surface plate heat exchanger provides pre-cooling the initial flue gas and simultaneous heating cooled and dried flue gas in order to prevent possible condensation of residual moisture in the flues and chimney. Heat recovery unit (gas-air surface plate heat exchanger) constructive single device is conventionally divided into two parts: cooler (dry zone) and the condenser. Air is fed to the condenser by

* Corresponding author: vic@tpu.ru

fan unit. Additional exhauster compensates aerodynamic resistance of the gas path in the heat exchanger and condenser.

2 Mathematical model

The mathematical model requires to preset the heat exchangers geometry. The heat exchanger and the heat recovery unit are compiled of the heat exchange modules (packages). The sheet length, width and thickness, the thermal conductivity coefficient, the number of sheets and the clearance between them are determined for each package. The number of packages and heat-carrier passages are set for each heat exchanger. The package geometry can be different for heat exchanger and heat recovery unit, however, for high-power installations (more than 1 *MW*) it is advisable to use the unified packages. So, the unified package has been developed for power installation 10 *MW* (flue gas flow 33 m^3/s), consisting of 72 sheets, size 2500×1250 *mm* and weight about 1.2 *tons* [4]. The number of packages for such a flue gas flow amounted to 16 for the heat exchanger and 24 for the heat recovery unit.

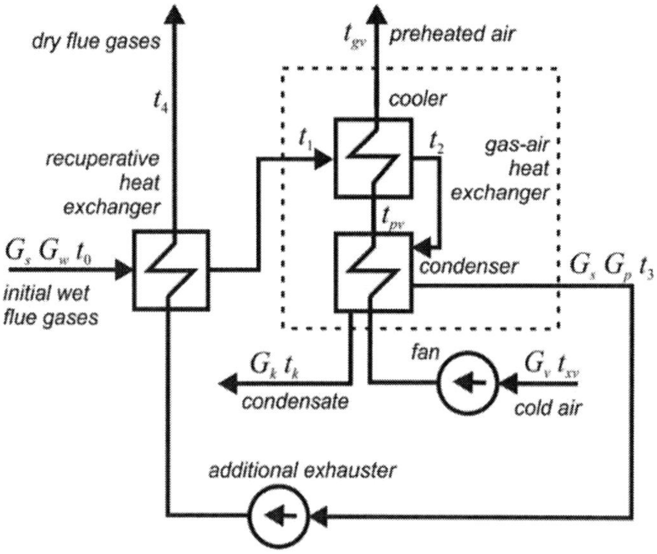

Fig. 1. The technology of flue gas deep heat recovery.

The input data for the mathematical model are the cold air temperature t_{xv} and flue gas mass flow consisting of a dry flue gas G_s and water vapor G_w with the temperature t_0. The flue gas outlet temperature t_2 from cooler assumes to be equal to the water vapor dew point at the initial humidity. Other parameters are calculated by successive approximations. A block diagram of the model structure is shown in Figure 2.

Block 1 contains the source data input and setting the initial values of the unknown parameters in the first approximation. This block also calculates the heat transfer surface area in all the elements F_p, F_t and the equivalent diameter of passage sections in the heat exchangers. The flue gas temperature at the condenser outlet t_3 is defined in a first approximation as a minimum (close to 0°C for negative air temperatures).

Block 2 is designed to calculate the flue gas composition G_s, G_w and expected condensate flow G_k. The heat transfer coefficient k_{tp} for heat exchanger is set in the first approximation, based on the gas flow velocity, and will be further refined by the simple iterations method.

Block 3 calculates the thermal balance of the recuperative heat exchanger. The log means temperature difference is defined by as a first approximation and refined by the dichotomy method.

Block 4 calculates a heat transfer process in the recuperative heat exchanger and finds the corrected value (next approximation) of heat transfer coefficient k_{tp}. If the relative error of the next approximation exceeds 0.1 % then there is a return to block 3. Otherwise, it is considered that heater has been conditionally calculated at a specified flue gas temperature at the outlet of the condenser t_3.

Block 5 calculates the thermal balance in the cooler and condenser to determine the air temperatures t_{gv} and t_{pv}. The heat transfer coefficient k_{tox} for cooler is found from heat transfer equations and then the dry zone area can be calculated.

Block 6 calculates the condensation zone heat transfer process. From the air flow side the heat transfer coefficient depends on the Reynolds number. The main difficulty is determining the heat transfer coefficient α_1 from a wet flue gas to the heat exchange surface with the water vapor condensation. Most of the known methods are based on experimental data processing and offer the dependences in rather narrow limits changes in the physical parameters. Applicability of a calculation method for functionally similar heat exchangers requires analysis and study [6]. Reviewed model allows to choose different methods to calculate the heat transfer coefficient from wet flue gas to the heat exchange surface.

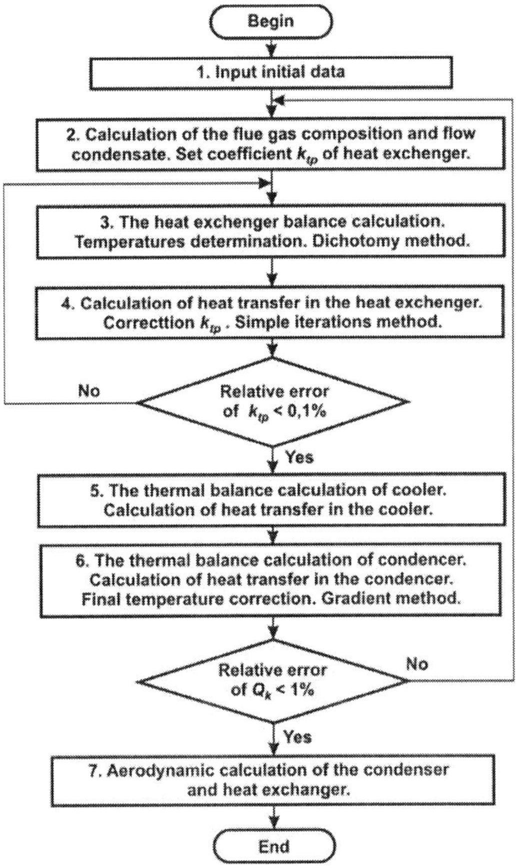

Fig. 2. A block diagram of the mathematical model structure.

Condenser power capacity Q_k is calculated by the heat transfer equation. If it is less the same power calculated from the balance equations, the final flue gas temperature at the condenser outlet t_3 is adjusted upwards by 0.01 °C, or downward by the same amount. Then return to the block 2. The algorithm prevents the transition to negative temperatures with the message about the condensate freezing possibility at the given parameters. The outer loop terminates when the thermal power capacity relative error lower 1 %. This cycle implements the gradient method. Other methods do not give a sustainable solution because the results are high sensitive to parameters value change.

Block 7 completes the necessary calculations. The pressure losses of air and flue gas are calculated for the heat recovery unit and the heat exchanger.

3 Results

The developed mathematical model is implemented as a software application that allows to choose the optimal geometric parameters of the installations for various capacities and verify their work.

A number of installations were designed by the calculations results [7]. The calculations were performed for the average temperature of the winter months in Siberia. At lower temperatures, the heat recovery will be more deep and the final flue gas temperature at the condenser outlet will decrease. The initial flue gas temperature is sufficiently high, resulting in a large dry zone in the heat recovery unit, reaching up to 45 % of heat exchange area. This decreases the condensation zone and reduces the heat recovery depth.

When installing the heat recovery unit is expedient to reduce the flue gas temperature at the boiler outlet to 100 °C by embedding additional surface for heating water in the boiler flue. This will increase the recovery depth and will reduce the final flue gas temperature below 20 °C.

4 Conclusion

A mathematical model of the condensing plate-type heat recovery unit using air as a heated medium contains the necessary heat balance equations and the heat transfer equation with selection different calculation methods. Equation systems are solved by various numerical methods that provide acceptable result accuracy.

The software application that implements a mathematical model allows to design the different installations and carry out verification calculations. The application will be useful to design organizations in the energy industry.

References

1. A. Kudinov, V. Antonov, Y. Alekseev, Thermal Engineering, **47**, 65 (2000).
2. K. Jeong, M. Kessen, H. Bilirgen, E. Levy, International Journal of Heat and Mass Transfer, **53**, 2361 (2010).
3. A. Garjaev, E. Tseplyaeva, Sur l'efficacite energetique, **1**, 187 (2003).
4. V. V. Bespalov, Power engineering of Tatarstan, **38**, 32 (2015).
5. V. V. Bespalov, L. Beljaev, D. Melnikov, MATEC Web of Conferences, **37**, 01009 (2015)
6. V. V. Bespalov, V. I. Bespalov, D. Melnikov, EPJ Web of Conferences, **110**, 01007 (2016)
7. V. V. Bespalov, Power engineering of Tatarstan, **42**, 39 (2016)

DOI: 10.1051/matecconf/20179101004

PROMISING DIRECTION OF PERFECTION OF THE UTILIZATION COMBINE CYCLE GAS TURBINE UNITS

Albina I. Gabdullina[1], *Nikolay* N.Galashov[1,*], *Svyatoslav* A. Tsibulskiy[1], *Denis* V. Melnikov[1], *Ilya* A. Asanov[1], and *Alexander* S. Kiselev[1]

[1]Tomsk Polytechnic University, 634050, Tomsk, Russia

Abstract. Issues of improving the efficiency of combined cycle gas turbines (CCGT) recovery type have been presented. Efficiency gas turbine plant reaches values of 45 % due to rise in temperature to a gas turbine to 1700 °C. Modern technologies for improving the cooling gas turbine components and reducing the excess air ratio leads to a further increase of the efficiency by 1-2 %. Based on research conducted at the Tomsk Polytechnic University, it shows that the CCGT efficiency can be increased by 2-3 % in the winter time due to the use of organic Rankine cycle, low-boiling substances, and air-cooled condensers (ACC). It is necessary to apply the waste heat recovery with condensation of water vapor from the flue gas, it will enhance the efficiency of the CCGT by 2-3 % to increase the efficiency of the heat recovery steam boiler (HRSB) to 10-12 %. Replacing electric pumps gas turbine engine (GTE) helps to reduce electricity consumption for auxiliary needs CCGT by 0.5-1.5 %. At the same time the heat of flue gas turbine engine may be useful used in HRSB, thus will increase the capacity and efficiency of the steam turbine.

1 Introduction

Combined cycle gas turbine (CCGT) unit of utilization type in the foreseeable future will be the main power generating units when operating on natural gas. This is due to: high efficiency, which has already reached more than 60 %; low cost of construction and commissioning of a small period of operation, which is about 1.5-2 times lower than that for steam-turbine power units of supercritical pressure; low consumption of process water, which is three times lower per unit of power output than in the steam-turbine thermal power plants (TPP); less impact on the environment due to lower emissions [1] into the atmosphere and heat and water; greater maneuverability when compared to steam turbine plants.

Actual aspects is implementation of CCGT in the Russian TPP. This is due to the fact that for the generation of electricity in the CCGT is currently in use and 70 % of natural gas in the next 30 years the proportion of combustible gas to generate electricity if and fall as a

*Corresponding author: gal@tpu.ru

result of substitution of coal, it is quite immaterial. In this regard, actual is matter of finding ways to improve these units.

The most complete technological issues of the operate CCGT utilization type described in [2, 3].

In utilizing CCGT gas turbine cycle and steam turbine cycle are linked through the HRSB and if there is no additional fuel combustion, the CCGT efficiency is determined by the equation [3]

$$\eta_{CCGT} = \eta_{GTU} + (1-\eta_{GTU})\eta_{HRSB}\cdot\eta_{STU}, \qquad (1)$$

where: η_{GTU} – efficiency of the gas turbine unit (GTU); η_{KY} – efficiency of the HRSB; η_{STU} – efficiency of the steam turbine unit (STU).

As the basis for the work of CCGT are turbine and steam turbine cycles are the following ways to increase their efficiency:

1. Improvement of the cycle and equipment of the gas turbine unit;
2. Improving cycle and equipment of the steam turbine;
3. Improving the processes in the HRSB;
4. Reducing the cost of electricity for own needs.

Consider the most promising ways of increasing the efficiency of the CCGT in the basis of the analysis of foreign and Russian scientific literature.

2 Improvement of the cycle and equipment of the gas turbine unit

Electrical efficiency of the gas turbine unit is determined by the product of the thermal, electro-mechanical and internal relative efficiency. Internal relative and electro-mechanical efficiency are determined by manufacturing gas turbine perfection, the compressor and electric generator and is currently close to its maximum value. Therefore, the main way of improving the GTU is to increase the thermal efficiency.

In modern CCGT is used gas turbine Brayton cycle with inlet and outlet of heat at a constant pressure. The thermal efficiency of this cycle is determined by inlet and outlet temperatures of heat. The temperature of heat input is interconnected with inlet temperature and during its growth increases. The main way to increase the thermal efficiency is to increase the heat input temperature. The temperature of heat input is determined by the material from which made the combustion chamber and the gas turbine, as well as the way of their cooling. Currently reliably mastered heat input temperature is 1500 °C, while powerful GTU in the company's General Electric, Siemens, Alstom received an electrical efficiency of GTU 40 [4, 5]. Experimental plants already operating successfully at 1600 °C. Expected electrical efficiency of GTU up to 43 %, GTU are being developed at a temperature of 1700 °C heat input that will provide efficiency up to 45 % [4].

For reliable modern cooling turbine components GTU operate with high excess air ratio, which leads to higher costs in the power the compressor.

Improving the efficiency of gas turbine technology was conducted mainly in the interests of the aviation industry and is to increase the initial temperature of the gases at the turbine inlet by cooling the gases washed by these parts of cyclic air GTU. Due to the complex control of water and vapor phase changes, and also because of the limited temperatures of the cooler suggested to excluded applying of water cooling of the turbine components. However, the cooling steam is developed and successfully applied by firms General Electric (GE) and Mitsubishi (MHI). At full cooling steam nozzle devices of the first stages of the gas turbine after the gas temperature are reduced by about 45 °C, while air cooling temperature drops to 100-120 °C. If the temperature of the gas in front of the rotor blades is maintained the same as in the air-cooling, the temperature of the gases in the combustion chamber and nitrogen oxide emissions will be significantly less [6, 7].

At the steam cooling decreased air bleed from the compressor for cooling the blades GT, reduced load on the compressor. The heat which gets in GT steam, the steam turbine power increases.

As a result, the efficiency of combined cycle power plant can be increased to 2 %. At the same time applying cooling air required [8, 9].

3 Improving cycle and equipment of the steam turbine

Steam turbine unit operates based on Rankine cycle. The main ways of improving the steam turbine plant are: increase the initial parameters; reheating steam; regenerative heating of feed water [10]; lowering the temperature of heat removal in the condenser; improving the flow part of the turbine.

Currently, the design flow of the turbine is used 3D modeling, which allowed to bring the upper limit of efficiency [11].

The initial temperature of the cycle of the steam turbine plant is determined by the temperature at the outlet of the gas turbine, which, depending on the type of turbine is in the range 450-640 °C. These conditions define the initial temperature of the steam at 420-610 °C. Increasing the initial pressure has little effect on improving efficiency, therefore to reduce capital investments and increase agility CCGT applied pressure is not higher than 16 MPa.

Intermediate superheat steam is mainly used for allowable moisture of steam at the outlet of the turbine.

The regenerative heating of feed water at CCGT does not increase efficiency, therefore does not apply.

The only way to increase the efficiency of the steam turbine plant is to lower the temperature of heat removal in the condenser, but the application of water-cooled condensers lower condensation temperature below 20 °C uneconomical. To address this shortcoming in the regions with low temperatures outside air is used air-cooled condenser (ACC), which allow at reduce winter heat removal the temperature to near 0 °C and below. But by using water as the working fluid the cycle it is possible to freeze and destruction of the condenser. Therefore, [12] suggested in a steam turbine installation to use the binary cycle, where the upper cycle of works on the water, and the bottom – organic Rankine cycle low boiling substance (LBS), which allows to reliably remove heat in the lower cycle at temperatures below 0 °C. Conducted in [13] research on the ACC and CCGT on different LBS showed that in the winter time can be 2-3 % increase CCGT efficiency.

4 Improving the processes in the HRSB

HRSG is the connecting link between the gas turbine cycle and steam turbine cycle. In equation (1) can be seen that the efficiency of HRSB significantly affect the second part, which defines the role of the steam turbine plant CCGT efficiency. The efficiency of HRSB is defined input and outlet temperatures of the flue gases. Inlet temperature of the gas of the HRSB is determined by a gas turbine and within 450-640 °C. To prevent condensation and corrosion metal gas path gases outlet from the HRSB temperature is maintained above 100 °C. As a result, the efficiency of HRSB does not exceed 86 %. Therefore, mainly by increasing the efficiency of HRSB is used of the technology operate boilers with a decrease in gas temperature below the point of Russia and the condensation of water vapor from the flue gas. Therefore, mainly by increasing the efficiency of HRSB technology is the use of boilers with a decrease in gas temperature below the dew point and condensation of water vapor from the flue gas. As shown in [14] the use of this technology allows in the increase the efficiency of 10-12 % HRSB, resulting in 2-3 % increase CCGT efficiency. Thus obtained from gas condensate can be effectively used at thermal power plants. In addition at water heat recovery from the flue gases is captured a significant portion of pollutants, which reduces pollution.

5 Reducing the cost of electricity for own needs

The electricity consumption for its own needs at the CCGT account for 1.5-2.5 %, with the main share comprises the cost of pumping water electric pumps. By changing the electric pump turbine engine (GTE), you can reduce the cost of electricity for own needs by 0.5-1.5 At the same time the heat of flue gas turbine engine can be useful in HRSB, which will increase the capacity and the efficiency of steam turbine plant.

6 Conclusions

1. The main ways of increasing the efficiency of GTU is to increase the initial temperature and decrease air flow for cooling. As a result in the foreseeable future GTU efficiency can reach 45-47 %.
2. The main ways of increasing the efficiency of steam turbine plant is to increase the initial temperature of the steam, which can only be achieved in the application of GTU with high inlet temperatures and heat removal, as well as reducing heat removal temperature as a result of the ACC and binary cycle with organic Rankine cycle in the lower stage. This will allow the winter to increase the CCPP the efficiency of by 2-3 %.
3. In order to increase the efficiency of HRSB is necessary decrease exhaust gas temperature below the dew point, which will increase the efficiency of HRSB 10-12 %, and CCGT the efficiency of by 2-3 %.
4. Replacement of steam turbine plant electric pumps gas turbine engine can reduce the cost of electricity for own needs CCGT by 0.5-1.5 %.

The research was realized with financial support of Minobrnauki of Russia in framework of FTP "Research and development in prior direction of scientific-technological complex of Russia in 2014-2020 years", unique R&D identifier RFMEFI58114X0001.

References

1. D. Gvozdyakov, V. Gubin, S. Shvab, A. Tanishev, MATEC Web of Conferences **72**, 01037 (2016)
2. R. Kehlofer, B. Rukes, F. Hannemann, F. Stirnimann, *Combined-Cycle Gas & Steam Turbine Power Plants* (PennWell Corporation, Tulsa, 2009)
3. A.D. Trukhniy, *Parogazovyye ustanovki elektrostantsiy* (MEI, Moscow, 2013) [in Russia]
4. G.G. Olkhovskiy, Therm. Eng. **2**, 10 (2013)
5. D. Robb, Turbomachinery Intern. **2**, 24 (2011)
6. G.G. Olkhovsky, Therm. Eng. **7**, 8 (2016)
7. S.C. Gulen, Gas Turbine World **44**, 26 (2014)
8. S.C. Gulen, Gas Turbine World **44**, 33 (2014)

9. K. Savina, A. Abramovskih, V. Haymin, S. Lavrinenko, MATEC Web of Conferences **72**, 01042 (2016)
10. K. Larionov, D. Gvozdjakov, A. Zenkov, V. Zaytsev, EPJ Web of Conferences **110**, 01034 (2016)
11. V. F. Kasilov, The series of siemens SST-200 – SST-900 Siemens in Russia, Therm. Eng. **62**, 240 (2015)
12. N. Galashov, Power Technology and Engineering **48**, 458 (2015)
13. N. Galashov, S. Tsibulskiy, V. Gubin, T. Serova, MATEC Web of Conferences **37**, 01021 (2015)
14. N. Galashov, S. Tsibulskiy, A. Kiselev, MATEC Web of Conferences **72**, 01029 (2016)

MATEC Web of Conferences **91**, 01005 (2017) DOI: 10.1051/matecconf/20179101005
Smart Grids 2016

NUMERICAL ESTIMATION OF THE FORMATION PROCESS OF ANTHROPOGENIC PRECIPITATION IN THE ATMOSPHERE

Dmitriy V. Gvozdyakov[1,*], *Vladimir* E. Gubin[1], *Alexandr* A. Gromov[2], *Anton* O. Tanishev[1], and *Svetlana* A. Shvab[1]

[1]Tomsk Polytechnic University, 634050, Tomsk, Russia
[2]Nürnberg University of Technology Georg-Simon-Ohm

Abstract. The processes of condensation of sulfur trioxide SO_3 and water vapor H_2O. Sulfuric anhydride is formed in flues of thermal power plants (TPP) by partial oxidation of SO_2 (up to 5 % of the total SO_2) from the combustion of high-sulfur fuels, and belongs to a class of mild-hazard products. Sulfuric anhydride in the interaction with water vapor, which refers to greenhouse gases, under certain conditions, it forms sulfuric acid.

1 Introduction

One of the factors of the negative impact of thermal power plants (TPP) on the environment is the possible formation of acidic precipitation in areas immediately adjacent to the TPP [1, 9]. Various pollutants released to the environment from the combustion of fuels [1].

In real practice the atmosphere always contains water vapor of natural origin [2]. In addition, a significant increase of the water vapor concentration in the air is possible in areas adjacent to the power plants by emissions of H_2O due to the work of the heat engineering equipment of the station [1, 3]. The analysis of processes of sulfuric acid drops formation in the air upon jointly proceeding processes of condensation of SO_3 and H_2O vapors on the condensation nuclei surface is of interest.

According to [1] it was assumed that the formation of sulfuric acid drops is the result of condensation of SO_3 and water vapor at the surface of the "condensation nuclei" – microscopic drops of water. The speed of the thermodynamic process significantly depends on the temperature [4]. Accordingly the reliability of the simulation results of condensation process is determined by the accuracy of the condensation surface temperature determination.

2 Study technique

The mechanism of the sulfuric acid formation [1] is based on cooperation, where sulfurous anhydride SO_3 is absorbed by water vapor.

*Corresponding author: dim2003@tpu.ru

$$SO_3 + H_2O \rightarrow H_2SO_4 + Q \quad (1)$$

The process of transfer of energy, momentum and mass in these circumstances is described by the system of non-stationary partial differential equations [5]. It is considered that the temperature and concentration on the left border of the solution domain (chimney of TPP) do not depend on time. Unsteady two-dimensional equations of mixed convection in the approximation of the Boussinesq approximation [6] are used to model the changes in the main sought for functions.

Calculation of the condensation rate was conducted using the formula [4]:

$$W_i^K = \beta \cdot \frac{P_A - P_{0i} \cdot \exp\left(-\frac{E_i}{R \cdot T}\right)}{\sqrt{\frac{2 \cdot \pi \cdot R \cdot T}{M}}}, \quad (2)$$

where P_A – the atmospheric pressure, Pa; P_{0i} – the preexponent, Pa; E_i – the activation energy of process of condensation, J/kg; R – universal gas constant, J/(mol·K); M –molar mass, kg/mol; β – the coefficient of condensation.

To solve the formulated boundary problem it was used the algorithm [7], developed for the decision of tasks of conjugate heat transfer in areas with a local energy source. Assessment of reliability of the obtained results was performed by verification of difference scheme conservative similar to [8].

The value of the coefficient of condensation of water vapor was varied in the range from 0.05 to 0.4. Below are the results of numerical simulation of growth of sulfuric acid drops in the condensation process of sulfur dioxide and water vapor, calculated at a wind speed of 5 m/s in summer.

Figure 1 shows the changes of the sulfuric acid drop sizes, depending on the time at various initial sizes of condensation nuclei.

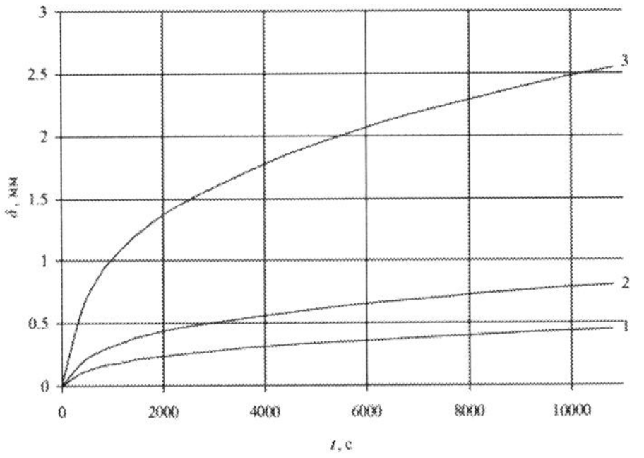

a

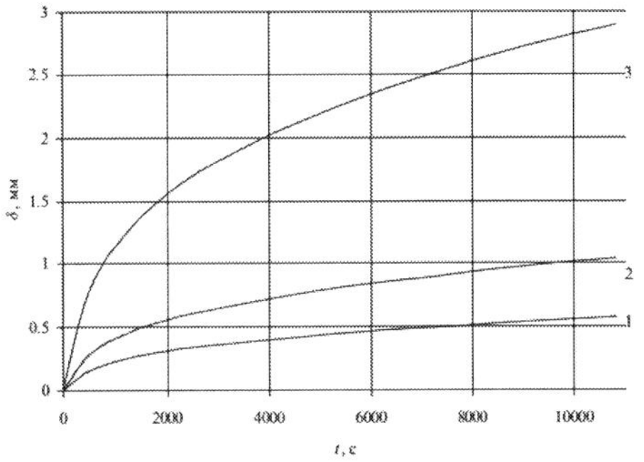

b

DOI: 10.1051/matecconf/20179101005

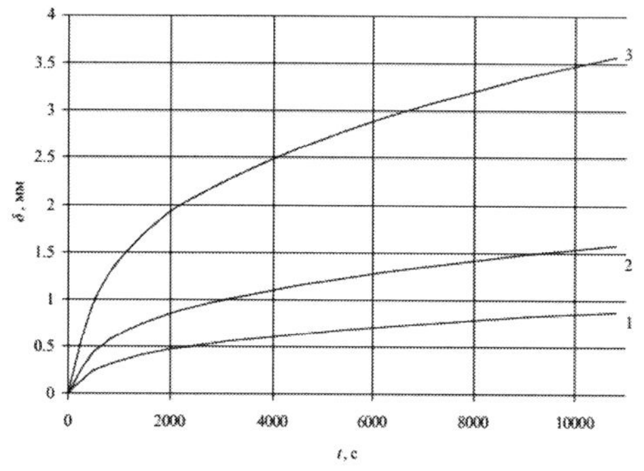

c

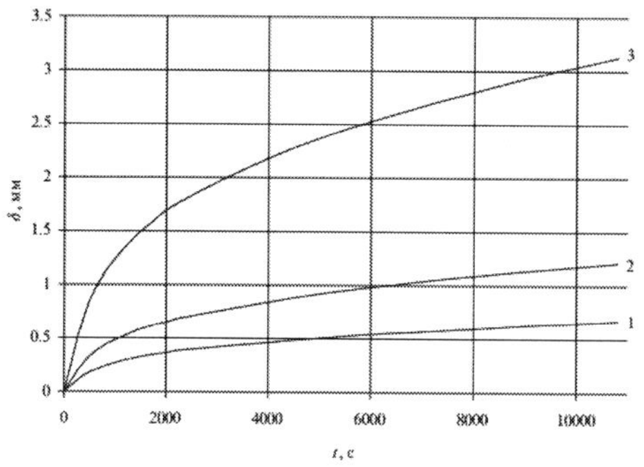

d

Fig. 1. Changes of the sulfuric acid drops size (1 – δ_0=0.8·10^{-6} m; 2 – δ_0=1.0·10^{-6} m; 3 – δ_0=5.0·10^{-6} m; δ_0 – the initial size of the condensation nuclei): a) β=0.05; b) β=0.1; c) β=0.2; d) β=0.4.

The results (Fig. 1) of numerical investigations show that when the coefficient of condensation β=0.05 it is possible to form the sedimentary acid precipitation already through 1800 s since the beginning of the process on the condensation nuclei surface δ_0=5.0·10^{-6} m. Further course of the simultaneous condensation process of sulfur dioxide and water vapor promotes the formation of sulfuric acid drops with sizes to 2.5·10^{-3} m during the analyzed time interval.

It was established that formation of acid precipitation occurs on the condensation nuclei surface at δ_0=1.0·10^{-6} m during 9000 and 7200 s, respectively, in the range of changes of the condensation coefficient from 0.1 to 0.2. For the considered period of time, the formation of sulfuric acid drops in the airspace adjacent to the power plants, with size from 1.0·10^{-3} m to 1.2·10^{-3} m is possible. Numerical evaluation of the particle border growth on the condensation nuclei surface at δ_0=5.0·10^{-6} m illustrates the possibility of the sulfuric acid sedimentary drops formation during the 1800 s and 600 s since the beginning of the simultaneous introduction process of sulfur dioxide and water vapor at the surface of the "embryo".

3 Conclusion

The formation of sulfuric acid drops in the atmosphere of the Earth occurs more intensive on the condensation nuclei surface with initial size from 1.0·10^{-6} m to 5.0·10^{-6} m when the coefficient of condensation β=0.05 as the researches have shown.

The work has been supported by the Russian Federal focused program "Research and development on priority orientation of science and technology in Russia 2014–2020", unique identifier 'Applied Research and Experimental Developments' project RFMEFI58114X0001.

References

1. Ju. A. Izrajel', I. M. Nazarov, A. Ja. Pressman, *Acid rains* (Gidrometeoizdat, Leningrad, 1989) [in Russian]
2. D. V. Gvozdyakov, Web of Conference **82**, 01026 (2015)
3. A. K. Vnykov, *Protection of atmosphere from emissions of energy facilities* (Energoatomizdat, Moskva, 1992) [in Russian]

4. A. D. Labuncov, *Physical fundamentals of Energy* (MEI Publ., Moskva, 2000) [in Russian]
5. P. Rouch, *Computational fluid dynamics* (Moskva, World, 1980) [in Russian]
6. Paskonov, V.I. Polezhaev, L.A. Chudov, *Numerical simulation of processes of heat and mass transfer* (Nauka, Moskva, 1984) [in Russian]
7. G. V. Kuznetsov, J. Eng. Phys. **79,** 56 (2006)
8. G. V. Kuznetsov, J. Eng. Thermophys. **18**, 162 (2009)
9. K. V. Slyusarskiy, M. K. Bejsekov, J. V. Marysheva, Y. Ja Rakov, MATEC Web of Conferences, **37**, 01053

MATEC Web of Conferences **91**, 01006 (2017) DOI: 10.1051/matecconf/20179101006
Smart Grids 2016

ANALYTICAL ESTIMATION OF THE DISTRIBUTION OF THERMAL NEUTRON FLOW

Dmitriy V. Gvozdyakov[1,*], *Aleksandr* S. Matveev [1], *Anton* O. Tanishev[1], and *Svetlana* A. Shvab[1]

[1]Tomsk Polytechnic University, 634050, Tomsk, Russia

Abstract. The neutron's evaluation of impact on the internal reflector as an example of the plate reactor in single-group approximation. As a result there were obtained wave equations analytical solutions and graphic solution of the formulated problem.

1 Introduction

The distribution of critical size or fuel critical concentration (mass) is the main task of the nuclear reactors theory. Consideration of the critical problem was limited of classical forms reactors (sphere, infinite cylinder and plate) without neutron reflector and with it.

Estimation of internal reflector considerate only limiting case of its diffusivity: for vacuum and black body (the diffusion equation is eliminates).

More than half a century of nuclear reactor operating experience identified the reasonability using internal and external reflectors [1, 6], also it's shown by previously researches [2, 3].

E.g., in research reactor IRT-T using not only external but also internal reflectors can lead to increase of the average density of the neutron flow [4]. Also we choosing beryllium as the material of the reflector cause it increase of the average density of the neutron flow. In general case flow distribution is carried out by numerical methods in multi-group approximation.

2 Mathematical problem and study technique

The distribution of the neutron flux in the core, the inner and outer reflectors in the single-group approximation is usually described by the following wave equations:

$$\frac{d_2\Phi_1}{dx^2} + \chi_1^2 \cdot \Phi_1 = 0, \qquad (1)$$

$$\frac{d_2\Phi_2}{dx^2} - \chi_2^2 \cdot \Phi_2 = 0, \qquad (2)$$

* Corresponding author:dim2003@tpu.ru

DOI: 10.1051/matecconf/20179101006

$$\frac{d_2\Phi_3}{dx^2} - \chi_3^2 \cdot \Phi_3 = 0. \tag{3}$$

Here χ_1, χ_2, χ_3 are the material parameters of the core and reflectors.

Figure 1 is showing schematic distribution of thermal neutron flux in single-group approximation.

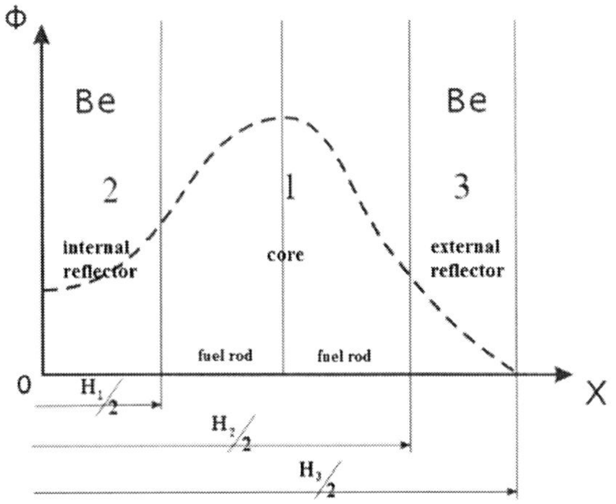

Fig. 1. The distribution of thermal neutron flow in single-group approximation.

Using the following conditions we will find the equations solution (1), (2), (3):

1. $x = 0$, $\nabla\Phi_2(0) = 0$, (4)

2. $x = \frac{H_1}{2}$, $\Phi_2\left(\frac{H_1}{2}\right) = \Phi_1\left(\frac{H_1}{2}\right)$, (5)

3. $x = \frac{H_1}{2}$, $D_2\nabla\Phi_2\left(\frac{H_1}{2}\right) = D_1\nabla\Phi_1\left(\frac{H_1}{2}\right)$, (6)

4. $x = \frac{H_2}{2}$, $\Phi_1\left(\frac{H_2}{2}\right) = \Phi_3\left(\frac{H_2}{2}\right)$, (7)

5. $x = \frac{H_2}{2}$, $D_1\nabla\Phi_1\left(\frac{H_2}{2}\right) = D_3\nabla\Phi_3\left(\frac{H_2}{2}\right)$, (8)

6. $x = \frac{H_3}{2} = \frac{H_2}{2} + \delta_{эф}$, $\Phi_3\left(\frac{H_3}{2}\right) = 0$. (9)

Here D_i is diffusion coefficient.

Application of the external reflector reduces loss of neutrons and saves core equal to effective additive [3].

$$\delta_{эф} = \frac{1}{\chi_1} arctg\left[\frac{D_1\chi_1}{D_3\chi_3} th\left(\chi_3 T\right)\right].$$

Now introducing an effective boundary condition:

 DOI: 10.1051/matecconf/20179101006

7. $x=\frac{H_2}{2}+\delta_{эф}$, $\Phi_1\left(\frac{H_2}{2}+\delta_{эф}\right)=0.$ (10)

Solutions of wave equations (1), (2) are following:

– core in infinite plate form

$$\Phi_1(x)=c_1\cdot sin(\chi_1\cdot x)+c_2\cdot cos(\chi_1\cdot x). \quad (11)$$

– reflector in finite plate form

$$\Phi_2(x)=c_3\cdot sh(\chi_2\cdot x)+c_4\cdot ch(\chi_2\cdot x). \quad (12)$$

After mathematical simplifications, finding the constant of integration it was obtained the critical equation for core in infinite plate form with an internal reflector:

$$D_2\cdot\chi_2\cdot th\left(\chi_2\cdot\frac{H_1}{2}\right)=-D_1\cdot\chi_1\cdot ctg(X_{KP}), \quad (13)$$

where:

$$X_{KP}=\chi_1\cdot\left(\frac{H_3}{2}-\frac{H_1}{2}\right). \quad (14)$$

3 Experimental results

Knowing the thickness of external reflector and material parameters, you can define a critical value of core size.

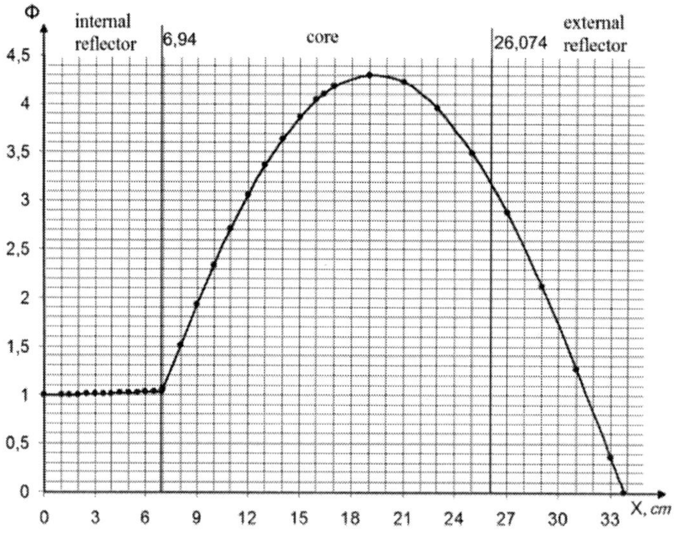

Fig. 2. The distribution of neutron flow for a core and for an internal reflector in single-group approximation.

Reflectors efficiency can be evaluated not only for saving active zone but also by the average thermal neutron flux ratio change in the core to its maximum value. For the reactor without reflector this value [5, 6] equals 0.637. If we have a beryllium reflector – it will be 0.939. In this research it's equal 0.82.

4 Conclusion

Research's result can be explained by the fact that single-group approximation isn't take into account the contribution to the accumulation of thermal neutrons in the reflector due to the slowing down of fast neutrons.

The work has been supported by the Russian Federal focused program "Research and development on priority orientation of science and technology in Russia 2014–2020", unique identifier 'Applied Research and Experimental Developments' project RFMEFI58114X0001.

References

1. G. G. Bartholomei, G. A. Bat', D. V. Baibakov, M. S. Alhutov, *Fundamentals of the theory and methods of calculation of nuclear reactors* (Energoatomizdat, Moskva, 1989) [in Russian]
2. D. V. Gvozdyakov, MATEC Web of Conferences **37**, 01024 (2015)
3. K. V. Tsyplenkov, A. V. Kuzmin, *The Influence of the Central reflector on the neutron flux distribution in the thermal reactor* (TPU Publishing house, Tomsk, 2005) [in Russian]
4. V. N. Parmon, A. D. Simonov, S. F. Sadykov, *Fizika goreniya vzriva* **51**, 46 (2015)
5. R. Murray, *Physics of nuclear reactors* (AI, Moskva, 1961) [in Russian]
6. D. V. Gvozdyakov, MATEC Web of Conferences **37**, 01024 (2015)

MATEC Web of Conferences **91**, 01007 (2017) DOI: 10.1051/matecconf/20179101007
Smart Grids 2016

INFLUENCE OF $NaNO_3$ AND $CuSO_4$ CATALYTIC ADDITIVES ON COAL OXIDATION PROCESS KINETIC DEPENDENCIES

Kirill B. Larionov[1,*], *Il'ya* V. Mishakov[1,2], *Alexandr* A. Gromov[3]and *Andrey* V. Zenkov[1]

[1]Tomsk Polytechnic University, 634050, Tomsk, Russia
[2]Boreskov Institute of Catalysis SB RAS
[3]Nürnberg University of Technology Georg-Simon-Ohm

Abstract. Experimental studies of bituminous coal and lignite oxidation were conducted with the addition of different nature catalytic additives: $NaNO_3$ and $CuSO_4$. The results showed that added mineral salts led to a noticeable decrease in the coals initial oxidation temperature and reaction acceleration at an early stage of the process.

1 Introduction

According to statistics [1], coal production occupies a leading position in the modern fuel and energy structure along with such conventional fuels as oil and gas. In turn, direct coal combustion in domestic and power-plant boilers is accompanied by a large amount of harmful substances release [2]. Today, the most eco-efficient technology of coal application is its gasification, as a result of which the generator gas is produced [3]. Active work on the development of various methods is conducted in the world as part of this technological process, aimed at improving the reliability and efficiency of gas plants. One possible solution is to use catalytic agents that improve kinetic characteristics of the process and increase the fuel application efficiency [4].

Most of the existing studies on catalytic gasification are directed to the use of alkali, transition and rare-earth metals oxides [5]. The main effect of using such catalytic additives is reduction of temperature at which fuel oxidation process starts. It is known [6], that the inclusion of catalytic additives based on various kinds of metal oxides in the gasification process is manifested most efficiently only at high temperatures (over 500 °C).

It is known [7] that the precursors of metal oxides, such as nitrates and sulfates, have strong oxidizing properties, low melting point (decomposition) and good solubility in water. It simplifies the process of their mixing with coal and improves communication of the "coal-catalyst" system by filling available pores of solid fuel [8].

The results of coal oxidation experimental studies with introduction of catalytic additives of different nature ($NaNO_3$ and $SuSO_4$) are presented in this paper. Also, an assessment of kinetic parameters changes is provided.

*Corresponding author: arryk@tpu.ru

DOI: 10.1051/matecconf/20179101007

2 Experimental setup and study technique

Bituminous coal from Osinnikovskiy coal deposit and lignite from Borodinskiy deposit were considered as initial solid fuel samples in this study. These coals are characterized by their morphology and the degree of metamorphism [9].

Coarse samples were milled in a ball-mill drum under identical conditions, and then sieved through a sieve with cells size of less than 80 microns.

Sieved coal powders were dried in an LF 25/350-VS1 drying oven (Loip, Russia) at 105 °C. Technical characteristics of these samples (Table 1) were determined by means of the existing techniques [10-13].

Table 1. Initial coal characteristics*.

Characteristic	Coal	
	Bituminous	**Lignite**
Ash content, mass %	7	5
Humidity, mass %	0,3	1,1
Volatiles, mass %	29	39
Fixed carbon, mass %	63,7	54,9

*Samples characteristics are given after their drying at 105°C.

The results in Table 1 show that bituminous coal, unlike lignite, has a higher content of connected carbon and mineral residue. In turn, the opposite can be said about the value of volatile substances output, due to different metamorphism degree of presented in the research coal types [10].

Microscopy of the researched samples (Figure 1) was performed via SEM JSM-6460LV device (JEOL, Japan) with focused ion beam to assess their morphological characteristics.

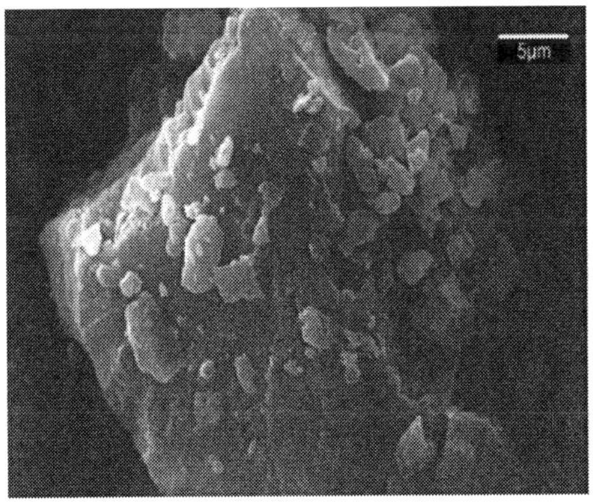

a

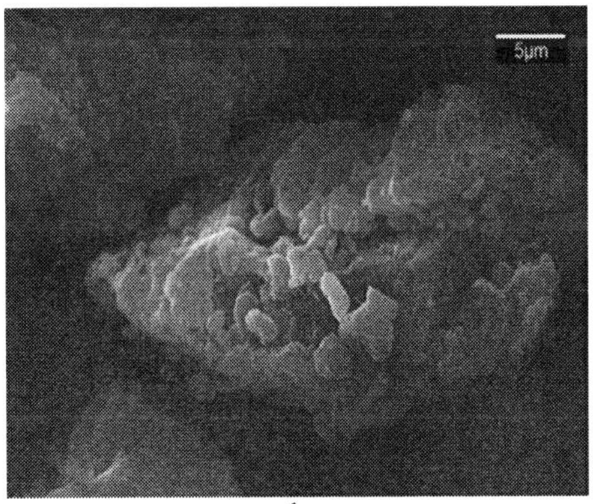

b

Fig. 1. Coal samples microphotograph with x3000 zoom (a – bituminous coal, b – lignite).

According to Figure 1, the bituminous coal particles have a dense oblong appearance. In turn, lignite particles have multi-faceted spherical shape and amorphous surface.

Next, obtained samples were mixed by capillary impregnation with catalytic additives $NaNO_3$ and $CuSO_4$, previously saluted in a water-alcoholic medium at a volumetric ratio of 1:1. The catalytic additive weight content was 5 % of the studied sample total weight (5 mass % of the catalytic additive and 95 mass % of coal).

The obtained slurry was dried in drying oven at 105 °C for 20 hours. After that, agglomerated samples were mixed in a stamp for 7 minutes to obtain a powder-shaped state. It is necessary to add that reference samples were prepared for comparative analysis of catalytic additives influence. This samples were exposed to identical preparation procedures, but mentioned above additives were not added.

A serial number was assigned to each sample, in order to facilitate conduction of subsequent experimental work and results analysis. List of the prepared samples is presented in Table 2.

Table 2. List of synthesized samples.

Sample name	**Serial number of the sample**
Bituminous coal	
Bituminous coal reference sample	1
Bituminous coal + 5% $NaNO_3$	2
Bituminous coal + 5% $CuSO_4$	3
Lignite	
Lignite reference sample	4
Lignite + 5% $NaNO_3$	5
Lignite + 5% $CuSO_4$	6

Determination of the kinetics of prepared coal samples oxidation was conducted with the use of Netzsch STA 449 °C Jupiter thermogravimetric analyzer (Germany). All experiments were carried out in similar conditions at a heating rate of 2.5 °C/min in a corundum crucible (sample weight of about 7 mg) in the temperature range of 50-600 °C. Mixture of air (60 ml/min) and nitrogen (10 ml/min) was used as an oxidizing medium.

3 Experimental results

Results of researched samples thermal analysis are presented in the form of TG and DTG curves in Fig. 2 and 3, respectively.

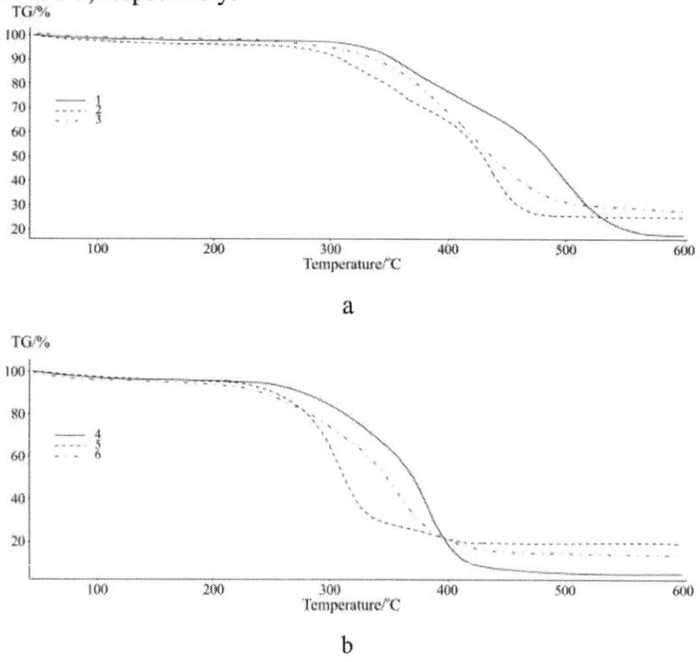

Fig. 2. Thermogravimetric curves (TG) of coal samples oxidation (a - bituminous coal samples, b - lignite samples).

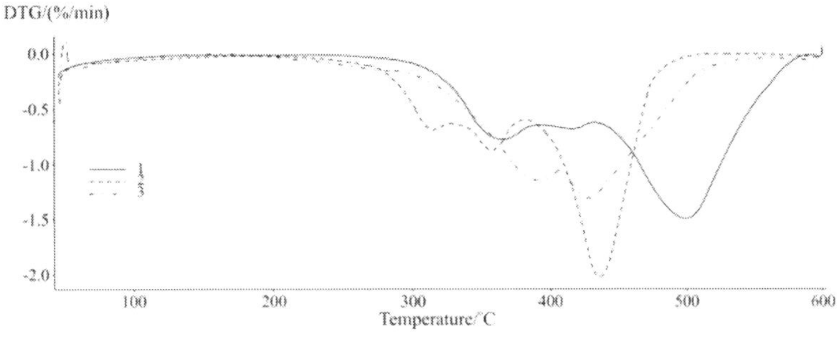

a

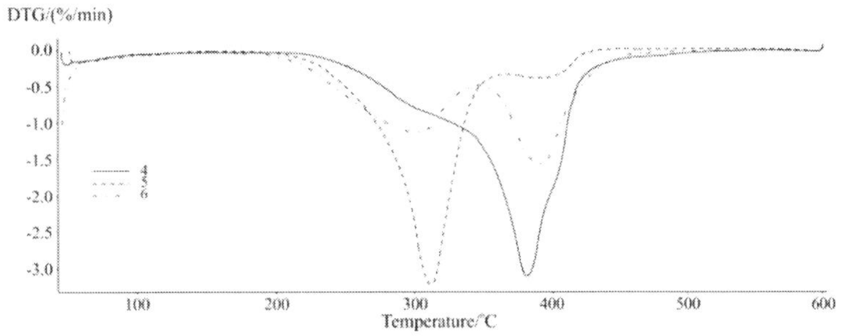

b

Fig. 3. Differential curves (DTG) of coal samples oxidation (a – bituminous coal samples, b – lignite samples).

According to Fig. 2, the temperature corresponding to 10 % in mass loss for initial samples 1 and 4 was 356 °C and 278 °C, respectively. For coals with addition of $NaNO_3$ (samples 2 and 5), this temperature is 338 °C and 244 °C (sample 5), and with the addition of $CuSO_4$ (samples 3 and 6) – 310 °C and 255 °C. Thus, the largest shift towards reduction of initial stage oxidation temperature is observed in samples 3 and 5, for which the change was 44 °C and 34 °C, respectively. Despite the fact that the addition of $NaNO_3$ has considerable catalytic effect at the initial stage of lignite oxidation process, the total reaction time in comparison with the original sample increased by 15 % because of residual carbon oxidation for both samples ends in the same temperature range (about 480 °C).

In turn, the greatest value of reaction maximum speed, determined from DTG data (Figure 3), corresponds to the samples with addition of $CuSO_4$. This figure rises by 37 % for bituminous coal and by 7 % for lignite. It should also be noted, that addition of catalytic additive is accompanied by increased reaction rate at an early stage of the oxidation. This is most clearly expressed for bituminous coal samples, which have lower content of longitudinal and transverse molecular bonds.

The effect of coal oxidation reaction rate increase (reduction of process start temperature) for samples with different catalytic additives is directly related to the decomposition of salts, which is observed at low temperatures area, with respect to the process of volatiles release. Oxide gas compounds, which act as transporters of oxygen from the oxidant to the oxidized substrate, are released as a result of salts decomposition. This process is accompanied by decrease of molecular bonding strength during intensified chain reaction, especially oxygen-containing groups, which leads to an earlier volatilization of volatile compounds and their subsequent ignition. In addition, synergy of this effect is reduction of diffusion resistance, because of coal structure thermal degradation, consisting of a large number of micro- and macropores.

DOI: 10.1051/matecconf/20179101007

4 Conclusion

The experimental studies have shown that the introduction of $NaNO_3$ and $CuSO_4$ mineral salts catalytic additives leads to a noticeable acceleration of the coal oxidation process as a result of oxide gas release at the decomposition stage. Released at an earlier stage of the process volatiles are split into light gases and enriched with carbon fine particles, making it easier to burn.

The work has been supported by the Russian Federal focused program "Research and development on priority orientation of science and technology in Russia 2014–2020", unique identifier 'Applied Research and Experimental Developments' project RFMEFI58114X0001.

References

1. *British Petroleum Statistical review of World Energy* (Pureprint Group Limited, Uckfield, 2014)
2. N. N. Galashov, S. A. Tsibulski, Power Technology and Engineering **48**, 459 (2015)
3. S. Jankovskiy, D. Luzhkovoj, K. Larionov, A. Matveeva, MATEC Web of Conferences **37,** 01026
4. V. N. Parmon, A. D. Simonov, V. A. Sadykov, Tihov S.F. Catalytic combustion: achievements and problems **51**, 143 (2015)
5. S. Yaman, S. Kucukbayrak, Thermochim. Acta **293**, 109 (1997)
6. M.J. Aho, J.P. Hämäläinen, J.L. Tummavuori, Fuel **70**, 1143 (1991)
7. U. F. Zavyalova, P. S. Barbashova, A. S. Lermontov, Kinetika i kataliz, **48**, 171 (2007) [in Russia]
8. I. V. Tokareva, I. V. Mishakov, D. V. Korneev, A. A. Vedyagin, K. S. Golokhvast,. Nanotechnologies in Russia, **10**, (2015) [in Russia]
9. K. V. Slyusarskiy, M. K. Bejsekov, J. V. Marysheva, Y. Ja Rakov, MATEC Web of Conferences, **37**, 01053 (2015)

THE INVERSE STATIONARY HEAT CONDUCTION PROBLEM FOR A CUBOID

Konstantin M.Yacenko[1,*], *Yri* Y.Rakov[1], and *Konstantin* V. Slyusarskiy[1]

[1]Tomsk Polytechnic University, 634050, Tomsk, Russia

Abstract. The heat conductivity coefficient is important characteristic which is used in various spheres. The original methodic for conductivity coefficient determination was proposed for samples in form of rectangular parallelepiped. The results of numerical solution of nonlinear heat conduction problem in heat conduction coefficient value range 0.04-5 W/(m K) with different sample relative size were presented.

1 Introduction

The heat conduction of substance is widespread phenomena and has a strong influence on various aspects of technological processes: from durability calculations to heat flux determination [1]. The heat conduction coefficient determines substance behavior in terms of heat conduction and extremely important for its correct description [2]. Existing experimental procedures for its estimation often have to handle with heat-transfer coefficient which is hard to measure directly. This causes significant error in heat conductivity coefficient final values [3]. To eliminate influence of this factor on measurements this procedure is realized in vacuum. To estimate exploitation parameters for measuring procedure for substances with heat conductivity value in range 0.04 – 5 W/(m K) the physical model of heat transfer was formulated and investigated.

2 Physical model of heat transfer

The test object is a model of a cuboid with the dimensions LX, LY, LZ. It is heated by the heat flux at one side but loses heat from all sides to the environment by radiation. Ambient temperature is constant and equal to T_{oc}. Integral emission coefficient ε and absorption coefficient of the sample surface A will take constant and coefficient of thermal conductivity λ does not depend on temperature [4].

* Corresponding author: brkmax@mail.ru

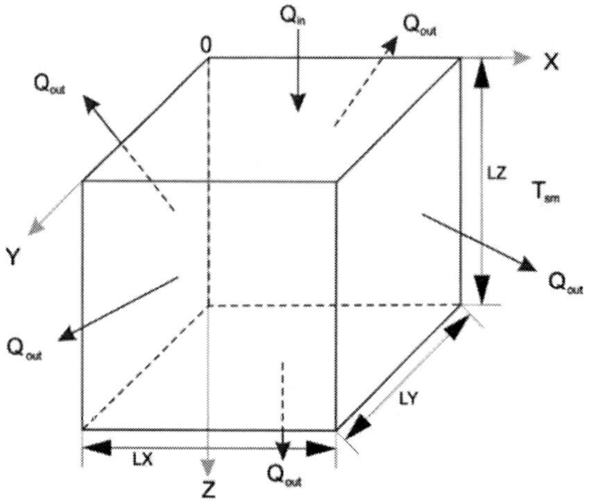

Fig. 1. The scheme of distribution of heat fluxes in the sample.

3 Mathematical model and decision methods

The mathematical dimensionless statement of the inverse stationary heat conduction problem for cuboid includes the heat equation (1), non-linear boundary conditions (2 - 7) and extreme conditions (8):

$$\frac{\partial^2 \tilde{T}}{\partial \tilde{x}^2}+\frac{\partial^2 \tilde{T}}{\partial \tilde{y}^2}+\frac{\partial^2 \tilde{T}}{\partial \tilde{z}^2}=0 \tag{1}$$

$$\left(\frac{\partial \tilde{T}}{\partial \tilde{z}}\right)_{\tilde{z}=0}=\pi_1\cdot(\tilde{T}^4_{\tilde{z}=0}-\tilde{T}^4_{oc})-1 \tag{2}$$

$$-\left(\frac{\partial \tilde{T}}{\partial \tilde{z}}\right)_{\tilde{z}=1}=\pi_1\cdot(\tilde{T}^4_{\tilde{z}=1}-\tilde{T}^4_{oc}) \tag{3}$$

$$\left(\frac{\partial \tilde{T}}{\partial \tilde{x}}\right)_{\pi_2=0}=\pi_1\cdot(\tilde{T}^4_{\tilde{x}=0}-\tilde{T}^4_{oc}) \tag{4}$$

$$-\left(\frac{\partial \tilde{T}}{\partial \tilde{x}}\right)_{\pi_2}=\pi_1\cdot(\tilde{T}^4_{\pi_2}-\tilde{T}^4_{oc}) \tag{5}$$

$$\left(\frac{\partial \tilde{T}}{\partial \tilde{y}}\right)_{\pi_3=0}=\pi_1\cdot(\tilde{T}^4_{\tilde{y}=0}-\tilde{T}^4_{oc}) \tag{6}$$

$$-\left(\frac{\partial \tilde{T}}{\partial \tilde{y}}\right)_{\pi_3}=\pi_1\cdot(\tilde{T}^4_{\pi_3}-\tilde{T}^4_{oc}) \tag{7}$$

$$\left|\pi_4-\int_{\tilde{F}}\left[\tilde{T}^4_{\tilde{z}=1}-\tilde{T}^4_{oc}\right]\partial\tilde{F}\right|\rightarrow \min \tag{8}$$

Here $\pi_1 = \dfrac{\varepsilon\sigma \cdot LZ^4 \cdot q_F^3 A^3}{\lambda^4}$, $\pi_2 = \dfrac{LX}{LZ}$, $\pi_3 = \dfrac{LY}{LZ}$, $\pi_4 = \dfrac{Q_{LZ}\lambda^4}{\varepsilon\sigma q_F^4 A^4 \cdot LZ^6}$; $\tilde{T} = T\Big/\left(\dfrac{q_F A \cdot LZ}{\lambda}\right)$, $\tilde{T}_{oc} = T_{oc}\Big/\left(\dfrac{q_F A \cdot LZ}{\lambda}\right)$, $\tilde{F} = \dfrac{F}{LZ^2}$, $\tilde{x} = \dfrac{X}{LZ}$, $\tilde{y} = \dfrac{Y}{LZ}$, $\tilde{z} = \dfrac{Z}{LZ}$; Q_{LZ} is integral heat flux from the bottom surface of the sample.

As follows from the mathematical statement of the inverse problem, the solution of the direct problem of heat conduction in different numerical values of coefficient of the thermal conductivity is necessary to calculate the minimum of objective function (2.8).

Numerical methods are used to solve nonlinear problems to get a table of approximate values. The finite difference method was chosen for this task.

Finite difference equations were solved by a method of fixed-point iteration and iterative method of successive over-relaxation (accelerated Liebmann method). The dichotomy method was used to determine the coefficient of thermal conductivity in the inverse problem [5].

4 Results and discussion

Figure 2 shows the dependence between the dimensionless parameters $\pi_1, \pi_2, \pi_3, \pi_4$ for the estimation of the geometric dimensions of the sample and the physical parameters of the experiment. It is necessary to determine the coefficient of thermal conductivity in the estimated range of its values.

The value π_1 should be in the range of 0.001–0.01 when the relative sizes of the sample $\pi_2 = \pi_3$ corresponding to the range from to 1 to 10. In this case, temperature changes in the sample will not exceed 100 K and use of mathematical models (1–8) becomes possible.

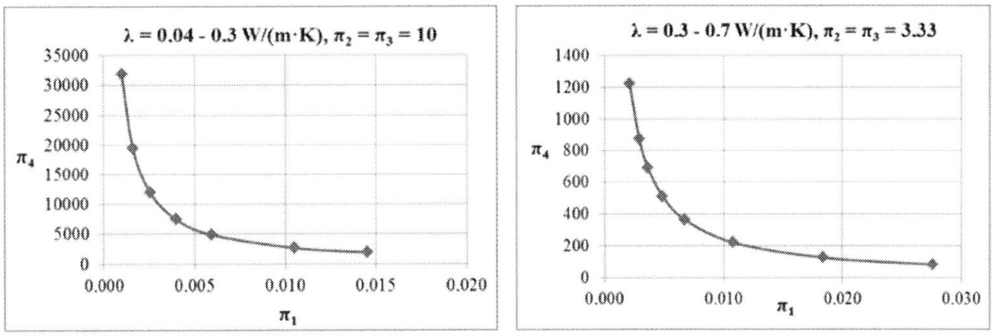

a b

 DOI: 10.1051/matecconf/20179101008

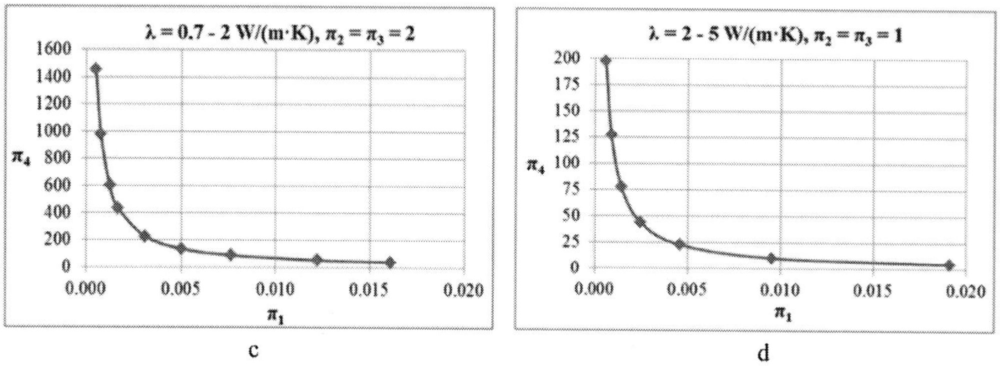

Fig. 2. The dependence between π_4 and π_1 : a – λ=0.04-0.3 W/(m·K), $\pi_2 = \pi_3 = 10$; b – λ=0.3-0.7 W/(m·K), $\pi_2 = \pi_3 = 3.33$; c – λ=0.7-2 W/(m·K), $\pi_2 = \pi_3 = 2$; a – λ=2-5 W/(m·K), $\pi_2 = \pi_3 = 1$.

5 Conclusion

The physical and mathematical model for thermal conductivity coefficient determination was formed for cubic sample. The optimal parameters for sample size and thermal flows were calculated in λ range 0.04-5 W/(m K).

This work was supported by Russian Science Foundation (project No. 16-19-10316).

References

1. K.V. Slyusarskiy, M.K. Bejsekov, J.V. Marysheva, Y.J. Rakov, MATEC Web Conf. **37**, 01053 (2015)
2. T.E. Shoup, *Applied numerical methods for microcomputers* (Higher School, Moskow, 1984)
3. E.P. Mikhalev, Yu. Ya. Rakov, 9 Int Sci Pract Conf Stud Post-Grad Young Sci. **1**, 56 (2003)
4. A.S. Zavorin, A.V. Kuzmin, Y.Y. Rakov, *Methods for determination the condensed matter thermal conductivity: tutorial* (Publishing house TPU, Tomsk, 2009)
5. D.D. McCracken, W.S. Dorn, *Numerical methods and Fortran programming with application in engineering and science* (Business Machines Corporation: Wiley international edition, 1965)

ECONOMIC EFFICIENCY OF A GAS-TURBINE TOPPING FOR STEAM REHEATING AT HEATING TURBO-INSTALLATIONS

Olga Romashova[1*], *Leonid* Belyaev[1], *Aleksandr* Tubolev[1], and *Egor* Skrebatun[1]

[1]Tomsk Polytechnic University, 634050, Tomsk, Russia

Abstract. The article considers the ways to improve the operating efficiency of conventional combined heat power plants using gas-turbine topping scheme allowing heat utilization of flue gases for steam reheating in the waste-heat steam boiler. The calculating analysis was performed for a steam-turbine power generating unit with a turbine of the T-110/120-130 brand name intended for various modes of operation.

1 Introduction

At the present stage of power engineering development the replacement of steam power cycle by combined one is one of the most efficient ways to improve performance of gas turbine thermal power stations. The coefficient of efficiency of combined cycle technologies is about 60-63 % and allows saving 10 % of fuel.

This work considers the ways to improve the operating efficiency of conventional combined heat power plants using gas-turbine topping scheme allowing utilizing heat of flue gases for steam reheating in the waste-heat steam boiler (WHB). The aim of this research is to estimate thermal effectiveness of steam reheating input in the basic cycle of a steam-turbine plant intended for various modes of operation of cogeneration turbine.

It is well-known that the use of conventional steam reheating in the combined heat power plant scheme leads to the increase of ventilating steam passage in the low-pressure section and thermal efficiency loss when operating in accordance with thermal schedule. Therefore, all cogeneration turbines with subcritical parameters of direct steam are designed without steam reheating. However, the rejection to use steam reheating leads to the decrease of thermal efficiency in the condensing mode on 3-4 % due to the decrease of steam specific work in the turbine and does not allow the turbines of T-110-130, T-175-130 types competing with condensing units with similar initial parameters (K-210-130, K-160-130) during the summer.

2 Statement of problem and limitations

The calculating analysis was performed for a steam-turbine power generating unit with a turbine of the T-110/120-130 type included in the steam-gas plant of heat recovery type.

* Corresponding author: roma@tpu.ru

The schematic diagram of a steam-gas plant is illustrated in Figure 1. Steam wasted in a high-pressure cylinder of the T-110 turbine flows in the waste-heat steam boiler of a gas turbine. The remained portion of gas heat is utilized in low-pressure and high-pressure regeneration systems of a steam turbine (FWGH – feed water gas heater, CGH – condensate gas heater). T,Q-diagram of the waste-heat steam boiler is shown in Figure 2.

Calculations were made for several modes of operation of the T-110-130 turbine: condensing mode by maximal possible volume steam flow rate in the condenser (mode I); in accordance with heating schedule with nominal heat supply from the turbine extraction (mode II); heat-extraction mode with single-stage preheating of heating-system water at minimal steam pressure in the controlled extraction and limited steam consumption at head of a steam-turbine (mode III).

Operation parameters of a gas-turbine part were defined from condensing mode and are invariable in heat-extraction modes. Temperature differences for some certain surfaces of the waste-heat steam boiler were estimated on [1]; waste-gas temperatures are accepted as θ_w^{WHB} = 100 °C. In all modes the initial and final steam conditions are invariable P_0 = 12.8 MPa, t_0=555 °C, P_k=0.0057 MPa. The condensing mode guaranteed by the manufacturing plant was taken as the design condition (initial parameters) for the T-110 turbine [2].

The following limitations were considered in the course of investigation [2, 3]:

- volumetric steam flow in the condenser $G_k \cdot v_k$ must not exceed the maximal-possible value, which defines geometrical dimensions of the final turbine stage. The input of steam reheating causes the increase of specific steam volume after steam reheating, which requires the reduction of mass steam flow consumption alongside the air-gas channel;
- pressure drop in chambers of controlled extraction leads to the electric power limitations (and main steam rate) due to the increase of bending stress in the blades of pre-extraction stages;
- vent steam flow for cooling of the low-pressure section by turbine operation in accordance to heating schedule depends on the temperature of steam coolant. The input of steam reheating leads to increase of steam temperature in the chambers of heat extraction and under other equal conditions requires the increase of vent steam discharge.

In addition, to ensure the turbine reliability by the input of steam reheating the strength calculation of the medium pressure section as well as the replacement of the material (steel) of air-gas channel components are required. The authors do not solve this problem in the framework of this work.

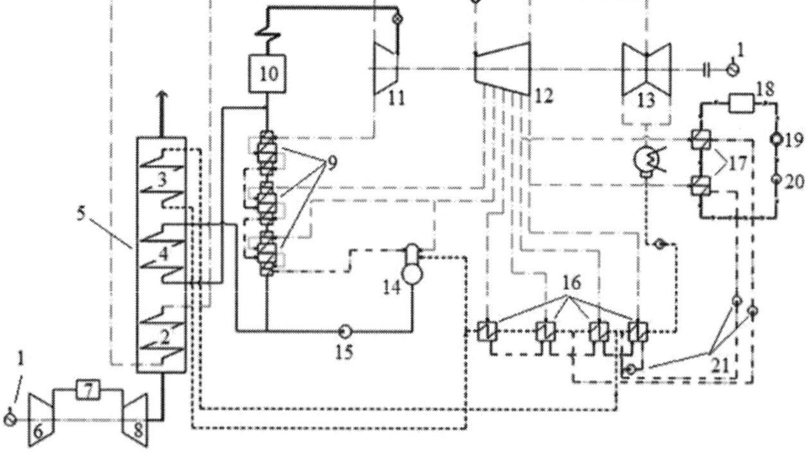

DOI: 10.1051/matecconf/20179101009

Fig. 1. Schematic thermal diagram of the steam-gas plant: 1 – electric generator; 2 – steam reheater; 3 – gas condensate heater; 4 – gas feed water heater; 5 – waste-heat steam boiler; 6 – compressor; 7 – combustion chamber; 8 – gas turbine ; 9 – high pressure heater; 10 – boiler; 11 – high pressure cylinder; 12 – middle pressure cylinder; 13 – low pressure cylinder; 14 – deaerator; 15 – feed-pump; 16 – low- pressure heater; 17 – heating-system water heater; 18 – peak water-heating boiler; 19 – heat consumer; 20 – heating water pump; 21 – drainline pump.

3 Efficiency calculation of steam reheating input in the thermal scheme of turbo-installation of T-110/120-130 type

The calculation of thermal scheme on condensing mode was made by the variation of steam temperature in the hot leg of steam reheating. Constant volume steam flow in the condenser

$$G_k \cdot v_k = const \tag{1}$$

is considered as the basic limit in this mode. The maximal steam flow rate from the turbine in the basic scheme is assumed to be equal to G_k = 301 t/h. The results of calculation of condensing mode are shown in Figure 2. The most efficient is the gas-turbine topping with the temperature of steam reheating which is equal to live steam temperature 555 °C.

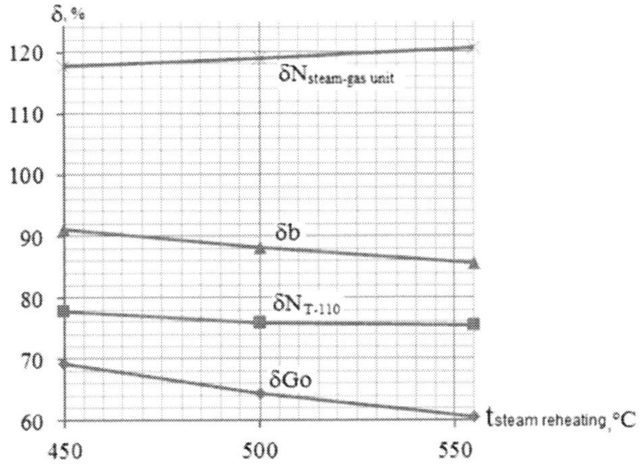

Fig. 2. The results of the calculation of the condensing mode of the T-110-130 turbine.

Therefore, steam consumption and its temperature in the inlet in WHB in the investigations of heat-extraction mode are taken from this scheme. Steam reheating till temperature of 555 °C leads to the reduction of specific reference fuel consumption on power generation on 14.3 %. The total capacity of the combined cycle gas turbine increases by 20 % compared with the initial scheme, in spite of considerable electric power decrease (about 25 %) of a steam turbine.

The impact of steam reheating on the indices of plant operation in heat-extraction modes depends on the parameters of the heat supply system and consumer's load. For the mode II calculations were made by thermal load of extraction Q_T=203 megawatt and in the range from 0.098 MPa to 0.25 MPa of steam variation in the upper controlled extraction. Steam temperature after steam reheating is calculated in iteration cycle depending on steam consumption and steam temperature in the outlet of a high pressure cylinder and its rate is 450-470 °C. Due to the decreased temperature of steam reheating compared with condensing mode and lower ratio of traditional regeneration of feed water in high and low

 DOI: 10.1051/matecconf/20179101009

pressure heaters, the volumetric steam consumption alongside the air-gas channel in these modes is lower than maximal possible values.

The studies show that the steam reheating by the operation of T-110-130 turbine on the thermal schedule decreases the electric power of a steam turbine on 11-13 % compared with its reference value due to the decrease of steam consumption in regenerative extraction by the increase of total capacity of the steam-gas plant on 39-44 % and almost invariable thermal efficiency.

For the mode III in accordance to [3] the electric power of a steam turbine is limited by the value N_E=80 MW, thermal load is equal to Q_T=100 MW by the pressure drop till minimal value of 0.03 MPa in the lower heat extraction. Steam consumption on turbine is calculated for the initial basic scheme without steam reheating and is constant after the introduction of steam reheating (100 kg /sec.). Since the main steam rate is considerably lower than maximal rate on turbine, steam temperature after steam reheating reaches its nominal value and gas reheating of feed water fully displaces the regeneration high pressure system (high pressure reheaters are disconnected) by the increase of feed water temperature. The highest electric power increment compared with the initial scheme was obtained in this mode. The capacity of a steam turbine increases from 80 MW to 100 MW, specific reference fuel consumption on power generation decreases till 25 % in spite of the increase of steam flow in the condenser.

Figure 3 and 4 illustrate the main results of calculations of considered modes.

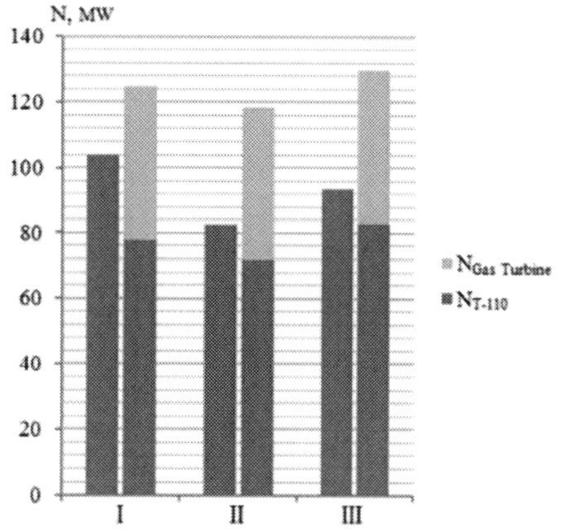

Fig. 3. The variation of the T-110-130 turbine power and steam-gas plant relative to the reference conditions of a steam turbine.

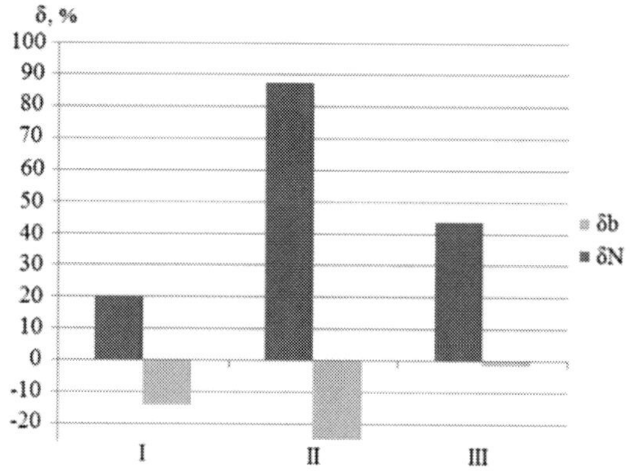

Fig. 4. The curves of electric power increment and fuel saving for a steam-gas plant relative to the reference conditions of T-110-130.

4 Conclusion

To sum up:

- in summer period when turbine of the T-110-130 type operates in the condensing mode the gas-turbine topping for steam reheating in the waste-heat steam boiler is able to reduce specific equivalent fuel consumption almost on 15 % by simultaneous increase of electric power and steam turbine output on 25 %
- the use of steam reheating by T-110-130 operation in accordance with the heating schedule with nominal heat output to the external consumers does not cause any changes of thermal effectiveness of the plant, however, the electric capacity of this plant can be increased on 39-44 % compared with the initial design of the combined-cycle plant depending on the pressure in extraction;

- maximal effect of the reheating input at cogeneration turbine becomes apparent for heat-extraction mode with minimal pressure in controlled extraction by limitation of steam flow at head of turbine. Electric power in a steam turbine recovers till rated power by the reduced flow consumption of the direct steam, while the total capacity of steam and gas part doubles compared to the initial scheme by the increase of thermal efficiency on 25 %.
- The introduction of steam reheating leads to the increase of reliability of a steam generating unit as well as live steam pipes and steam inlet elements of the turbine due to the consumption reduction of direct steam.

References

1. Tsanev S.V. *Gas-turbine and steam-turbine installations of thermal power plants: Textbook for institutions of higher education* (Publishing House MEI, Moscow, 2006)
2. Trukhniy A.D. *Heat steam turbines and turbine plants: [textbook for institutions of higher education* (Publishing House MEI, Moscow, 2006)
3. Instruction manual of heating unit of TG-8 brand name of Tomsk Regional Power Plant-2

MATEC Web of Conferences **91**, 01010 (2017) DOI: 10.1051/matecconf/20179101010
Smart Grids 2016

INDICATORS THE TIME PRODUCTION CYCLE TURBINE

*Lyudmila*V. Savostyanova[1,*] and *Valeriy*V. Litvak[1]

[1]Tomsk Polytechnic University, 364050, Tomsk, Russia

Abstract. A study of repair documentation turbines with a capacity of 25-150 MW and in operation for at least 20 years old to obtain statistical data on the indicators the time production cycle turbines. Keyed indicators differ from those that are used to select the duration of the overhaul time at the moment. The stated approach allows to predict an individual resource turbines on the analysis of the calculations carried out using the repair documentation.

1 Introduction

The purpose of work is to provide a method for forecasting the individual re-resources turbines, spent significant periods of time and the justification of the recommendations of on the volume and terms of repair and maintenance of turbines, preserving a certain number of defects in those or other sites, and that accumulate due to significant periods operation starts and the number of turbines [1].

Information about repair and operational characteristics extracted from the documents submitted by power plants, which employ condensation, heat supply and backpressure turbine capacity from 25 to 180 MW. The paper presents the materials carried out studies of the effect of the production cycle characteristics of an individual resource turbine T-100-130 TMZ.

2 The study

The study of individual life, that is, the total time from the moment the object control its technical condition before it transitions to the limit state, the test turbine is of considerable interest because fleet renewal is going very slowly. Existing turbines have significant wear and periodic scheduled maintenance do not provide a complete replacement of worn nodes and continue to carry the load.

The production cycle turbine is called a calendar duration of the maintenance period, from start to work after the end of the previous overhaul before the end of the subsequent planned overhaul.

* Corresponding author: savost@tpu.ru

The time during which the turbine plant has a load is a major techno logical process, and regardless of the load, will be referred to hereinafter - operating time. "Reserve" turbine mode is provided for cases when turbine for operations schedule is waiting and ready to receive the load (after the start-up operations). "Repair" mode provides for the implementation of planned and unplanned repairs.

For the turbine T-100-130 average cycle time was 20761 hours (865 days), and the structure is shown in Figure 1.

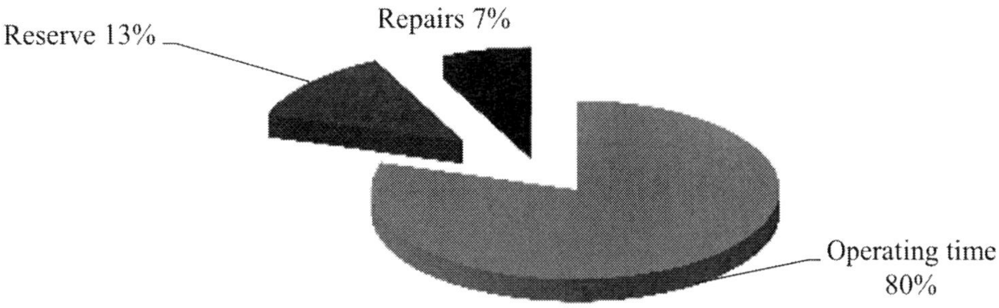

Fig. 1. Structure of the production cycle turbine T-100-130 TMZ.

The structure of the production cycle turbine units operating at different power plants, practically does not differ from each other. This is due to the unity of repair and maintenance of policies and coincidence of dispatching management approaches [2].

Interest in the study of the production cycle steam turbine equipment structure is based on the fact that the ratio of operating time, and the duration of the repair reserve determine the basic characteristics of reliability - the probability uptime, uptime, time to failure, the assigned resource availability and others.

In this regard, it is important to know the extent to which the production cycle parameters change during the life of the turbine [3]. Since turbine T-100-130 TMZ, was commissioned in 1967 for two different decades showed different characteristics of the production cycle in relation to the average calendar pro-duration cycle (Table 1).

Table 1. Characteristics of industrial turbine T 100-130 cycles for two different decades.

Index	**The period from 1968 to 1978**		**The period from 1989 to 1999**	
	Number of hours	**%**	**Number of hours**	**%**
Operating time total	84089	–	71344	–
The average operating time per cycle	28029	85	14268	74
The average duration of the repair	2337	7	2356	12
The average duration of the reserve	2549	8	2652	14

These changes are caused not only by the state of turbine performance, but also the known changes in management and market reforms in the energy sector.

The average duration 20761 h turbine cycles and repair downtime – 1495 hours. This means that the individual characteristics of the turbine - components with defects, the rate of growth of defects, the duration of the turnaround time, corrective maintenance programs and other factors play a more significant role in ensuring the efficiency, than park. It detects

DOI: 10.1051/matecconf/20179101010

and confirms the fact that defects arise and grow in the turbine sites differently [3]. This means that the period of the rise of the defect is mean time between failures.

Therefore, it is advisable to form a power information database for repair and maintenance services – analogue of the "Diagnostic cards" and the repair form. This electronic document allows more informed decisions about extending the operating turbine. On the duration of the overhaul time turbines, among others, at the rank, seriously affects the number of stops and starts [4]. Figure 2 shows the distribution number of starts turbine T-100-130 from 1978 to 2011.

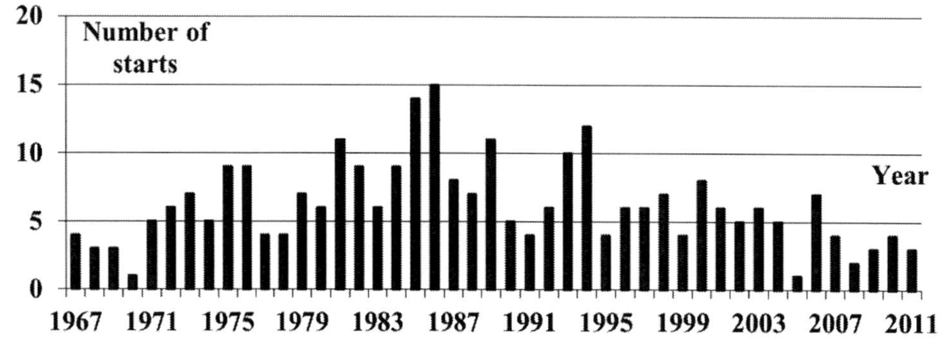

Fig. 2. Distribution of the turbine starts T-100-130, which has an operating time 316 478 h.

The average number of starts per year for the operation of the turbine period was 6. With an average cycle time 20761 h average number of starts per cycle is 13. In the assessment of individual performance reliability turbine should consider not only the achievements, but also the number of starts. Individual statistical characteristics can be easily converted into the characteristics of reliability, in this case, the calculated reliability indices for the period of 1 year [1]:

- probability of failure of 0.6;
- availability factor of 0.91;
- probable uptime after a repair 13302 hours.

On the basis of information on the repair and maintenance history for each turbine can be created "diagnostic map" in the form of an electronic database on which you can restore genuine individual maintenance portrait. This makes it possible to shape the flow of data across nodes turbine, type of defects, defect rates rise, time and other technological features. Accordingly, the reliability of the characteristics can be calculated for the individual components and turbine units. In the first place – the likely uptime of each node.

The repair history of each turbine can be recreated at each plant based on the forms of repair, defective statements, customized documents in the form of an information base.

3 Conclusion

1. This paper discusses the characteristics of the production cycle turbine, which are very dependent on the quality and quantity of repairs, as well as the consumer demand for heat and electricity in a calendar year.
2. It was established experimentally that the average time between overhauls turbine was 15369 hours, while the regulatory overhaul life for the considered turbine is 34000 hours [5].
3. Reduction of disposable resource turbine associated with a long service life and approaches the limit of the individual nodes, requires re-turn to the individual

DOI: 10.1051/matecconf/20179101010

planning repair procedures. This can be done only on the basis of the repair history for the entire period of operation.

References

1. L.V. Savostyanova, V. V. Litvak, Bulletin of the Tomsk Polytechnic University 321, 11 (2012)
2. V.V. Litvak, V.F. Panin. *The reliability of heat and power equipment and ecological environment around the TPP* (Publishing house of the YTL, Tomsk, 2009) [in Russia]
3. Rodin V.N., Sharapov A.G., Murmanskiy B.E. et al. *Repair of steam turbines* (Publishing House of the Ural State Technical University, Ekaterinburg, 2002)
4. Rezina V.F., Gladshtein V.I., Avrutsky G.D. *Increased resource continuously operating turbines* (Publishing House MEI, Moscow, 2007)
5. *SO 34.04.181-2003. Rules of the organization of maintenance and repair of equipment, buildings and facilities of power stations and networks* (Moscow, 2004) [in Russia]

PHYSICAL-MATHEMATICAL MODEL FOR FIXED-BED SOLID FUEL GASIFICATION PROCESS SIMULATION

Konstantin V. Slyusarskiy[1,*], *Alexander* G. Korotkikh[1], and *Ivan* V. Sorokin[1]

[1]Tomsk Polytechnic University, 634050, Tomsk, Russia

Abstract. Phycial-mathmatical model for fixed-bed coal gasification process simulation is proposed. The heterogeneous carbon oxidation chemical reactions were simulated via Arrhenius equation while homogeneous reactions in gas phase were calculated using Gibbs free energy minimization procedure. The syngas component concentration field and fuel conversion distribution as well as syngas final temperature and composition were defined for fixed bed gasification of T-grade coal of Kuznetskiy deposit. The optimal fuel residence time and gasifyer specific productivity were defined. The prevail reactions in oxidizing and reduction zones together with its height were defined.

1 Introduction

Solid fuel is one of major sources of primary energy in the world nowadays according to BP data [1] and provides nearly 30 % of world generation. While energy consumption is expected to increase by 35 % to 2035 year the share of coal generation is expected to be significant and has a value close to 25 % [1]. Such decrease is seem to appear due to economic and ecological issues connected to coal usage like carbon, sulfur and nitrogen oxide emissions, lower efficiency compared to combined cycle for natural gas and solid particles in flue gases [2].

The solid fuel gasification may solve both these problems. Syngas may be used in combined cycle. Gasification process features allow to decrease nitrogen oxide formation intensity and apply modern gas-cleaning technics [3] which are inefficient with existing direct combustion technology.

Gasification – process of turning fixed carbon from coal into combustible gas – include many chemical reactions and physical processes. That makes it quite difficult to define process outcome at different expoitational conditions. To resolve this problem the physical-mathematical model with macrokinetic limitations was proposed. Based on it the gasification process characteristics were calculated for T-grade bituminous coal of Kuznetskiy deposit.

2 Material characterization

* Corresponding author: konstantinsv@tpu.ru

The T-grade bituminous coal was used to simulate fixed bed coal gasification process. The fuel proximate analysis was carried out using ISO 17246:2010 [4]. The moisture, ash, volatile matter and fixed carbon content values were defined. Ultimate analysis was carried out via Euro EA 3000 CHNOS analyzer. Results presented in Table 1.

Table 1. Proximate and ultimate analysis results for T-grade bituminous coal.

Proximate analysis, wt. %				Ultimate analysis, wt. %				
Moisture	**Ash**	**Volatiles**	**Fixed carbon**	**C**	**H**	**N**	**S**	**O**
3	21	20	66	68	4	3	<1	4

Kinetic characteristics of heterogeneous carbon oxidation reactions for T-grade mark coal were taken from literature data [5]. The frequency factor and activation energy values for reactions of carbon interreacting with oxygen and carbon dioxide as the most intense reactions are given in table 2.

Table 2. Kinetic constant values for heterogeneous chemical reactions.

Химическая реакция	**$C+O_2=CO_2$**	**$C+CO_2=2CO$**
Pre-exponential factor A_0 , 1/c	211	71230
Energy of activation E_a , кДж/моль	21	156

3 Model statement

The fixed bed counter-flow gasification process was modeled. Reactor chamber was presented as a set of consequently connected cylindrical domains with equal height and fuel residence time. The temperature distribution in each domain was considered to be uniform. The rate of heterogeneous chemical reactions were simulated via Arrhenius equation. Considering fuel residence time in each domain to be marked as τ, the total amount of turned into gas phase carbon could be calculated as:

$$w_i = A_0 \exp\left(\frac{-E_a}{R \cdot T_i}\right)(1-X_i)\tau[OX] \tag{1}$$

there R – gas constant, J/(mole K); T_i – temperature in domain i, K; X_i – fuel conversion value; τ – fuel residence time, s; $[OX]$ – oxidizer molar concentration. Conversion value could be calculated as:

$$X_i = \frac{m_i - m_a}{m_0 - m_a} = X_{i-1} - w_i \tag{2}$$

here m_i – fuel mass after i^{th} domain, kg; m_a – ash mass, kg; m_0 – initial fuel mass, kg.

Because next conversion value depends on previous, to increase calculation preciseness the equation 1 was integrated by time:

$$\int_{X_i}^{X_{i+1}} dX = \int_{t_i}^{t_{i+1}} A_0 \exp\left(\frac{-E_a}{R \cdot T}\right)(1-X)dt \tag{3}$$

there t is time moment, s.

Considering $t_{i+1} - t_i = \tau$ analytical solution of equation 3 will take the following form:

$$X_{i+1} = 1 - (1 - X_i)\exp\left(-A_0 \exp\left(\frac{-E_a}{R \cdot T}\right)\tau\right) \tag{4}$$

The gas-phase composition was defined by Gibbs free energy minimization procedure with assumption that gaseous mixture in each domain exists in equilibrium by following equation [6]:

$$G = RT\sum_{i=1}^{m} n_i \left((\mu_i / RT)_i + \ln\left(n_i / \sum_{i=1} n_i \right) \right), \tag{5}$$

there n_i – molar amount of ith substance, mole; μ_i – molar potential of ith substance, J/mole, which could be defined as follows:

$$\mu_i = h_i - s_i T + RT\ln(p / p_0), \tag{6}$$

there h_i, s_i – enthalpy and entropy of formation of ith substance at temperature T, J/mole, J/(mole K); p_0 and p – ambient and current pressure in system, Pa.

To follow material balance of system with N substances and M types of atoms the possible solution domain was limited by following set of equations:

$$\sum_{i \in N} a_{ij} n_i = b_j, \ j \in M, \ n_i > 0, \tag{7}$$

there a_{ij} – stoichiometric coefficient (number of *j*-type atoms in molecule *i*); b_j – total amount of *j*-type atom in system, mole.

The Gibbs free energy minimization procedure was realized using Lagrange multipliers via steepest descent methods. As long as presence of components with close to zero content has a negative impact on calculation preciseness the number of taken into account substances was decreased to six following: H_2O, CO, CO_2, H_2, CH_4 and O_2. Nitrogen content in air and inorganic matter content in fuel were considered to be chemically passive and were taken into account only in equation of thermal balance.

To define thermal effect of turning system from initial condition into final the Hess law was used:

$$\Delta H_r = \sum n_i^{pr} H_{fi}^{pr} - \sum n_i^{f} H_{fi}^{f}, \tag{8}$$

there H_{fi} – enthalpy of formation of ith substance, J/mole; «*pr*» and «*f*» indexes refer to reaction products and feed, respectively .

Thermal effect of process could be defined as follows:

$$Q = \Delta H_r + \sum H_i^{f} m_i - \sum H_i^{pr} m_i, \tag{9}$$

there H_i – i^{th} substance heat of formation, J/kg; m_i – mass of ith substance, kg.

3 Simulation results

Using proposed model the fixed bed gasification process characteristics were defined for T-grade bituminous coal. The initial conditions were 400 °C for gaseous feed and 20 °C for fuel. The coal/steam/air ratio was 1/1/1.5 according to [7]. The residence time of fuel in gasifyer was chosen to supply final conversion 0.9 and was defined to equal to 0.15 min. The temperature and conversion fields were obtained (fig. 1) as well as substance concentration distribution (fig. 2) over reactor height.

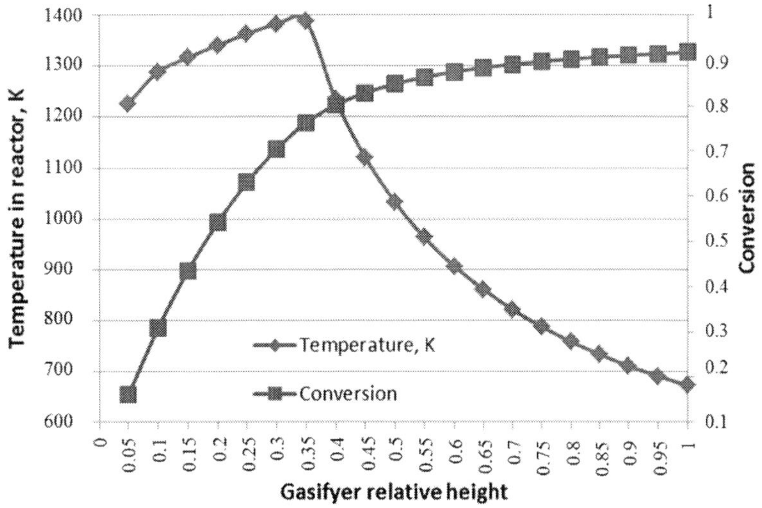

Fig. 1. Syngas temperature and fuel conversion distribution over reactor height.

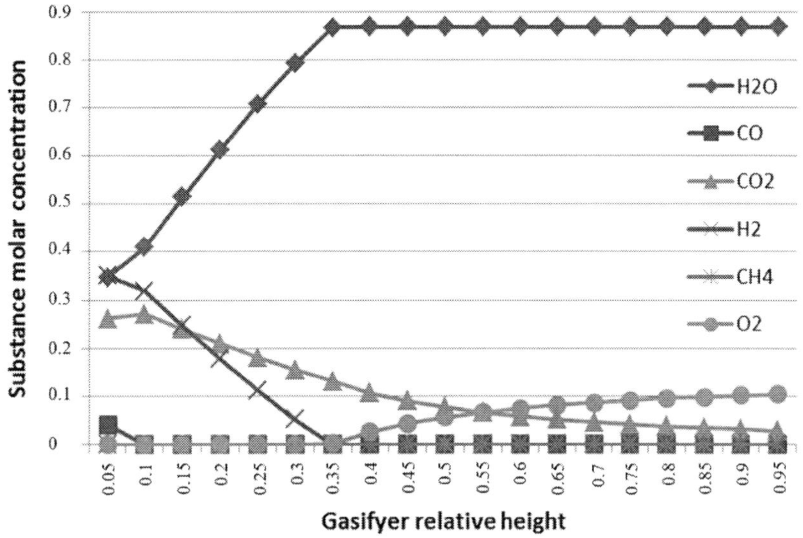

Fig. 2. Syngas component molar concentration over reactor height.

Graph analysis allows to clearly distinguish two reaction zones. The first is characterized by carbon complete oxidation reaction prevailing:

$$C+O_2=CO_2. \quad (1)$$

Other substances remain inactive – their concentration does not change. It begins at the 0.35 of gasifyer relative height and lasts till the end of process. It is characterized by rapidly increasing temperature and slowly increasing conversion growth rate. It is connected to high temperatures and low conversion values at the beginning of this zone while the average temperature in this phase is quite low. The syngas calorific value does not change and stays equal to zero.

The second phase is connected to water-gas shift reaction:

$$CO+H_2O=CO_2+H_2. \quad (2)$$

As long as this chemical reaction endothermic the temperature decreases during this stage. The carbon dioxide content is keep growing together with hydrogen content. At the very end of this stage the signs of carbon heterogeneous oxidation by water appears:

$$C+H_2O=CO+H_2. \quad (3)$$

This resulted into more intense temperature drop and calorific value increasing. The methane content values over the reaction zone was small and could be not taken into account in further calculations.

4 Conclusion

Equilibrium thermodynamics and heterogeneous chemical kinetics combination were used to simulate counter-flow fixed bed gasification process for T-grade bituminous coal of Kuznetskiy deposit. The optimal fuel residence time was defined as well as syngas composition and final temperature. Based on temperature distribution and syngas concentration fields over the reactor height the oxidation and reduction zones were distinguished. Prevailing chemical reaction on each step were defined.

References

1. *British Petroleum Statistical review of World Energy* (Pureprint Group Limited, Uckfield, (2016)
2. A. Zenkov, S. Yankovsky, A. Matveeva, S. Lavrinenko, A. Gromov, MATC Web of Conf. **72**, 01131 (2016)
3. K. Larionov, I. Korjashov, S. Jankovskij, V. Gubin, A. Zenkov, MATEC Web of Conf. **37**, 01030 (2015)
4. ISO 17246:2010 (2010)
5. A. Korotkikh, K. Slyusarskiy, A. Ditts, Rus. J. of Phys. Chem. B. **10**, 576 (2016)
6. D. Svishchev, A. Kozlov, I. Donskoy, A. Ryzhkov, Fuel **168**, 91 (2016)
7. A. Korotkikh, K. Slyusarskiy, I. Sorokin, MATC Web of Conf. **72**, 01104 (2016)

RESEARCH OF MECHANICAL TREATMENT INFLUENCE ON RHEOLOGICAL PROPERTIES OF COAL-WATER FUEL BASED ON LOW-GRADE 3B COAL

A. A. Tolokolnikov [1], S. A. Yankovsky[1,*], V. E. Gubin V.E.[1], A. V. Zenkov [1] and A. S. Zaitsev[1]

[1]Tomsk Polytechnic University, 634050, Tomsk, Russia

Abstract. Experimental study of mechanical treatment effect of rotary flow modulation device on coal-water fuel rheological properties were conducted. The slurry was prepared on the basis of low-grade 3B coal from Balakhtinskoe deposit.

1 Introduction

Energy Development Strategy of Russia [1] involves the widespread use of solid fuels in energy production by direct combustion at thermal power plants. The main consumer of high-grade coal is the steel industry due to its technological features at the expense of other energy objects [2-5].

It is known [2] that the variety of thermal power stations operate on non-projected fuel, often with low thermal parameters, which leads to serious problems, especially during peak load of boilers [2-5].

At nowadays, there is a prospective technology of low-grade coal application in energy sector [6], so called coal-water fuel [6]. This technology represents composite dispersed system consisting of a solid phase in the form of finely divided coal, liquid medium (water) and plasticizing agent [6]. Research in this area is urgent due to the fact that there is no scientifically sound procedural framework describing physical and chemical processes in CWF.

2 Experimental setup and study technique

Experimental studies were conducted on 3B brown coal from "Balakhtinskoe" deposit.

The initial coal samples were sieved through a sieve with a mesh size of 80 microns after primary treatment. After that, the obtained material was milled in a ceramic drum of 2 liter volume with distilled water and grinding bodies for 1.5 and 9 hours. The weight ratio of researched coal and water are shown in Table 1. The obtained slurry was subjected to further treatment in a rotary flows modulation device (RFMD) after reaching a given

* Corresponding author: Jankovsky@tpu.ru

 DOI: 10.1051/matecconf/20179101012

milling time. RFMD was included in the CWF preparation scheme to improve its rheological properties. Density and dynamic viscosity were determined after the treatment by Brookfield viscometer. Experiments to determine fuel sedimentation stability after 24 hours were performed with the use of beakers.

The percent-mass concentrations of the examined CWF samples are presented in Table 2.

Table 2. Examined CWF compositions.

Sample name	Mass content of coal, %	Mass content of water, %
3B grade coal	48	52
	45	55
	42	58

3 Experimental results

As a result of a series of experiments, the dependencies presented in Figures 1-2 were obtained.

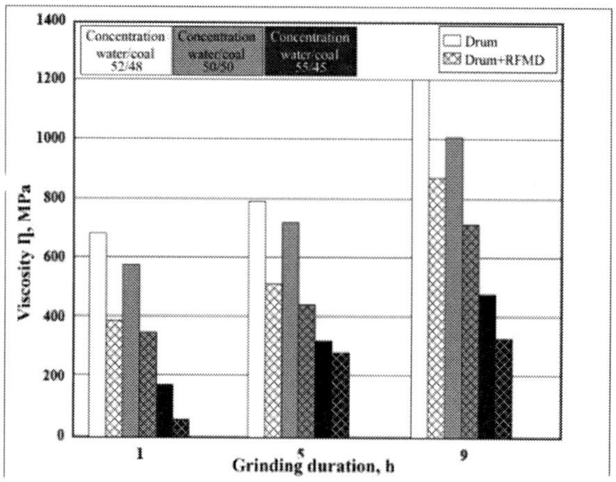

Fig. 1. Dependence of CWF viscosity based on 3B coal from grinding duration and concentration composition at a given sliding velocity of viscometer.

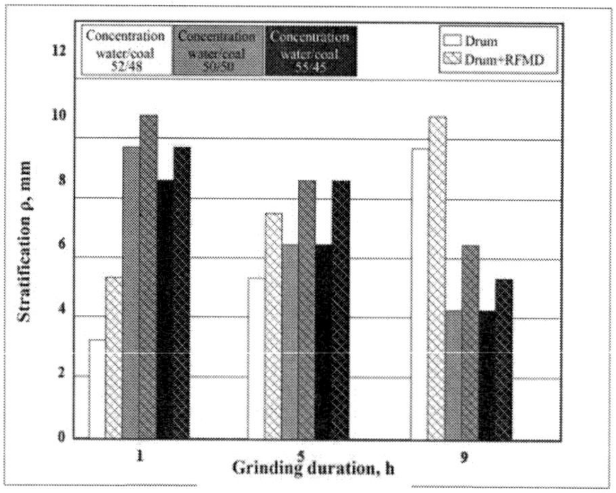

Fig. 2. Effect of grinding time on CWF sedimentation stability for 3B grade coal.

Analysis of the obtained results showed a decrease in samples viscosity due to the dispersion of the solid part in coal-water fuel because of the acoustic, hydrodynamic and mechanical impact of the rotary flows modulation device.

CWF sedimentation stability is worsened as a consequence of more active particulate solids precipitation than before mechanical treatment.

4 Conclusion

Results of the conducted research revealed that the use of rotary flows modulation device reduces coal-water fuel viscosity and increases its sedimentation stability.

The work has been supported by the Russian Federal focused program "Research and development on priority orientation of science and technology in Russia 2014–2020", unique identifier 'Applied Research and Experimental Developments' project RFMEFI58114X0001.

References

1. B. Vorobyev, Herald of the Russian Academy of Sciences **1**, 65 (2011)
2. T. Taylasheva, A. Gil, E. Vorontsova, Bulletin of the Tomsk Polytechnic University **327**, 128 (2016)
3. M.P. Baranova, V.A. Kulagin. *Fiziko-himicheskie osnovi polucheniya toplivnih vodougolnih suspenziy: monografiya* (Sibirskiy federalniy universitet, Krasnoyarsk, 2011) [in Russia]
4. A.G. Korotkikh, K.V. Slyusarskiy, MATEC Web of Conf. **23**, 01020 (2015)
5. S. Jankovskiy, D. Luzhkovoj, K. Larionov, A. Matveeva, MATEC Web of Conf. **37**, 01026 (2015)
6. A.V. Zenkov, Yankovsky, S.A. Matveeva, S.V. Lavrinenko, A.A. Gromov, MATEC Web of Conf. **72**, 01131 (2016)

MATEC Web of Conferences 91, 01013 (2017)
Smart Grids 2016
DOI: 10.1051/matecconf/20179101013

THE ANALYSIS OF THERMAL EFFICIENCY OF CCGT WITH BACK- PRESSURE VIA THE METHODS OF SMALL-DEFLECTION

Artem A. Uvarov[1] and *Aleksandra* M. Antonova[1,*]

[1]Tomsk Polytechnic University, 634050, Tomsk, Russia

Abstract. Presents the results of studying the influence of efficiency gas turbine and steam section of CCGT on the thermal efficiency of combined-cycle plant with back-pressure based on a powerful high-temperature gas turbine. To assess the impact of selected parameters was made analysis by method of small-deflection. The advantages of one-circuit combined-cycle in comparison with more complex schemes. According to the research results, have been made corresponding conclusions.

1 Introduction

Combined Cycle Gas Turbines (CCGT) that allow significant increase of efficiency and biological safety [1, 2] of thermal power engineering have been actively used in world's power engineering in recent years. Binary CCGTs of utilization type on the basis of modern high-temperature gas turbines (GT) proved to be the most efficient [3, 4]. These CCGTs surpass power units in terms of thermodynamic efficiency and ecological factors. The endeavors the increase CCGT efficiency call for the development of units on cogeneration steam cycle with back-pressure steam turbine (CCGT DH) applied [5, 6]. These units are the object of the given research.

2 Thermal efficiency of CCGT DH

In CCGT DH, utilization of heat of exhaust GT gasses in heat recovery steam generator (HRSG) is done on two segments of heating cycle, i.e. with the resulting overheated steam and with heating reverse DH system water in tail surface of HRSG. The heat of exhaust steam is utilized with its transition to the consumer of technological steam or district heating system water fed to the direct network.

The CCGT DH diagram is rather simple as there is no need to use several levels of steam pressure and steam reheating. Fig. 1 shows the diagram of CCGT-DH with district heater (DH) providing the customers with heat and hot water.

The increase of HRSG efficiency and the unit as a whole is realized through deep utilization of heat via decrease of temperature of exhaust gasses while transmitting heat to the reverse DH system water in gas-water district heater (GWDH). Temperature mode of

* Corresponding author: anton@tpu.ru

the heating system is a key one for the characteristics of CCGT DH steam section as the temperature of DH system water defines the pressure of exhaust steam at the exhaust of gas turbine

Thermal efficiency of CCGT DH can be assessed with so-called total efficiency of CCGT indicating the share of heat of fuel beneficially used in the unit. Beneficially used heat encompasses electric power of CCGT N_E and thermal power transmitted to heat consumers Q_{HC}.

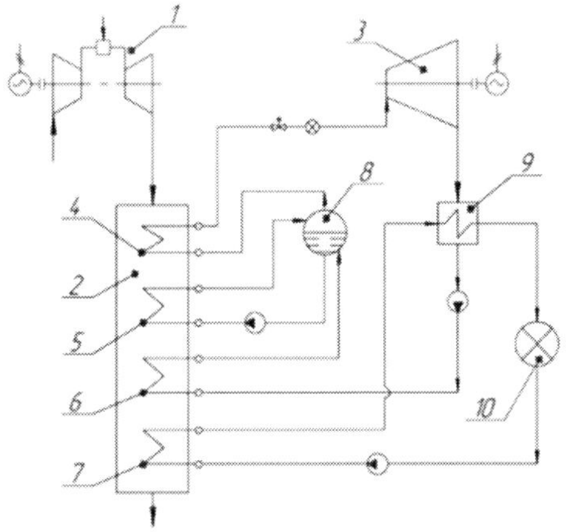

Fig. 1. The diagram of CCGT-DH with back-pressure steam turbine: 1 – gas turbine; 2 – heat recovery steam generator; 3 – steam turbine; 4 – superheater; 5 – evaporator; 6 – economizer; 7 – GWDH; 8 – drum; 9 – DH; 10 – heat consumer.

The total efficiency of CCGT:

$$\eta_{CCGT} = \frac{N_E + Q_{HC}}{Q_{CC}} = \frac{N_E^{GT} + N_E^{ST} + Q_{HC}}{Q_{CC}} = \frac{N_E^{GT}}{Q_{CC}} + \frac{N_E^{GT} + Q_{HC}}{Q_{CC}}. \tag{1}$$

in which Q_{CC} is heat fed from fuel to working fluid in the combustion chamber of gas turbine; N_E^{GT}, N_E^{ST} – are electric powers of gas and steam turbines.

The energy balance of GT suggests: $Q_{CC} = N_E^{GT} + Q_{HRSG}^{H} + \Delta Q_{EG}$, in which ΔQ_{EG} is loss of heat with exhaust gasses from the heat recovery steam generator; Q_{HRSG}^{H} is useful output of heat recovery steam generator. Loss of heat with outgoing gasses ΔQ_{EG} can be defined using the equation for efficiency of heat recovery steam generator:

$$\eta_{HRSG} = \frac{Q_{HRSG}^{H}}{Q_{CC} - N_E^{GT}} = \frac{Q_{CC} - N_E^{GT} - \Delta Q_{EG}}{Q_{CC} - N_E^{GT}} \tag{2}$$

in which denominator stands for heat of exhaust gasses of GT, from which they are fed to HRSG. Useful output of heat recovery steam generator equals:

$$Q_{HRSG}^{H} = Q_{CC} - N_E^{GT} - (1 - \eta_{HRSG})(Q_{CC} - N_E^{GT}) = (Q_{CC} - N_E^{GT}) \cdot \eta_{HRSG}. \tag{3}$$

On the other hand, useful output of heat recovery steam generator is consumed for two components: electric power of steam section of CCGT N_E^{ST} and thermal power fed to the heat consumer Q_{HC}. Hence, obtain the total efficiency:

$$\eta_{CCGT} = \frac{N_E^{GT}}{Q_{CC}} + \frac{N_E^{ST} + Q_{HC}}{Q_{CC}} = \eta_{GT} + \frac{Q_{CC} \cdot (1-\eta_{GT}) \cdot \eta_{HRSG}}{Q_{CC}} = \eta_{GT} + (1-\eta_{GT}) \cdot \eta_{HRSG}. \quad (4)$$

This formula lacks efficiency of steam section of CCGT (SC) or any other factor characterizing its economy, i.e. thermal efficiency of CCGT DH does not depend on SC economy; it functions with no losses in the cycle: condensation heat of exhaust water steam is completely used for covering the load of heat consumers.

3 The analysis of efficiency via method of small-deflection

Method of small-deflection [7, 8] allows obtaining the solution in general and numerical type by using the results of calculations of a certain basic variant. The method is based on equations linearization. As a result, the initial analytical equation is replaced with the equation with small-deflection of the arguments comprised. The obtained evident analytical dependence binds the changes of initial variables with the changes of function, which enables to analyze the changes of function with the help of coefficients of variable values effect.

Following the method of small-deflection, differentiate the equation of efficiency of CCGT DH (4):

$$d\eta_{CCGT} = \frac{\partial \eta_{CCGT}}{\partial \eta_{GT}} d\eta_{GT} + \frac{\partial \eta_{CCGT}}{\partial \eta_{HRSG}} d\eta_{HRSG}. \quad (5)$$

Here, partial derivatives are: $\frac{\partial \eta_{CCGT}}{\partial \eta_{GT}} = 1-\eta_{HRSG}$; $\frac{\partial \eta_{CCGT}}{\partial \eta_{HRSG}} = 1-\eta_{GT}$. Substituting partial derivations in (5), we obtain: $d\eta_{CCGT} = (1-\eta_{HRSG})d\eta_{GT} + (1-\eta_{GT})d\eta_{HRSG}$. Let's first consider finite increment and then relative deviations:

$$\frac{\Delta\eta_{CCGT}}{\eta_{CCGT}} = \frac{1}{\eta_{CCGT}}\left[(1-\eta_{HRSG})\eta_{GT}\frac{\Delta\eta_{GT}}{\eta_{GT}} + (1-\eta_{GT})\eta_{HRSG}\frac{\Delta\eta_{HRSG}}{\eta_{HRSG}}\right].$$

Considering the indication of coefficient at variable, we obtain:

$$\delta\eta_{CCGT} = K_1\delta\eta_{GT} + K_2\delta\eta_{HRSG}. \quad (6)$$

Here, coefficients of effect K_1 and K_2 are defined through values of "basic" variant, having index "0" in the formulae:

$$K_1 = \frac{(1-\eta_{HRSG0})\eta_{GT0}}{\eta_{CCGT0}};\ K_2 = \frac{(1-\eta_{GT0})\eta_{GT0}}{\eta_{CCGT0}}.$$

CCGT DH, shown in figure 1, takes Siemens SGT5-4000F gas turbine as a basis. In nominal conditions at ambient temperature of minus 40 °C, taken as a basic variant, gas turbine has the following characteristics: electric efficiency – 35.37 %; electric capacity – 247.6 MW; heat fed to the combustion chamber with fuel – 700.1 MW; exhaust gas temperature – 594.8 °C; exhaust gas flow – 644.9 kg/s.

For the steam section of CCGT, values of indices in basic variant are determined by the balance sheet [8] and given in table 1. For the given values of variables of basic variant of CCGT DH, coefficients are as follows: K_1=0.0687, K_2=0.6050.

Table 1. Basic values of steam section of CCGT DH and efficiency of the unit in basic variant.

Parameter, dimension	Value
Initial pressure of the steam, MPa	5.0
Initial temperature of the steam, °C	540
Electric capacity, MW	68.51
Thermal power, MW	302,9
Temperature of the direct heating water, °C	110
Efficiency of HRSG	0.828
Total efficiency of CCGT	0.885

Taking into account the resulting values, we obtain dependency of efficiency of the analyzed CCGT DH on the factors in the form of equation in small-deflection:

$$\delta\eta_{CCGT} = 0.0687 \cdot \delta\eta_{GT} + 0.6050 \cdot \delta\eta_{HRSG}. \tag{7}$$

The analyses of the equation shows that efficiency of using heat of gas turbine exhaust gasses in heat recovery steam generator has a much stronger impact on the efficiency of using heat of fuel than efficiency of gas turbine. With 1 % change of η_{HRSG}, efficiency of CCGT η_{CCGT} changes on 0.6050 %. The increase of efficiency of gas turbine on 1 % leads to the increase efficiency of CCGT on 0.0687 %. Temperature of exhaust gasses from HRSG has the defining value. In CCGT DH, this temperature definitely depends on the temperature of reverse DH system water.

4 Conclusion

According to the research results, we can draw the following conclusions.

1. A simple schemes of CCGT with back-pressure steam turbine provides high reliability of CCGT due to single circuit and lack of steam reheating.
2. Total efficiency of CCGT DH does not depend on economy of steam section, as SC functions with no losses in the cycle.
3. Thermal efficiency of CCGT DH is mostly affected by the completeness of using of heat of gas turbine exhaust gasses in heat recovery steam generator.
4. To increase the efficiency of heat recovery steam generator and CCGT DH as a whole, a decrease of temperature of exhaust gasses within sustainable limits is needed. This is defined by the valid choice of temperature mode of the heating system.

References

1. D.Gvozdyakov, V.Gubin, Chem. and Pet. Eng. **51**, 201 (2015)
2. A.V. Vorobiev, E.M. Dutova, A.S. Latyshev, MATEC Web of Conf. **72**, 01125 (2016)
3. G. Olkhovsky, V. Trushechkin, Therm. Power Plants **1**, 2 (2013)
4. A. Uvarov, A. Antonova, A. Vorobjev, MATEC Web of Conf. **37**, 01062 (2015)
5. A. Gushchin, Turbiny i dizeli **7**, 16 (2007) [in Russia]
6. A. Trukhniy, *Parogazovyye ustanovki elektrostantsiy* (MEI, Moskow, 2013) [in Russia]
7. G. Olkhovsky, *Power gas turbine plants* (Energoatomizdat, Moskow, 1985) [in Russia]
8. A.M. Antonova, A.V. Vorobiev, A.S. Orlov, MATEC Web of Conf. **23**, 01069 (2015)
9. K. Savina, A. Abramovskih, V. Haymin, S. Lavrinenko, MATEC Web of Conf. **72**, 01042 (2016)

CALCULATION OF THE THERMAL STATE OF THE GRAPHITE MODERATOR OF THE RBMK REACTOR

Alexander V. Vorobiev[1,*], *Vitaly* O. Dmitriev[1], *Ivan* G. Tkachenko[1], and *Alexander* S. Matveev[1]

[1]Tomsk Polytechnic University, 634050, Tomsk, Russia

Abstract. This work is devoted to study the temperature field of the graphite stack of the RBMK reactor. In work was analyzed the influence of contact pressure between the components of the masonry on the temperature of the graphite moderator.

1 Introduction

One of the factors that determine the reliability and safety of NPP with reactors water-graphite reactor is the reliability of graphite moderator (masonry), calculated for the whole lifetime of the reactor.

The graphite walls of the reactor RBMK-1000 has a cylindrical shape with a diameter of 11.8 m and height 7 m, made up of 2488 columns. In the holes of the columns there are pipes of the technological channels (TC) and CPS channels. Fuel assemblies are located inside the channels are cooled by water with a temperature of 250-280 °C.

During the reactor operation part (5-6 %) of the full thermal power of the reactor is highlighted in graphite masonry. This heat is removed to the pipes TC. To ensure thermal contact with the blocks on the pipe channels wearing split graphite rings (Fig. 1), and in the gaps between the pipes and blocks the purge gas is supplied (a standard composition of 10 % nitrogen and 90 % helium by volume).

The design and schematic solutions of reactor facilities RBMK allow for changes in the composition of the purge gas [1].

Temperature field in the graphite walls is determined primarily by the capacity of the reactor. The content of nitrogen X_N in the nitrogen-helium mixture (the composition of the purge gas) effects on the temperature of graphite.

The temperature of the graphite masonry is one of the main parameters limiting the operation of RBMK reactor. In practice, it is assumed that all modes of reactor operation the maximum temperature of the graphite should not exceed the values of threshold temperature 750-800 °C [1-3]. In addition, fluctuations in the average temperature of the graphite masonry (moderator) lead to corresponding changes in the reactivity of the reactor,

* Corresponding author: worob@tpu.ru

as these units are RBMK reactors offer a significant magnitude positive effect of reactivity on the temperature of the moderator [1].

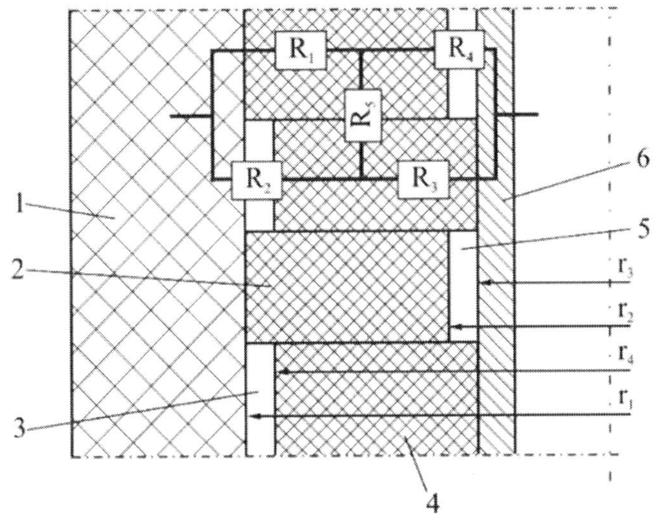

Fig. 1. The scheme of installation of the pipe TC in graphite masonry: 1 – graphite block; 2 – outer graphite ring (sleeve); 3, 5 – gas gaps; 4 - inner graphite ring (bushing); R_i - thermal resistance; r_i - is the radius.

Consequently, the definition of the thermal state of the graphite stack of the RBMK reactor is an urgent task.

Temperature field of masonry water-graphite reactor is three-dimensional. Program of numerical calculations of two-dimensional and three-dimensional temperature fields very difficult [4, 5]. Therefore, for an approximate estimation of the maximum temperature of the graphite is necessary to resort to various simplifications and assumptions.

2 The calculations and methodology of the study

In the method (3) thermal resistance of the bushings and the gas gaps can be likened to the electrical resistance. Making and simplifying the circuit of these resistances, it is possible to determine the value of the equivalent circuit, and hence the thermal resistance of the gap of the graphite block – pipe and technological channel (Fig. 1).

The reasons for this discrepancy may be unaccounted thermal resistance at the contact graphite block outer ring, outer ring and inner ring, the inner ring pipe of the TC (Fig. 1).

To clarify the source of differences between calculated and experimental data on the temperature of the graphite technique was refined by incorporating additional thermal contact resistances. To calculate the last used criteria equation from [6].

$$\alpha_k = \frac{\lambda_{mx} \cdot Y}{h_0} + 8 \cdot 10^3 \cdot \lambda_{av} \cdot \left(\frac{P_k}{3 \cdot \sigma_{str}} \cdot K \right)^{0,86} . \qquad (1)$$

Where λ_{mx} - thermal conductivity of gas mixture; Y =3.33 - the relative values of the gap; h_0 - the total roughness of the contacting surfaces; K - is the coefficient depending on h_0; P_k - contact pressure ; σ_{str} - the tensile strength of more plastic of the contacting materials.

By using this method were calculated the temperature of individual graphite block with variation of the composition of the purge mixture and the amount of contact pressure. The calculation results are presented in Fig. 3.

3 Experimental results

The results of the variant calculations of temperature separate the graphite block is considered the technique shown in Fig. 2.

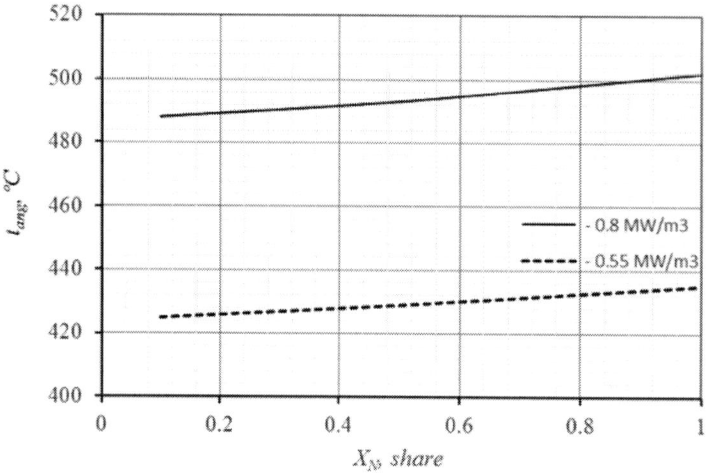

Fig. 2. The dependence of the maximum temperature of the graphite block (in the corner of graphite block) from the composition of the purge mixture excluding thermal contact resistance.

Content analysis Fig. 2 shows that the maximum (angular) temperature of the graphite block t_{ang} with increasing content of nitrogen in the purge mixture from 10 to 100 % increase by 10-15 °C. These figures differ significantly from the experimental data given in [2], where the mean value of the coefficient of the effect of mixture composition on the temperature of the graphite is estimated in the range of 0.9-1.2 °C/%.

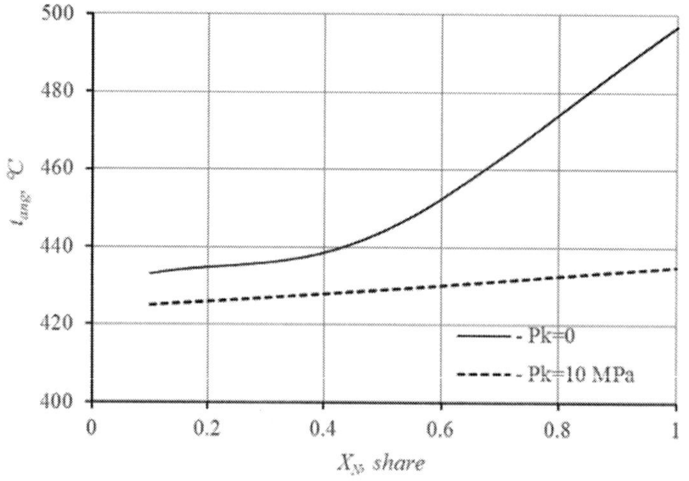

Fig. 3. The dependence of the maximum temperature of the graphite block (in the corner of graphite block) from the composition of the purge mixture at different values of contact pressure.

The analysis of graphs in Fig. 3 shows that the thermal contact resistance is estimated significantly (1.05-1.2 times) increase the temperature of the graphite moderator. It also increases significantly the degree of influence of nitrogen content in the mixture to the temperature level of the unit.

4 Conclusion

As a result of work were revised algorithm and program of calculation of the thermal state of the graphite moderator of the RBMK reactor, taking into account thermal resistance at the contact elements of the graphite stack and TC.

The use of this program allows to significantly improve the accuracy of engineering methods of calculation of the temperature of the graphite moderator of the RBMK reactor.

References

1. N. A. Dollezhal, *Channel nuclear power reactor* (Atomizdat, Moskow, 1980) [in Russia]
2. A. V. Vorob'ev, A. M. Antonova, Bulletin of the Tomsk Polytechnic University **322**, 182 (2013)
3. A.V. Vorobiev, A.M. Antonova, M.P. Vinogradov, EPJ Web of Conf. **110**, 01053 (2016)
4. A.M. Race, *Thermal-hydraulic calculation of the reactor RBMK-1000 reactors: textbook* (MEI, Moscow, 1982) [in Russia]
5. D.V. Gvozdyakov, A.O. Tanishev, S.A. Shvab, V.N. Martyshev, A.V. Kuzmin, MATEC Web of Conf., **37**, 01024 (2015)
6. Y. P. Shlykov, E. A. Ganin, S.N. Tsarevsky, *The thermal contact resistance* (Energy, Moscow, 1977) [in Russia]

EVALUATION OF ENGINEERING TRAINING COMPETENCES IN THE FIELD OF SMART SYSTEMS AS A FACTOR OF RUSSIAN INNOVATIVE DEVELOPMENT THROUGH THE EXAMPLE OF TOMSK POLYTECHNIC UNIVERSITY

Igor B. Ardashkin[1,*], and *Ekaterina* A. Kirsanova[1]

[1]Tomsk Polytechnic University, 634050, Tomsk, Russia

Abstract. The article dwells on Russian system of educational engineering training in the field of smart systems with a view to its innovative component through the example of National Research Tomsk Polytechnic University (TPU). The article highlights that the development of innovations has two significant components (technological-economic and social-humanistic). It is proved that only with equal combination of these two components in innovations it is possible to obtain maximum innovative development effect for economy of any country. It is concluded that the system of educational engineering training in TPU is misbalanced in terms of innovative development due to considerable technological-economic component dominance over social-humanistic one. The authors propose possible ways to overcome this misbalance of innovative development in the system of engineering education of TPU upgrading event block on staff and students motivational competences improvement.

1 Introduction

Smart systems are such systems which even being technologically sophisticated (with cutting-edge technologies and advanced equipment) will not run without such components as cultural, social, economic and organizational. Their main task is to use technical systems as close as possible to human capacities and at the same time to exclude human participation in their functioning to the greatest possible extent. Their application is of universal nature and presupposes the ability of usage in highly expanded range of social areas of human life. But these systems will work only when society will understand the necessity of such developments and follow the rules of its application.

As a consequence, the appropriate system of educational training is needed for innovative economy as well as for the development and implementation of Smart systems.

* Corresponding author: ibardashkin@tpu.ru

As a result, the main purpose of the article is to evaluate present educational programs of engineering training (main competences as indicators of their effectiveness) through the example of Tomsk Polytechnic University with a view to their innovative component.

2 Results and discussion

National Research Tomsk Polytechnic University (TPU) is the first engineering university founded in Asian part of Russia, which has a great experience of training engineers. Currently according to internal ratings TPU is one of the leading Russian technical universities beyond Moscow and St. Petersburg [1]. It is necessary to clarify certain concepts before starting the evaluation of competencies of engineering education in the field of Smart systems as a factor of innovative development of Russia through the example of TPU.

In particular, it refers to such concepts as innovation, innovative development, innovativeness. It is important in order to clarify the topic of discussion and show the connection between educational engineering training, Smart systems, innovations and innovative development.

When Joseph Schumpeter in his "Theory of Economic Development" used the concept of "innovation", he wanted to emphasize the difference between the concepts of economic growth and economic development. The concept of "innovation", according to Schumpeter, is important to demonstrate the qualitative difference in the implementation of a process, in the functioning of a phenomenon. Schumpeter expressed five types of qualitative changes, demonstrating the nature of innovation [2].

Another author Peter Drucker believed that innovation is a characteristic of human mentality and his ability to control his own destiny and the destiny of individuals, organizations and enterprises. Innovation according to P. Drucker is not necessarily a characteristic of commercial entity, it can be used in the non-profit sector. P. Drucker wants to emphasize social and humanistic aspect of any innovation, illustrating such important components of innovation and innovative development as: the opportunity for human self-realization, tuning into creation and introduction of something new; the importance of social contacts, allowing to capture the mood and the needs of the present society, realizing them in the form of innovation (it would be much more difficult to make it in the situation of closed nature); technological component through which innovation is implemented and used; economic component aimed at keeping financial and any other efficiency innovations [3].

Thus, Peter Drucker wants to emphasize that technological characteristics of innovation alone (new technology, profit, organizational transformations) cannot be regarded as a significant indicator of innovative development, as we can often find in publications and speeches of local researchers, administrative officers, etc. Humanistic and social components of innovation, drawing on the understanding of Schumpeter, allow making some qualitative transformations in current changes, turning economic growth into economic development.

Nowadays this aspect is especially significant because the evaluation of innovative development of countries, enterprises, business vehicles and universities can differ widely. This difference can be seen very clearly in various ratings of innovative development of countries all over world. It is necessary to clarify that the differences in evaluation are mainly related to the coverage of basic innovation components: technological, economic, social and humanistic. Let us demonstrate this with the help of two ratings of global innovative development evaluation: The Global Innovation Index formed by World Intellectual Property Organization (WIPO) magazine in cooperation with Cornell University (USA) according to the methodology of STEAD Business School (France) and

DOI: 10.1051/matecconf/20179101015

The Bloomberg innovation index. According to these ratings for the previous 2015 year Russia ranked 49 and 14 positions for the development of innovation among other countries of the world. Taking into consideration that the difference between these positions is considerable, it can be concluded that different methods of calculation were used, since in one case we are talking about averages, and in the other case we are talking about the position, displaying relatively high levels of innovation.

In the rating of Global Innovation Index the methodology takes into account two groups of indicators: the resources and conditions for innovation (institutions, human capital and research, infrastructure, development of internal market; business development) and present practical results of innovation (technology development and the knowledge economy; the results of creative activity).

The Bloomberg innovation index methodology is based on 7 criteria: Research and Development intensity, productivity, concentration of high technology, concentration of research, production capacity, and efficiency of tertiary and patent activity.

If we compare these two ratings on the criteria of evaluation, it appears that innovation Bloomberg index rating is more of technological and economic nature, whereas the Global Innovation Index seeks tries to take into account socio-humanistic component of innovative development. Such criteria as institutions, human capital, infrastructure, results of creative activity, etc. involve not only technological and economic characteristics, but also the evaluation of conditions, relationships, human activity, which means that social and humanistic component of innovation, is taken into consideration.

If we refer to the experience of Tomsk Polytechnic University (TPU), it is possible to detect similar trends and in its functioning.

Innovative component is the most important trajectory of development of the University because the goal of the University is the establishment and development of TPU as a research university - one of the world leaders in the field of resource efficient technologies that solve the global problems of humanity on the path to sustainable development [4]. This goal is unreachable without innovative filling. The University makes great effort in the indicated direction. This also can be applied in educational and research component.

The emphasis is laid on innovative component of educational training. In recent years, curricula were annually adjusted, taking into account wishes of students and prospective employers with whom the leaders of main educational programs are trying to cooperate. These changes can be tracked through the documentation on educational areas of training in the last 5 years on the University website in "Education programs fund of TPU" and "Education" sections.

At different levels of training special emphasis is laid on the development of creative abilities, independence and initiative. In absolutely all areas of training the University provides teaching and research work, creative project – specific discipline in which students learn to carry out research. In addition, in each of the disciplines, which students learn, there is a focus on project-based approach, which allows not only absorbing some information and knowledge, but also producing their own information and learning how to use it for practical purposes.

The course "Creative workshop for engineers", which allows not only identifying the degree of human creativity, but also managing the development of creative abilities is introduced as an elective course for the development of creative and heuristic competences at the undergraduate level [5]. The course of "Scientific fundraising" is introduced as an elective one at the graduate level allowing undergraduates not only to learn how to organize, to conduct and to protect their own projects, but also to practically implement, simultaneously learning and getting adequate support from various funds.

By analogy, we can consider the scientific component. As a rule, an innovative trajectory of the University is expressed in the creation of innovative infrastructure of universities around the enterprises established with the participation of teachers, students, business representatives (the most famous zone of such enterprises - Silicon Valley at Stanford University in the US. Tomsk Polytechnic University has its own business incubator - a place which is specially designed for start-ups at the interface of science and business. This place has several small innovative enterprises, where active organizers implement and sell innovative developments. Only on the website of the University it is indicated 47 enterprises [6]. It is enough for half a million population of Tomsk, where there are five large state universities apart from TPU. Here is a list of products and services that they can provide to potential customers. This also is not the whole list of scientific events of innovative type, but it gives an understanding of the seriousness of the consideration of the University.

There are seven main reasons preventing successful innovative development of Tomsk Polytechnic University: not enough level of English proficiency of students and staff; low level of internationalization; concentration of "big" science in state academies; orientation of a substantial part of successful school graduates to capital universities; a small number of educational programs in English; infrastructure of the campus is not modern enough; "Technological stagnation" of 90s in Russia, accompanied by law demand for the engineering profession; barriers complicating the employment of foreign scientists in Russian universities [7]. Among these reasons, only two of them are caused by technological and economic aspects: insufficient infrastructure of the campus and "technological stagnation" of 90s in Russia, accompanied by law demand for the engineering profession. The last reason is hardly included, as law demand for engineering profession - is the cause of social and humanistic nature. All other reasons related to socio-humanistic field and may be grouped as follows: the lack of qualified staff and the lack of employee motivation. It should be added that usually a lack of qualified personnel is a consequence of a lack of motivation. Motivation sphere – is a very "sophisticated" area which cannot be controlled by any quantitative indicators.

Smart systems act as a result of such activities, which let us estimate the completeness of representation of all aspects of developing innovations (technological and economic, social and humanistic). Only relatively equivalent representation of designated aspects of innovation enables successful development and application of any smart system. The fact that Tomsk Polytechnic University addresses to such systems as Smart Grid shows the correctness of chosen trajectory of innovative development, another thing is that without the emphasis on socio-humanistic aspect this development will be severely constrained and stopped. It was stated earlier by the authors [5, 8, 9, 10].

3 Conclusion

To conclude with, it is necessary to say that the evaluation of educational system of engineering training competencies in TPU (as well as in Russia as a whole) in the field of Smart systems is uneven, due to the fact that the technological and economic aspect are predominant in innovative development. It is expressed through the development and use of new technologies and products focus on commercial exploitation of innovations created in university. As a result, technical and economic success is neutralized by the fact that it does not lead to corresponding social transformation in minds and behavior of people, and therefore does not lead to effective use of development of society (e.g., the acquisition of modern medical equipment initiated by Russian government program did not improve the quality of health services in most cases, because the equipment was never used and was left in warehouses). Reconstruction of the balance between technological and economic, social and humanistic aspects of innovative development is the only trajectory to improve the quality of educational programs of TPU and other higher education institutions of Russian Federation, as well as the quality of life of our population. But to do this it is important to implement the restructuring of motivational aspects of university employee's activity, where corporate culture can become the most convenient tool.

The research is conducted with the financial support from Russian Humanitarian Science Foundation 16-16-70006

References

1. Reyting TPU [Online] http://tpu.ru/today/facts-numbers/ratings/
2. J. A. Schumpeter. *Teoriya economicheskogo razvitiya, Capitalizm, socialism I democratiya* (ECSPO, Moscow, 2007) [in Russia]
3. P. Drucker, *Management Challenges for the 21st Century* (Taylor & Francis Ltd., New York, 2001)
4. Plan meropriyatiy po realizatsii Programmy povesheniya konkurentosposobnosti ("dorozhnaya karta") federalnogo gosudarstvennogo avtonomnogo uchrezhdeniya vysshego obrazovaniya "Nationalnyy issledovatelskiy Tomskiy politechnicheskiy universitet" na 2013–2020 gody (2 etap – 2015–2016 gody) (Moscow, 2015) URL: http://tpu.ru/f/2365/dk_2015.pdf
5. I. Ardashkin, L. Korobeynikova, A. Popova, Status of social competencies of power engineers in the context of forming the concept of an intelligent network or smart grid, MATEC Web of Conf. **37**, 01003 (2015)
6. Malye innovazhionnye predpriyatiya. Corporativnyy portal Tomskogo politechnicheskogo universiteta (Tomsk, 2016) [Online] http://portal.tpu.ru/departments/otdel/bi/mip
7. Programmy povysheniya konkurentosposobnosti. Sayt TPU (Tomsk, 2016) URL: http://tpu.ru/today/today/programs/viu/
8. A. Chmykhalo, V. Khaliulina, M. Abushaeva, Innovative Power Systems and The Formation of The Creative Class in Russia, MATEC Web of Conf. **37**, 01016 (2015)
9. M. Makienko, N. Kurkan, A. Strelcowa, Philosophical approach to engineering education under the introduction of the smart grid concept in Russia, MATEC Web of Conf. **37**, 01031 (2015)
10. I. Ardashkin, A. Bykov, E. Kirsanova, The status of world outlook and environmental competencies in aerospace engineering training: the general and special, MATEC Web of Conf. **48**, 06004 (2016)

HUMAN RESOURCING CHALLENGES OF THE SMART GRIDS DEPLOYMENT PROJECTS IN RUSSIA

Alexander Y. Chmykhalo[1,*], *Vasilya* R. Khaliulina[1], and *Olga* V. Mikhailova[1]

[1] Tomsk Polytechnic University, 634050, Tomsk, Russia

Abstract. The article addresses the issue of the causes of human resources gaps for the implementation of projects for the deployment of intelligent electricity grids. The "focus group" research technique used in Tomsk Polytechnic University enabled to identify numerous barriers encountered by present-day engineering students in Russia on the path of their professional development. The barriers explored impede greatly the creation and introduction of state-of-the-art technological systems, Smart Grid being one of them. The elimination of those barriers will accelerate the accomplishment of the Smart Grid project.

1 Introduction

The current socio-economic situation in the world is characterized by intense competition in the field of innovations development among the countries. The governments of many countries are taking extra measures to engage the population in entrepreneurial activities and innovation development.

Many scientific papers tend to present the process of promoting innovation development as a complex problem, requiring an integrated approach for its analysis, according to which creativity, innovation and entrepreneurship make up an integrated entity, or "construct". That construct is viewed not only as an engine driving forward a country's socio-economic development, but also as a meta-competence to be addressed in the education system, particularly in the system of engineering education [1, 2]. How to integrate approaches aimed at training enterprising, creative-minded, innovation-oriented graduates into existing universities' curricula is at the core of the current education agendas in all the world's countries that face the urgency of social and technological modifications of their academic institutions for the purpose of their upgrading. Increasingly topical is this issue for the system of engineering education designed to simultaneously maintain advanced scientific and technical research and encourage the development of novel products and technologies to meet societal needs and priorities [3, 4]. Universities are becoming increasingly aware of the need to graduate engineers with a creative mindset, capable of designing new products and services, equipped with methods of recognizing market trends, detecting people's needs as well as evaluating social challenges [5].

* Corresponding author: ovmtom@tpu.ru

Under these conditions, there is a growing concern in many countries about the quality of engineering education. It has been reported that fewer young men choose science and engineering careers despite the increased labour market demand for experienced engineers and technicians since new technical systems have more and more application opportunities and require more “active feedback” from end-users [6]. The deployment and maintenance of those systems necessitate collaborative working, which, in its turn, requires a new set of professionals and engineers with augmented competences.

The most striking example of those novel technical systems launched in the world’s developed countries over the past decades is the intelligent power supply system - Smart Grid. Among the problems that the European Union countries have faced while implementing the program for Smart Grids deployment the most crucial one has turned out to be that of human resourcing. The “European Technology Platform Smart Grids. Strategic Deployment Document for Europe’s Electricity Networks of the Future” published in 2010 highlights that “engineering in the energy sector, electricity grids in particular, is seen by many as old-fashioned and “difficult” as it requires a high level of competence in mathematics, physics and other sciences. This discourages the potential new students from studying and pursuing a career in power engineering” [7].

In response to this claim, the aim of this study is to find out whether the idea of power engineering being ‘old-fashioned and difficult’ is shared by Russian students in the Tomsk Polytechnic University (TPU) (one of the leading institutions of engineering education in Russia) [8]. What are the students’ perceptions of the skills and competences the present-day engineer should possess? What, they believe, are the greatest barriers hampering the professional development of the engineer?

The challenge of providing the engineering personnel for the implementation of Smart Grids programs in Russia, in the city of Tomsk in particular, is of special importance to us, because, it was in Tomsk where the development of intellectual power systems in Russia was started. In 2011, the “Intellectual Power Supply Systems” research center was established to produce high-technology methods for heating, power supply and telecommunication. However, thus far, there has not been any noticeable breakthrough in these fields either on the territory of the Tomsk region, or any other region of Russia. The analysis of the current situation makes us suggest that one of the key barriers constraining the implementation of that, as well as other innovative projects, is human resourcing, namely the shortage of engineering graduates whose professional skills and competences are enough to adequately respond to the challenges of an innovation-driven economy [9].

The purpose of this paper is by using the “focus group” research technique to examine the perception held by Russian students of the content of an engineering major, of the most significant barriers impeding the professional engineering training in Russia.

The sampling frame for the study consisted of students enrolled in a Bachelor’s degree programme in “Electrical engineering and equipment” and in a Master’s degree programme in “Nuclear physics and nuclear energy technology at the Institute of Physics and Technology and the Institute of Power Engineering in TPU”. The survey and its analysis were implemented in 2016.

2 Results and discussion

For the purpose of the survey two sets of questions were offered to respondents. The first set questions were designed to explore the contemporary Russian students’ perception on the content of an engineering major, the distinctive attributes and abilities of engineers in comparison with other specialists. A second set of questions was aimed at detecting the main barriers having negative implications for the professional development of the present-day engineer. The survey enabled the respondents to convey their position on the issues under study. The list of essential qualitative characteristics given below represents the views shared by the majority of respondents:

1. The main objective of the professional activity of the engineer is to invent, design and produce new equipment, to devise and develop new materials and products.
2. The present-day engineer has to have good professional education. An engineering graduate has to be provided with a profound grounding in engineering disciplines, to have scientific research experience, to keep track of the latest science and technology advancements.
3. The engineer has to be able to use information technology and related applications to convey and retrieve information, to use presentation software to create, edit, and give digital representations of information about produced items, and deliver oral self-presentations.
4. The present-day engineer has to be able to use available and create new channels of retrieving information, to work cooperatively with others, as either a leader or a follower, on the project, participate in professional or other social groups and communities both nationally and internationally.
5. The present-day engineer aspires to be the best in his field and demonstrates a commitment to self-development and improvement of knowledge and expertise by self-education and by participating in relevant training and professional development programs.
6. The well-qualified engineer has to possess and use knowledge pertinent not only to the specific area of engineering practice or allied sciences, but also Humanities, Social Sciences, Economics.
7. Engineers are people who solve problems of others by means of technical devices. Engineers are capable of coming up with fresh, original, but workable ideas to address some routine problems.

The analysis of the results of the survey has made it possible to identify a few main barriers affecting various aspects of the present-day engineering training:

1) firstly, there are barriers connected with the content of engineering education in Russian universities:
   - the shortage of academic hours in engineering major subjects at a bachelor's degree level;
   - a big share of the engineering departments syllabi is taken up by humanitarian and socio-economic courses whose content does not always correlate with the major subjects courses and does not contribute to the formation of the requisite knowledge and skills in management and communication;
   - the constant growth of information to be processed and absorbed makes it imperative to apply an integrated interdisciplinary approach to curriculum development.
2) secondly, there are barriers connected with the organization and supervision of learning and research processes of engineering students:
   - with a student having just one research supervisor, there are risks of their adopting not only positive attributes of the latter, but also negative ones. There is a need for students' being involved in diverse scientific societies and communities at the department, university and inter-university levels. The sharing of scientific information within those communities has to be ensured.
   - good engineers can be prepared only by good engineers. There is a lack of, though a great demand for effective highly-qualified university teachers;
   - a poor level of the organization of students professional practice at enterprises, which looks more like an orientation field-trip. Employers are not motivated to offer really challenging workplacement experiences, as a consequence, graduates

don't get a real insight into the modern enterprise and find themselves neglected by employers;
  - rarely do industry's leading enterprises offer work experience opportunities, so there is no chance for students to familiarize with their best practices;
  - students have low motivation for research and inventions development, as those activities do not affect directly their employment and traineeship opportunities;
3) thirdly, there are barriers connected with the shortcomings of the country's educational system in general:
  - the level of school education is stated to have declined, which entails having fewer young people aspiring to engineering training;
  - there is no logical and content continuity between the programs of baccalaureate and Master's degree levels;
  - there is no explicit strategy for the development of engineering industries and engineering education in the country, as a result, there is a great number of misguided school-leavers making future career decisions;
4) fourthly, there are cultural and value barriers:
  - the notably lower social and economic status of the engineer in the modern society (a low prestige of a profession with the population);
  - the public is reluctant to accept changes initiated by engineers, so there is no due concern in the society about involving young people in engineering, which discourages them from pursuing engineering careers;
  - the society doesn't have a clear idea of the role of young specialists (engineers) in its economic, political and social life, lacks explicit value imperatives with regard to young people;
  - with the government participating greatly in the national economy, political feasibility is a crucial guiding force for the state bureaucracy in decision-making, thus innovations are being neglected;

The analysis of our survey results makes it possible for us to argue that they correlate in many aspects with other findings addressing the quality of engineering education in Russia, and in TPU in particular. Those findings cover the statistical data on the employment of Bachelor's programs graduates, the number of innovative enterprises launched by TPU graduates, income earned from vending invention patents granted to TPU students, lecturers, engineers [8], etc.

3 Conclusion

The data analysis of the deployment plans for electricity networks of the future (Smart Grids) in the European Union countries clearly demonstrates that providing human resources (engineers, technicians and craft personnel) for the project is a crucial factor in its successful completion. The results of the survey conducted among TPU students acknowledge the topicality of the human resourcing challenge in Russia. The numerous barriers impeding the process of training competent, creative, innovation-minded engineers have been identified. The impact of those barriers is tangible even in the universities regarded as the clear leaders of power engineering education in Russia.

Thus, we can argue that human resourcing is a major obstacle to the implementation of the smart grids deployment project in the Tomsk region, and Russia as a whole. For the future intelligent electricity grids to be introduced it is urgent to undertake efforts and activities to eliminate the identified and explored barriers.

This work was carried out with the financial support from the grant of the Russian Humanitarian Science Foundation (RHSF) (project No. 16-16-70006).

References

1. M. Edwards-Schachter, A. García-Granero, M. Sánchez-Barrioluengoa, H. Quesada-Pineda, N. Amara, Thinking Skills and Creativity **16**, 27 (2015)
2. B. Ardashkin, L. A. Korobeynikova, A. V. Popova, Status of social competencies of power engineers in the context of forming the concept of an intelligent network or smart grid, MATEC Web of Conf. **37**, 01003 (2015)
3. M. A. Makienko, N. V. Kurkan, A. A. Strelcowa, Philosophical approach to engineering education under the introduction of the smart grid concept in Russia, MATEC Web of Conf. **37**, 01031 (2015)
4. D. K. Gattie, N. K. Kellam, J. R. Schramski, J. Walther, EJEE **36**, 521 (2011)
5. M. H. Morris, J. W. Webb, J. Fu, S. Singhal, JSBM **51**, 352 (2013)
6. R. Freeman, National Bureau of Economic Research, Working Paper 11457, (2005)
7. Smart Grids European Technology Platform. URL: http://www.smartgrids.eu/
8. A.Y. Chmykhalo, O. V. Mikhailova, N. A. Vtorushin, R. A. Bykov, Innovation development in Russia: the value aspect, SHS Web of Conf., **28** (2016)
9. A.Y. Chmykhalo, V. R. Khaliulina, M. E. Abushaeva, Innovative Power Systems and The Formation of The Creative Class in Russia, MATEC Web of Conf. **37**, 01016 (2015)

ABOUT EXTRA-CURRICULAR ACTIVITIES OF THE UNIVERSITY GRADUATE – STUDENT'S VIEWPOINT

Veronica Ivanova[1,*], *Kseniya* Mertins[1] ,and *Semen* Kaftasev[1]

[1]Tomsk Polytechnic University, 634050, Tomsk, Russia

Abstract. Extra-curricular activity of modern youth, including university students, is considered to reduce. Students are thought to demonstrate consumer's attitude, unwillingness to participate in any social activity, and as a result, the level of soft skills acquired by university students is decreasing. The study of social activity of 100 Bachelor graduates from TPU, including the program "Electrical engineering", indicated the role of the university in creation of conditions for competence development during training of students according to individual educational trajectory and development of their social (extra-curricular) activity.

1 Introduction

Modern society brings up the problem of development of social activity of university graduates able to design and build a new environmental reality, to eliminate outdated stereotypes of activity, make innovative changes in different spheres of social life.

According to [1], some researchers concerned with understanding of the universities training process as a system that integrates academic and extra-curricular activities. Many universities around the world offer extra-curricular activities, from sports to training or professional activities. One of the main objectives and concerns of colleges and universities is to reach an appropriate employability level for their students. But employability is a complex and vague concept, that goes from identifying personal and academic students skills to other educational outcomes as demographic data affecting graduate employment. [2]

These kinds of skills are more likely to be developed in non-formal education in youth organisations. [3]

In particular, the professional competence formation in a future specialist during practical training is interconnected with the educational projects system introduction, which includes: integration of personal, professional, financial, technical and educational resources; competent specialists innovative training in the specific environment; the scientific projects adaptation to the specifics of practical and innovative learning processes.

* Corresponding author: kurator@tpu.ru

Based upon a literature review and analysis of best practices, it seems almost self-evident that student engagement, including extracurricular activities, has a positive impact on student academic performance. [4]

The university system of training of engineers must take into account all professional aspects of a modern specialist and identify not only the subject and activity structure of the education, but also the significant conditions of a specialist's professional activity, the absence of which makes this professional activity complicated [5] .

2 MATERIALS AND METHODS

2.1 Objectives of the research

Hypothesis. Extra-curricular activity enhances soft skill development. Participation in extra-curricular activity increases the level of soft skills among graduates.

Objective: to determine the basic forms of student social activity, to design and conduct a survey of students, to identify the competencies acquired by the university graduates as a result of their participation in different types of student activities, as well as to find out the role of the university in soft skill development.

2.2 Theoretical and empirical methods

The following research methods were used to achieve the objectives: theoretical - analysis of scientific and pedagogical literature on research, modeling method; empirical method through observation, pedagogical experiment, testing, questioning, conversation; mathematical methods of data processing.

3 Results and discussion

To study the social activity of TPU bachelor graduates, a questionnaire was designed with 19 questions related to the types of activities and the conditions that enhance the development of soft and hard skills. The survey was conducted among 100 bachelor graduates, including students trained in "Electrical engineering".

3.1 The results of the survey conducted among TPU students

According to the survey, 79 % of TPU bachelor graduates were engaged in extra-curricular activities, and 57 % of respondents consider themselves successful people. This indicates that they gained sufficient basis for start up of their professional activity or social development. The categories of the success highlighted by the respondents were graduation from the top university, occupational work, and skills and abilities gained.

The survey revealed that 70 % of respondents participated in scientific conferences, 62 % of graduates have experience in writing research papers and have their papers published, 22 and 13 % of respondents conducted research and developed their own engineering solutions, respectively. The study results showed that 6 % of the respondents were engaged in student coucelling and 30 % of bachelor graduates participated in creative life of the university, 29 % of graduates were active in sports, and17 % of respondents were involved in volunteer activities.

To answer the question about the most significant achievement in student's life, the graduates identified: participation in scientific conferences (10 %) (3 % of them became winners), publication in periodicals (including those in English (6 %), self-engineered products (4 %), including those patented (3 %), scholarships (including 2 % of those received the scholarship of the President of the Russian Federation), volunteering, creativity, athletic performance, practical training and internship (9 %). However, 2 % of respondents emphasized the importance of obtaining a diploma of higher education as a whole. More than 50% of master's students plan to be involved in various student activities while doing their Master's degree program that can be attributed to previous positive experience, the cue from successful projects of group mates, the influence of the supervisor and the need for soft skills development. The survey results show more than 2-fold increase in students' desire to be engaged in scientific activity (23 %), to develop self-engineered products (8 %), to have their research published (9 %), to actively participate in creative activity (9 %), sports (9 %), self-government (3 %) and counselling and mentoring (2 %). In addition, master's students must successfully combine studies, research and work (implementation of their own projects, fulfillment of orders and engineering development and etc.), gain experience and develop soft and hard skills.

The authors also tried to identify the factors that motivate students to student activities. Most of the respondents (56 %) indicate self-development, 54 % of respondents lay emphasis on the work during their studies and 52 % of respondents emphasize the opportunity to participate in internships in companies and participation in conferences and other events in Russia and Europe. The survey showed that 32 % of respondents work out a portfolio to submit it to the employer. Internal motivators like a letter of appreciation or a diploma from the University administration were chosen by 43 %. The factor of being a member of a team was reported by 31% of respondents, and access to corporate events was mentioned by 25 % of respondents. The scholarship as a stimulating bonus was found important by 93 % of respondents.

More than 70 % of TPU graduates assert that the university created the conditions for soft skill development. The respondents said that student activities help to develop communication skills (63 %), self-presentation skills (46 %) and teamwork skills (46 %), as well skills to establish business relations (45 %), to participate in projects (35 %), skills in gaining experience (portfolio) (39 %) and talent development (29 %).

While participating in student activities, future graduates are oriented towards the requirements of employers. These include work experience (9 %), English proficiency (11 %), ability to work as a team (14 %), readiness for self-development and self-education (4 %), the ability to communicate (4 %), self-presentation (5 %), orientation to knowledge, including the basics of electrical power engineering (4 %), literacy (2 %), the ability to give scientific credence to the proposed technical solution (development of a new power supply device and etc.) (2 %), implementation of modern software tools (CAD) as well as personal qualities: punctuality, attentiveness, creative approach to work, ability to work individually, ability to manage a project or work within the framework of the project, responsibility, organizational skills, analysis of scientific and texts (information), development of scientific and engineering texts (information). None of the respondents focused on the requirements for proficiency in modern computer technologies that can be attributed to sufficient ICT level of the graduates (more than 44 % of respondents according to their own estimates).

The university, in turn, helped 95% of TPU graduate respondents develop the following soft skills: team work (60 %), the ability to make self-presentation (49 %), the ability to present their developments (39 %), the ability to work with information (64 %), creativity (26 %), leadership (30 %), the use of knowledge to manage complex engineering activities (8 %), skills in written and oral communication in Russian (34 %) and English (5 %), exercitation of legal and cultural aspects, knowledge of issues related to safety and health (17 %), awareness of the university's corporate culture (4 %) and organization (11%).

The survey showed that 95 % of respondents acquired competences, but at the same time they do not admit their involvement in extracurricular activities. Thus, we can conclude that soft skills are developed among more than 25 % of students as a result of learning activities (active learning and interactive techniques).

According to the survey results, the University organizes the events involving employers to identify the requirements for future graduates in the form of job fairs (68 %), provides participation in network projects (14 %), organizes professional (71%) and research (41%) training, invites leading representatives of enterprises and companies to deliver lectures (25 %), holds summer employer schools (14 %), roundtables (27 %) and scholarship programs (22 %). In addition, competence of graduates can be identified through a survey, poll, employer interviewing (14 %), and company open days (42 %). Therefore, it is necessary to develop cooperation with companies, including that to identify key soft skills.

The analysis of the events in which students are encouraged to participate by university, showed that the percentage of the participants of university and city levels is higher than that of the students participating in the events inside the campus. The number of the students involved in the activities at Russian and international level is actively increasing.

Conclusion

The results of the survey confirmed the viability of pedagogical maintenance of student social activity as a system. The study revealed that the level of soft skills competencies of those participating in student activities is higher by more than 40 %, than that of the students not involved in student activities. The level of soft skills development must be considered to build individual educational trajectories of students.

Students who participate in extracurricular activities generally benefit from the many opportunities afforded them. Social activity of students during training in a bachelor' degree program is a basic prerequisite for the graduate to get a sense of success to hold this position in postgraduate study.

This work is funded within the framework of realization of Strategic Programme on National Research Tomsk Polytechnic University Competitiveness Enhancement in the Group of Top-Level World Research and Academic Institutions.

References

1. I. Rudenko, N. Fedorova, I. Muskhanova, S. Alieva, B. Kasumova, Z. Gadaborsheva, E. Gunina, K. Tumarov, J.of Sust. Dev. **8**, 6 (2015)
2. C. del Campo, B. Navallas, M-del-M Camacho-Minano, The Int. J of Man. Educ. **14** 263 (2016)
3. S. Miles, J. of Y. Stu **10** (3) (2007)
4. J. Wang , J. Shivele, *Research project*. URL: http://www.csus.edu/oir/research%20projects/student%20activity%20report%202009.pdf (2016)
5. I. B. Ardashkin, L. A. Korobeynikova, and A.V. Popova, Mat Web of Con. **37**, 01003 (2015)

MATEC Web of Conferences **91**, 01018 (2017) DOI: 10.1051/matecconf/20179101018
Smart Grids 2016

STUDY OF THE CONTROL SYSTEM WITH THE SMITH PREDICTOR AND AUTOTUNING ALGORITHM AT-2 FOR THERMAL CONTROL OBJECTS WITH THE CONSTANT SPEED ACTUATOR

Viktor F. Kuzishchin[1,*], Elena I. Merzlikina[1], and *Hoang* Van Va[1]

[1]National Research University Moscow Power Engineering Institute, 123458, Moscow, Russia

Abstract. The application of the Smith predictor in the control system with a constant speed actuator is considered. In order to make it possible to use the Smith predictor a PID-controller with autotuning is used, the autotuning provides the control object model using a self-oscillation mode. The control object has a second order transfer function with a transport delay. The performance of the system was analyzed on an experimental workbench with the industrial constant speed actuator and a physical model of the control object. Recommendations are given, how to correct the PID-law parameters in order to improve the system performance.

1 Introduction

The Smith predictor can be used in order to improve the performance of a control system with a control object with time delay [1, 2]. For effective operation of the Smith predictor an exact model of the control object is necessary, the model must be renewed periodically, and it can be a problem in industrial control systems.

When the Smith predictor is used it is also necessary to correct the controller parameters in order to use the possibility to improve the operation speed of the system, this possibility is provided by the delay elimination regarding to the controller signal. It is necessary to estimate the performance of the system when disturbances influence the control object.

In order to obtain the control object model it is offered to include an autotuning (AT) module into the controller software. The control system with a constant speed actuator widely used in power engineering is considered, therefore it is recommended to use so called AT-2 [3] which provides self-oscillation mode in order to determine the control object mathematical model by means of frequency methods and to calculate the PID-law parameters later.

The aim of the present work is to analyze the performance of the control system with the predictor, PID-controller, constant speed actuator and AT-2, and to give

* Corresponding author: MerzlikinaYI@mpei.ru

DOI: 10.1051/matecconf/20179101018

recommendations how to correct the PID-law parameters in order to use all possibilities of the control system with the predictor.

2 Experimental workbench and autotuning algorithm

The study was carried out on an experimental workbench (see Figure 1) containing a physical model of the control object (it is an electrical heater connected with a laboratory adjustable autotransformer), a constant speed actuator, a programmable controller (Owen PLC 154) and a personal computer.

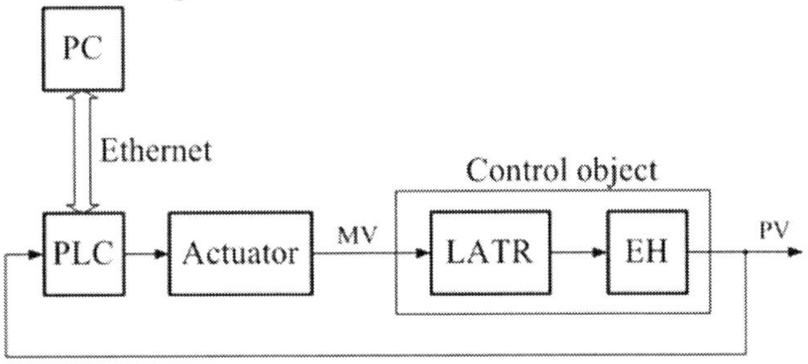

Fig. 1. The structure of the control system. PC – personal computer, PLC – programmable logic controller, LATR – laboratory autotransformer, EH – electric heater.

The control action on the object was carried out by means of the heater supply voltage change by the actuator connected with the slider of the autotransformer. The full stroke time for the actuator was 30 seconds.

Programming environment CoDeSys 2.3 was installed on the computer, connected with the controller through Ethernet, CoDeSys allowed to program the controller, obtain, process and archive information about the system operation. In the controller software there was an autotuning module (AT-2) which calculated the PID-law parameters and determined the control object model parameters. The control object model has the second order transfer function with transport delay

$$W_{ob}(s)=\frac{K_{ob}\cdot\exp(-s\cdot\tau)}{(T_1\cdot s+1)\cdot(T_2\cdot s+1)}, \tag{1}$$

where K_{ob} – transfer constant; T_1, T_2 – time constants; τ – delay time.

The optimal parameters of the PID-law are calculated using the object model obtained in order to minimize the integral from the control error module, the frequency oscillation index M=1.4. AT-2 is used in the software of different controllers, for example, made by Owen company. The PID-law used has the following transfer function.

$$W_{PID}(s)=K_P\cdot\left[1+\frac{1}{T_I\cdot s}+T_D\cdot s\cdot\frac{1}{(T_f\cdot s+1)^2}\right], \tag{2}$$

where: K_P, T_I, T_D, T_f – the transfer constant, time constants of integration, differentiation and filtering; $T_f = T_D/8$.

Transient processes in the control system when the AT-2 is operating are given in Figure 2. When $t \approx 400$ s the operation of AT-2 finishes and the system begins to work in the control mode.

The AT-2 calculated the following parameters:

– The object model parameters: K_{ob}=183.5; T_I=13.3 s, T_2=240.4 s; τ=12.6 s;

 DOI: 10.1051/matecconf/20179101018

- The PID-law parameters: K_P=6.36; T_I=58.6 s; T_D=13.3 s.

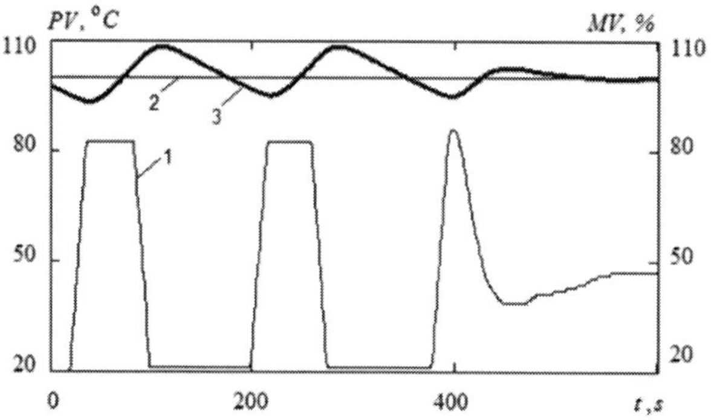

Fig. 2. Transient processes when AT-2 is operating. 1 – the signal from the actuator, *MV*, %; 2 – the setpoint signal, *SP*, °C; 3 – the control object output (the heater temperature), °C.

The block diagram of the control system with the Smith predictor is given in Figure 3. The predictor is built on the basis of the control object model with two outputs, one of them has the transport delay and the other does not, the control object output is also used in the predictor. This structure was used because the control system was based on the programmable controller using event-driven libraries.

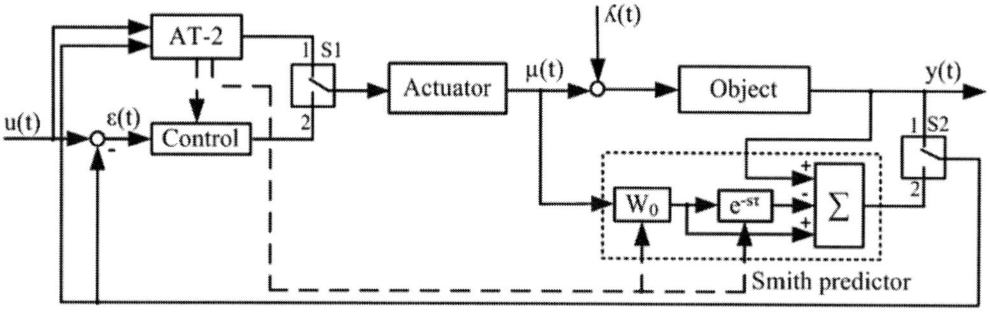

Fig. 3. Block diagram of the control system with the Smith predictor.

3 Experimental results

The performance of the control system was estimated by the transient processes when the setpoint was changed and when the step disturbance *ΔMV* influenced the control object. The step disturbance was created moving the LATR slider in the manual mode. The setpoint was changed by 5°C (from 100 to 105°C), the disturbance *ΔMV* was 20%.

The following variants of the control system parameters were considered:

- The PID-law parameters calculated by the AT-2, no predictor;
- The PID-law parameters calculated by the AT-2, the Smith predictor included;
- The Smith predictor included, the calculated PID-law parameters were corrected using the correction coefficient k=2 (K_P was doubled).

The transient processes when the setpoint was changed are given in Figure 4.

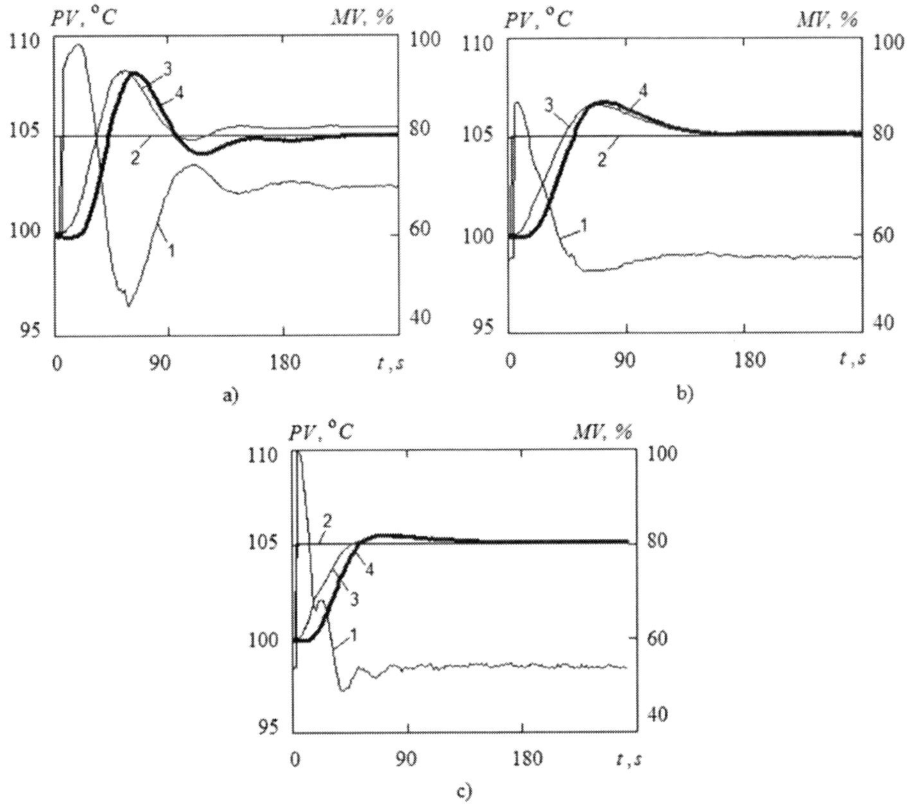

Fig. 4. The transient processes when the setpoint was changed: a)no Smith predictor; b) the Smith predictor is operating, the calculated PID-law parameters are used; c) the correction coefficient k=2.0. 1 – *MV*; 2 – *SP*; 3 – the output of the predictor; 4 – *PV*.

The control system performance can be characterize by the overshoot $\delta = PV_{max} - SP$, the settling time T_{cntr} and the damping degree. The process with k=2.0 (see Figure 4c) can be considered the best. The overshoot and the settling time for this case are considerably lower than for the process where the PID-law parameters calculated by AT-2 are used (see Figures 4a and 4b).

The transient processes for the case when the step disturbance acted on the control object are given in Figure 5.

 DOI: 10.1051/matecconf/20179101018

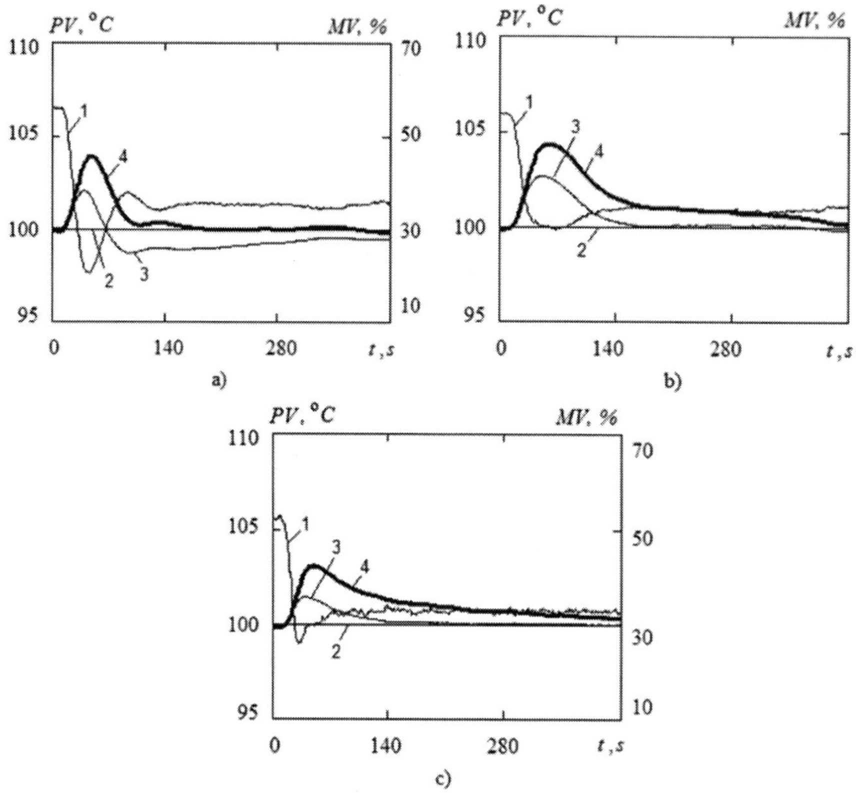

Fig. 5. The transient processes when the step disturbance acted on the control object: a)no Smith predictor; b) the Smith predictor is operating, the calculated PID-law parameters are used; c) the correction coefficient *k*=2.0. 1 – *MV*; 2 – *SP*; 3 – the output of the predictor; 4 – *PV*.

4 Conclusion

According to the experimental results given above the following conclusions can be made:

- When the correction coefficient *k*=2 the overshoot is minimal, though the settling time is rather long.
- When the correction coefficient is higher than two, the transient process becomes oscillatory, it is inadmissible.
- In this case it is unsuitable to correct the time constants of the PID-law.
- The use of the Smith predictor in the control system with a thermal control object with time delay is suitable when the setpoint changes often during the operation of the system. For this case it is recommended to use the correction coefficient *k*=2 and to correct the transfer constant calculated by the AT-2.
- If the setpoint is mainly constant and remarkable disturbances influence the control object, the effectiveness of the Smith predictor is doubtful.

References

1. Ren Honge, Cao Xuehai, Guo Jifeng, Advanced Science and Technology Letters **77**, 101 (2014)
2. V. V. Hoang, V. F. Kuzishchin, E. I. Merzlikina, Reports of MMTT-28 **8**, 197 (2015)
3. V. F. Kuzishchin, V. S. Tsarev, Thermal engineering **61**, 281 (2014)

PURIFICATION OF GAS EMISSIONS FROM THERMAL POWER PLANTS BY MEANS OF APPARATUS WITH JET-BUBBLING CONTACT DEVICES

Ilnur N. Madyshev[1,*], *Oksana* S. Dmitrieva[2], and *Andrey* V. Dmitriev[2]

[1]Kazan National Research Technological University, 420016 Kazan, Russia
[2]Kazan State Power Engineering University, 420066 Kazan, Russia

Abstract. A jet-bubble contact device is designed to enhance the operation efficiency of heat and mass exchange units. New devices should be introduced for the contact of gas and liquid, which can provide the minimum entrainment of liquid by the gas flow and have a high phase contact area and relatively low hydraulic resistance. Results of study of a resistance of the dry layer of contact elements at design parameters of jet-bubbling contact devices have been presented. Influence of average gas consumption rate on the specific hydraulic resistance of jet-bubbling contact devices has been analyzed.

1 Introduction

According to the data of The Energy Research Institute (The Russian Academy of Sciences), they expect the industrial growth and energy consumption by 35-70 %, depending on the scenario of the electric-power industry development, if there is a significant substitution of oil and gas by renewable resources and coal [1]. This situation implies the development of effective purification methods of gas emissions from thermal power plants (TPP), operating on heavy fuels.

The problem solution of catching the sulphur dioxide and nitrogen oxides out of the flue gases from the thermal power plants is connected with the processing of large volumes of gas emissions, exceeding in some cases million cubic meters per hour. Used for the industrial purposes, the absorption apparatus are eliminated for the purification of gases, because they have low gas throughput rate (1.5-2.0 m/s). The increase in their diameter leads to an increase of gas and liquid non-uniformity along the cross section of the apparatus and the significant reduction of the purification efficiency [2].

For the effective purification of industrial large-capacity gas emissions, the use of vortex type devices is very promising, as they have a high throughput rate, intensive interaction of phases and a wide range of stable operation [3-5]. However, significant disadvantages of vortex type devices are high specific energy costs due to high hydraulic resistance.

[*] Corresponding author: ilnyr_91@mail.ru

2 Jet-bubbling contact device

To eliminate these disadvantages the authors have developed the jet-bubbling contact device [6], shown in Fig. 1. This device consists of drain cups (down comers), with vertical walls, required to maintain the liquid level inside them. In each row the drain cups are interconnected by means of rods and their top ends are open, while bottom ends (the bottom) have sets of holes, required for the liquid drain on top of the below drain cups.

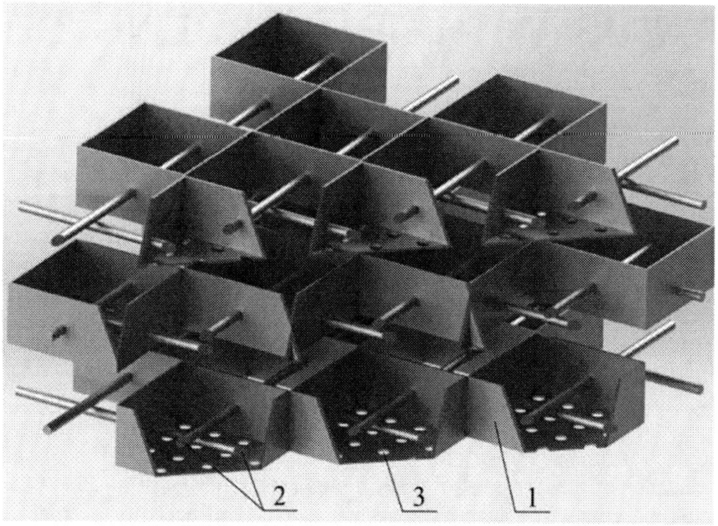

Fig. 1. Jet-bubbling contact device: 1 – drain cup; 2 – rods; 3 – holes.

The jet-bubbling contact device, developed by the authors, operates as follows. The liquid, through the plurality of holes, made in the bottom of the drain cups, is dispersed in the form of jets on top of the below drain cup. Thus, the liquid level in these drain cups, is maintained by the vertical walls of the drain cups. The drain cups are arranged horizontally chequerwise and form a tray. The below tray has an offset of drain cups, forming their vertical chequerwise location. For this reason, the gas, coming from the bottom under the plate, takes on a zigzag movement.

When moving, the liquid jets break up into a multitude of drops. They collide with the liquid surface, inside of the drain cups, and splash every which way. Thus, there is a developed, continuously renewing contact surface of phases, which is determined by the presence of relatively small gas bubbles in the liquid layer and droplets departing from the surface. In addition, the upward gas flow is in contact with the falling drops or liquid jets and, thus, appears the second contact area of gas and liquid.

When the distances between the drain cups are on the same level and equal to the width of the drain cup, it provides stable, uniform gas flow and leads to the reduction of hydraulic resistance of the proposed jet-bubbling contact device. Thus, formation of the original interaction between gas and liquid allows stimulating heat and mass exchange processes both in liquid and in gas phases at relatively simple apparatus design.

3 Calculation procedure and results

The purpose of the numerical studies is to determine the hydraulic resistance of the proposed contact devices. For that purpose, all-gas flows through the layer of contact elements were computer simulated via ANSYS software application. In this case, the average gas velocity was changing from 1 to 3 m/s, the width of the contact elements – from 0.04 m to 0.1 m.

The resistance of the dry layer of contact elements can be determined by a modify Darcy-Weisbach equation:

$$\Delta P_{dry} = \xi \frac{H}{b} \frac{\rho_G W_G^2}{2}, \tag{1}$$

where ξ – resistance coefficient; H – height of the layer of contact elements, m; b – width of the contact element, m; ρ_G – gas specific gravity, кg/m^3; W_G – average gas consumption rate, m/s.

It should be noted that the resistance coefficient is almost independent of the Reynolds criterion for gas (if Re_G = 1.5×10^3–1.8×10^4) and of the width of the contact element. Therefore, in engineering calculations it can be taken equal to 9.6, while the relative error does not exceed ±5.2 %.

Fig. 2 and Fig. 3 illustrate the dependence of the specific hydraulic resistance of jet-bubbling contact devices on the average gas consumption rate and width of the contact element. Fig. 2 shows that the impact of the specific hydraulic resistance on the width of the contact elements is most pronounced at high gas velocities.

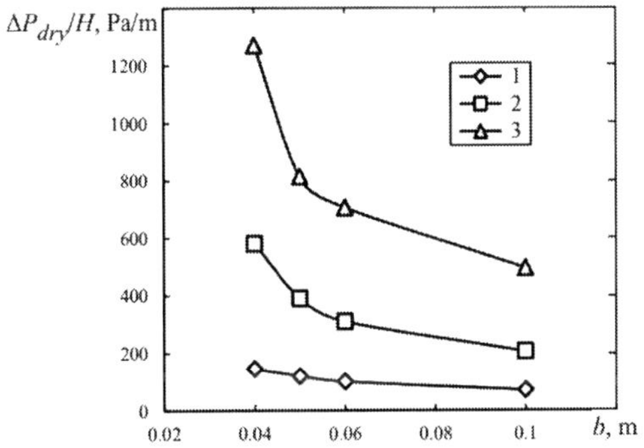

Fig. 2. The specific hydraulic resistance of jet-bubbling contact devices, depending on the width of the contact element at different average gas consumption rate W_G, m/s: 1 – 1; 2 – 2; 3 – 3.

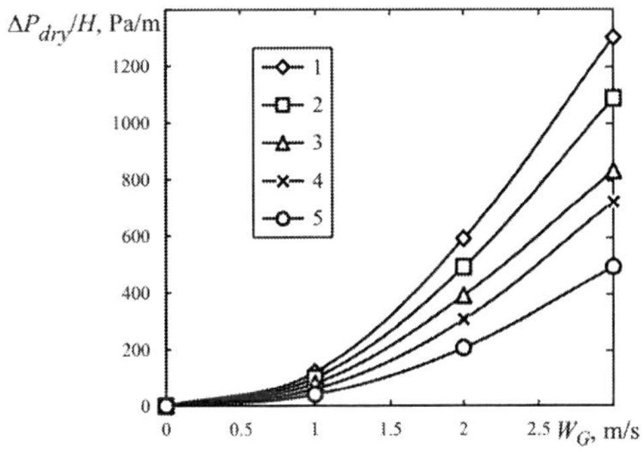

Fig. 3. The specific hydraulic resistance of jet-bubbling contact devices, depending on average gas consumption rate at different width of the contact element b, m: 1 – 0.04; 2 – 0.05; 3 – 0.06; 4 – 0.075; 5 – 0.1.

Wherein, the increase of the cell width of a contact device leads to the increased average velocity of a fluid flow, which is associated with increasing height of the land which has the greatest speed.

Analyzing the diagrams, it can be concluded that the increase in the width of the contact elements and reduction of the average gas consumption rate leads to a significant reduction of the hydraulic resistance of the proposed contact devices.

4 Conclusion

According to the results, the following conclusions can be made:

- if the distance between the drain cups at one level is chosen corresponding to the width of the drain cups, a constant section for the gas flow is provided, which leads to the decrease in the hydraulic resistance of the suggested jet-bubbling contact device,

- with the increase of the width of elements the average velocity of a flow under any conditions increases,
- developed delivery devices use energy of the liquid, flowing downwards, which reduces the flow resistance for the vapour phase (gas phase),
- thus, it can be stated that the use of the proposed contact devices for purification of gas emissions from thermal power plants, is promising, as they can provide good quality purification at high loads with relatively low energy costs.

The reported study was funded by RFBR, according to the research project No. 16-38-60081 mol_a_dk.

References

1. A. A. Makarov, Science & technology forecasts and problems of the power engineering development in the Russian Federation up to *2030* **79**, 206 (2009)
2. A. I. Skoblo, Y. K. Molokanov, A. I. Vladimirov, V. A. Shelkunov, *Processes and devices for oil & gas processing and petrochemical industry* (Gubkin Russian State University of Oil and Gas, Moscow, 2007).
3. L. R. Zijatdinova, A. N. Nikolaev, Ecology and industry of Russia, **3**, 23 (2009)
4. O. S. Makusheva, A. V. Dmitriev, N. A Nikolaev, Chem. Pet. Eng. **46**, 330 (2010)
5. N. A. Voinov, O. P. Zhukova, N. A. Nikolaev, Theor. Found. Chem. Eng. **44**, 213 (2010)
6. I. N. Madyshev, O. S. Dmitrieva, A. V. Dmitriev, A. N. Nikolaev, Chem. Pet. Eng. **51**, 383 (2015)

ETHICAL IMPLICATIONS TO THE ASSESSMENT OF THE SMART GRID TECHNOLOGY IN RUSSIA

Marina A. Makienko [1,*], *Natalia* V. Kurkan[1], and *Ekaterina* E. Emelyanenko [1]

[1]Tomsk Polytechnic University, 634050 Tomsk, Russia

Abstract. This paper brings up the issue of staff training for professional development of the Smart Grid technology and for use of its elements by customers in households. It was revealed that the considerable part of the respondents was not familiar with the definition of Smart Grid. That required the development of communication skills by energy engineering students and their social activity as well. The reasons mentioned make actual the following elements of engineering education: social responsibility, stress resistance, ability to forecast the future.

1 Introduction

Machines have been involved into the human life from the olden times initially intended to ease the rough labor as well as to make the environment available for living and provide the security of human life. It should be pointed out that in modern world, the new functions of machines start prevailing such as the ability to improve the human nature, change the social reality and the ways of communications, and provide the conditions for humanistic values. The authors interpret the humanistic values within the concept of the sustainable development formulated from the Brundtland Report of 1987 [1]. In modern view, the social dimension is an important part of the sustainable development which involves personality orientation, the stability of social and cultural systems, fair distribution of economic and social benefits, reducing the number of human conflicts [2]. The Smart Grid technology is no exception, it is subject to the analysis against the sustainable development background. The special aspect in adoption of the complex systems such as the Smart Grid is the uncertainty about the full range of the potential hazards when implementing such technology because the social and technological risks are revealed after a long period only. The research presents the components of the ethical evaluation which can be used to analyze the requirements considered when adopting the Smart Grid as well as the problems of the ethical competence development in training the engineering undergraduates.

2 Results and discussion

* Corresponding author: mma1252@tpu.ru

A review of papers enables us to identify the Smart Grid concept constituents as follows: the power engineering strategy aimed at the use of alternative sources of energy, energy-efficient technology as well as at the active involvement of clients into production, consuming and distributing the energy; the development strategy of the companies which produce the equipment for power industry; the development concept for power engineering companies, which is based on innovations and the responsible participation of a consumer. In Russia and other countries, the Smart Grid concept was stimulated by the government. In 2009, the government of the Russian Federation approved the Russia's Energy Development Strategy for the period until 2030. The developed strategy is oriented towards the transition to the new generation energy, the implementation of new technology, the efficient use of conventional and new energy sources [3]. This paper provides the directions and the strategy implementation techniques. The analysis of those points suggested that the economic and technical conditions as well as the results of the strategy adoption are primarily emphasized in Russia. The human resourcing of the power industry is, above all, the main social condition: the development of professional training, corporate staff training, the creation of well-equipped and secure workplaces. However, there is a problematic issue in this paper, in our opinion. This is the poor analysis of the cultural and intellectual constituents of the Smart Grid innovation according to the sustainable development concept. The extended introduction of this technology as the Smart City [4] allows us to consider this process as the change of the social reality.

To name this process, modern authors often use the term *social experiment* [5, 6]. An experiment as a research method supposes a researcher creating the controlled conditions to test a hypothesis. In view of this, there is no difference between a man-operated experiment and an experiment which involves material systems – the experiment is artificial to some extend as a researcher plans and monitors the situation that does not exist in the real world. However, there is a possibility of the irreversible effects during the experiment, which prevents a researcher from stopping the experiment or its negative impact. This requires some extra conditions to conduct the experiment as well as to adopt new technologies in society. The modern researchers propose to forecast risks and negative effects at the early stages of the technology adoption to neutralize the unintended consequences. So, the Responsible Research and Innovation approach is being developed today [7]. Innovation is always related to planning the future because the present has not provided yet any algorithms to carry out the innovation as well as any patterns for interaction of social institutions and groups both with each other and with the adopted technology. The lack of interaction models for the new technology results in necessity of the principle that provides the framework for the evaluation of innovations. For that end, the precautionary approach based on the sustainable development concept is often used. The core value is to meet the needs of population and society to the greatest possible extend with preventing the negative effects. The similar emphasis on the expectation takes place in Europe (the Horizon 2020 program) and in the US (National Nanotechnology Initiative, NNI) as well.

The next technique to analyse the new technology adoption, which is recognized among the modern researchers, is based on bioethics principles [8] – keeping the test person informed, the positive impact on society, undiminished security of the test person, human dignity, professional and responsible behavior of the test conductor. For the first time, the principles of bioethics were set up in the Nuremberg Code after the Nuremberg trials were over in 1947 [9]. In this paper, the principle of the informed consent is analysed. The informed consent for engineering purposes was studied in the paper by Mike W. Martin, Roland Schinzinger [10]. The principle of the informed consent is proven and to some extend, obvious for a modern medical experiment. It supposes that a participant of the experiment is aware of the test purpose, methods, duration and possible negative effects. It is necessary to extrapolate the informed consent in the sphere of engineering because the technology impacts and changes the human life to the same extend as the medicine. There is no doubt that the implementation of this principle can be complicated by following issues: the consumer awareness and the activity of the engineer. A consumer is entitled to the information provided as well as to the information which they are interested in, such as resources, the economic feasibility, the impact on environment, the production management, etc.

The next aspect to be considered when discussing the informed consent principle is the risk forecasting. A necessary basis for forecasting in the current practice is the system of values, on which a decision about the possible risks is made, and the concept of the future considered as the ground for planning the present. A conventional element of the decision making within the risk theory is an attempt to strike the possible balance between the economic benefit and the social-ethical responsibility. The economic benefit is apparent and of general importance, however, the ethical component should be analysed. The ethical issues of the technology implementation such as human security, preserving the personal and cultural identity, privacy right, development of traditional forms for preserving the sociality are the topical problems for our future. Hence, it is necessary to place greater emphasis both on individual values, which are of general importance as well, and on social values which will determine the life of humanity in the future.

The constituents of the ethical evaluation mentioned above are interrelated – the engineering inventions are adopted in the society considering the precautionary principle. The technology adoption is an experiment of a specific nature – a social experiment which is conducted to serve mankind. To determine, whether the invention is really helpful or not the informed consent is applied. In our opinion, these principles, despite some constraints, can be applied for analysing the Smart Grid technology in a social context. The sustainable development involves preserving the planet with the conditions and the amount of resources that provide the fulfilling life for our descendants. The requirement for the sustainable development is the combination of the economic development, environmental system, social development and traditional values, which are unique for every culture. Our understanding is that the Smart Grid fully complies to the sustainable development concept.

The Smart Grid provides the opportunity to use the alternative sources of energy which enhance the chances to preserve the nonrenewable sources for the next generations. While the renewable energy sources reduce the negative impact on environment, they increase the consumer responsibility. Being simultaneously a producer, a consumer has the opportunity to cut down the energy costs and to supply the rest to other consumers. The adoption of this technology positively impacts different groups, which interests are affected by the technology adoption: it meets demands of each consumer, the potential for technological risks is reduced which is the essential advantage for all nations; the smart control system provides the realtime data about the equipment status reducing the labor costs of the system operator; the need in reconstruction of the workstation provides the safe working conditions for human resources; the Smart Grid designing stimulates the interrelated fields of science and technology, so, the positive working environment and facilities for research and development are provided; the reduction of production costs due to the cutting down the energy costs provides the positive effects on business [11].

In spite of the positive aspects of the Smart Grid adoption, it should be noted that the ordinary consumers make the significant part of the mentioned groups, which can benefit from using this technology. This part, as was earlier discussed [12], is poorly aware of the Smart Grid itself as well as of possible positive and negative impacts of the technology adoption (this is the proven fact for Russia). So, the adoption requires that an engineer interacts with the society and clarifies possible impacts of the technology and supports this way the informed consent [13]. That is the aspect which strengthens the social constituent of the engineer's activity in the modern world, and this means changing the goals of the engineering education declared in the Learning for the future concept [14]. That document noted that the education is a condition for the sustainable development as the appropriate world view, knowledge and understanding of problems and solutions are formed through the education. It is interesting that the document formulated the requirements to the academic staff who make students develop the necessary competences in the field of the sustainable development.

According to that paper, every teacher, irrespective of their specialty and training, should develop themselves four groups of competences: they should teach the principles of communication between different cultures, generations, social groups, people and environment; they should teach considering the students' life experience and solving the urgent problems; they should help to coordinate the different visions of the future, shape the view of the sustainable development; they should understand the impact of the present mechanisms on the future. These competences are obviously to be developed by students. Targeting this problem is a topical issue in Russia. Tomsk Polytechnic University currently provides the action plan aimed at the development of the competences required. This process is performed with the change of the social and humanity training the engineering students. According to the new concept, the training of social and humanity competences is to be resulted in the qualifications of a graduate who should be able to make the professional decisions with the principles of social responsibility, meet social challenges and demands, be an efficient worker both in national and international teams [15]. The concept adoption is complicated by the fact that a system of certain values should be its final goal. So, it should be noted that the modern society makes two conflicting requirements to the engineering training: it is necessary to shape the worldview on the sustainable development basis, on the one hand, but the intention and ability for the continuous production, on the other hand. The issue whether the combination of these requirements is possible in the training process as well as in the worldview of a modern person remains an urgent problem when developing the necessary competences of the perspective engineers.

3 Conclusion

The adoption of new technology in a society can be considered as a social experiment as it provides the transformation of both the material and the social realities as well as the system of values. The very nature of the social experiment is in the fact that it is hardly possible to forecast its procedure and results and to stop the experimental execution as well. Moreover, the participants of a social experiment should be large groups of people, who rarely consider themselves as the experiment parts, and the risks taken by the participating groups in case of the unfortunate occurrence. The specified characteristics focus the researchers on the ethical requirements which are to be met: the informed consent, the positive impact on social life, the safety of the test person, human dignity, professional and responsible behavior of the test conductor. The informed consent should be a condition for the adoption of the global technology such as the Smart Grid and stimulate its expansion among populations. Implementing this principle is possible on condition of the changing the goals of the engineering education according to the sustainable development. However, there is a conflict of the values in the modern society i.e. the simultaneous commitment to consumption and saving. The next research will provide the possible solutions of this conflict in the engineering education.

This research was conducted with the support from Russian Humanitarian Science Foundation (RHSF) 16-16-70006.

References

1. Report of the World Commission on Environment and Development: Our Common Future.
2. Li M. National Institutions for Sustainable Development: A preliminary review of the institution literature. United Nations Department of Economic and Social Affairs, (2012)
3. Russia's Energy Development Strategy for the period up to 2030. Moscow, 174 (2010)
4. The Smart Cities Council. URL: http://smartcitiescouncil.com/
5. N. Doorn, S. Spruit, Z. Robaey, Science and Engineering Ethics, **3**, 22 (2016)
6. C. Ansell, What is a «Democratic Experiment?», **2**, 9 (2012)
7. A. Grunwald, Enterprise and Work Innovation Studies, 7 (2011)
8. Ibo van de Poel, Science and Engineering Ethics, **3**, 22 (2016)
9. Der Nuernberg Kodex 1947. URL: http://www.ippnw-nuernberg.de/aktivitaet2_1.html
10. Martin M, Schinzinger R., *Introduction to Engineering Ethics* (McGraw-Hill Publ., 2000)
11. B.B. Kobets, I.O. Volkova. *Innovatsionnoye razvitiye electroenergetiki na base contseptsii Smart Grid* (Moscow, Energy Publ., 2010) [in Russian]
12. Marina A. Makienko, Natalia V. Kurkan, Anna A. Strelcowa, Philosophical approach to engineering education under the introduction of the smart grid concept in Russia, MATEC Web of Conf., 37, 01031 (2015)
13. I. B. Ardashkin, L. A. Korobeynikova, A. V. Popova, Status of social competencies of power engineers in the context of forming the concept of an intelligent network or smart grid, MATEC Web of Conf, **37**, 01003 (2015)
14. *Learning for the future. Competences in Education for Substainable Development*, (2012)
15. A. Chmykhalo, V. Khaliulina, M. Abushaeva, Innovative Power Systems and The Formation of The Creative Class in Russia, MATEC Web of Conf., **37**, 01016 (2015)

MATEC Web of Conferences **91**, 01021 (2017) DOI: 10.1051/matecconf/20179101021
Smart Grids 2016

METHOD OF CALCULATION OF CURRENT OF THE GROUND FAULTS IN THE PARALLEL OVERHEAD TRANSMISSION LINES 110-220 KV

Alexandr V. Ukraintsev[1,2,*],and *Ivan* V. Nagay[1,2]

[1] Platov South State Federal Polytechnic University (NPI), 346428 Novocherkassk, Russia
[2]AvtomaticaDon, 347900 Taganrog, Russia

Abstract. The mutual induction between the phase wires of the different overhead lines which situate close to each other cause unbalanced redistribution of currents in the line wires. This leads to emergence of the out-of-balance zero sequence current which affects negatively on the sensitivity of the zero-sequence current protection. It is impossible to estimate such out-of-balance current by the means of the typical calculation programs for short circuit currents. This paper describes the method of "virtual" lines for an extra correction of the values of zero-sequence currents during the current ground faults happening in the overhead lines 110-220 kV. There is an example of using this method for three parallel overhead lines 220 kV passing close to each other.

1 Introduction

In the multiended overhead lines 110-220 kV which include multiwire overhead parallel line (OHPL) and OHPL with single or several branches there is a problem of taking in account the zero-sequence current while adjusting sensitive stages of zero-sequence current protection[1]. This problem exists because of the high out-of-balance currents in the lines as much as in normal operating conditions and during the remote asymmetrical short-circuits which happen by the reason of mutual induction between phase wires of one line or two separate lines. In these circumstances the values of currents and the distance between the wires play an important role [2]. In the much loaded OHPL such influence might cause zero-sequence currents of several dozens of amperes [3]. Out-of balance currents are approximately the same as the currents in the set points of sensitive functions (zones) of zero-sequence current protection. The set points are put out functioning to avoid their false work and the reliability of backup is reduced.

The zero-consequence out-of-balance currents result the great difference between the real and calculated short-circuit currents of single-phase short circuits while using typical calculation programs [4].Finally, there is a requirement to accept set points of protection with a large safety factor to exclude non- selective operation, but sensitivity of protection is reduced.

* Corresponding author: xelandr@mail.ru

2 Experimental setup and study technique

To consider the influence of the mutual induction of the triple-circuit transmission line on the phase currents flowing in it while fault regime the part of a network represented in the Fig. 1 has been taken for analysis. It has three parallel OHPL: W1, W2, and W3.

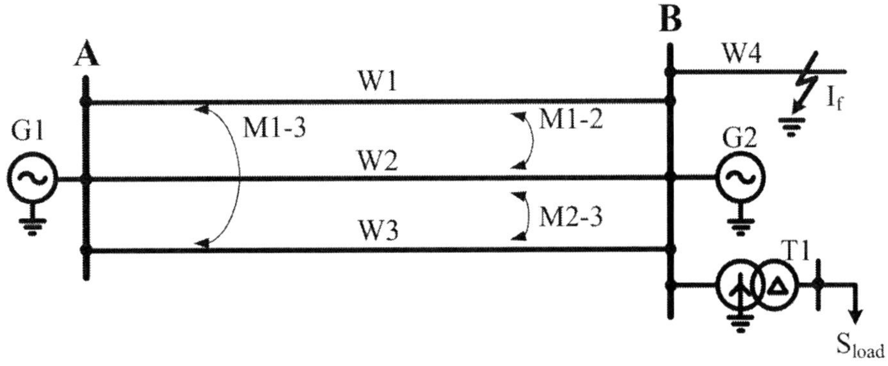

Fig. 1. Network scheme.

The section consists of three OHPL which go close to each on the same routingand have the start points at one substation and the end points at another substation. This makes it possible to consider them as triple circuit line. All the lines have a metal lightning arrester grounded on each tower. Every line is divided for 7 segments where the distance between the wires is different.

According to the analysis implemented with the program of calculation of the short-circuit currents in electric Power Networks RTKZ 2.03 [5], for unbalanced conditions caused by ground short circuit of one of the phases of considered OHPL,there is no required accuracy in values of short-circuit currents because they do not correspond to those which were registered by recording emergency instrument at the fault moment. The essential of the main obstacle to achieve the needed result is the absence of an opportunity to set mutual induction between separate phases and impossibility to take into consideration the affection of grounded wire hawsers and unbalanced conditions of the load. These limits can be eliminated by the application of extra methods for more precise definition of the parameters of conditions.

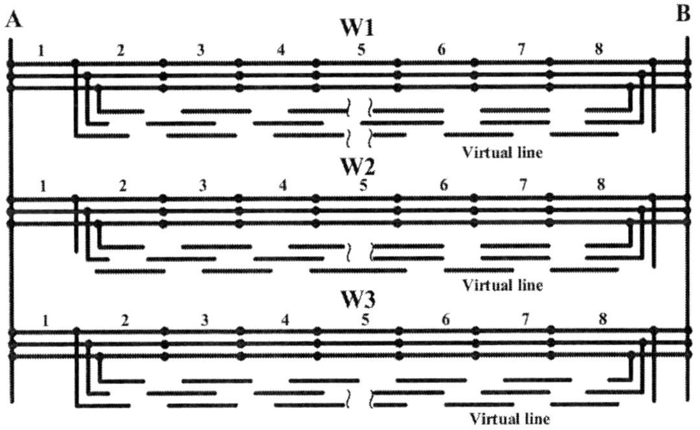

Fig. 2. An explaining scheme of the lines having three "virtual" ones to simulate an asymmetry.

The inherent and mutual resistances of the equivalent circuit have been calculated during the research based on the geometrical parameters of the lines.

It is offered to set the "virtual" lines to simulate the asymmetry caused by the mutual induction between not only the separate circuits (equivalent wires of circuits) but separate phase wire of the single line. In Figure 2 there is an explaining scheme of the lines having three "virtual" ones.

The resistances of "virtual" lines are calculated by the following method. For example, let us assume that $I_A > I_B > I_C$. In this case it is needed to set two parallel lines with the simulation of the open-phase mode in the line №1 (the failure of the phases B and C) and in the line №2 (failure of the phases A and C). The resistance of an extra line №1 z_{wi1} in the base of the main line can be defined the following way:

$$z_{wi1} = \frac{I_{wiA}}{I_{wiA} - I_{wiC}} \cdot z_{wi}, \tag{1}$$

$I_{wiA,B,C}$ - phase currents of the main line W_i

$$z_{wi2} = \frac{I_{wiA}}{I_{wiA} - I_{wiB}} \cdot z_{wi}, \tag{2}$$

z_{wi2} - the resistance of an extra line №2.

There also can be the mutual induction between the main circuit W_i and "virtual" lines W_{i1} and W_{i2}.

Thus, two "virtual" lines must be set in parallel of each line.

There is another way of setting the "virtual" lines according to it one "virtual" line is set in parallel to basic line. The resistances of the "virtual" lines are defined by the means of the following method:

The phase with the most current $I_{wi\max}$ is defined;

Then the resistance of the "virtual" line z_{wi1} connected to this phase is defined according to the following formula:

$$z_{wi1}^{(1)} = \frac{I_{wi\max}}{I_{wi}^{0}} z_{wi}^{(1)}, \qquad z_{wi1}^{(0)} \approx z_{wi1}^{(1)} k, \tag{3}$$

$z_{wi1}^{(1)}$, $z_{wi1}^{(0)}$ -resistance of the direct and zero sequence of the "virtual" lines;

$z_{wi}^{(1)}$ - resistance of the direct sequence of the main line; k=3÷4.

3. Two phases with the minimal current are switched off.

The currents have been calculated using the method confirm its efficiency because the error of the calculation is not more than 10 % meanwhile without using the method described above it might be about 20-30 %.

3 Conclusion

Thus, the proposed method to define the parameters of the "virtual" lines with the program RKTZ 2.03 which allows to calculate the modes of the network having numerous longitudinal-transversal asymmetries and the load lets to simulate the affect of the line's own mutual induction between the wires "phase-to-phase" and "phase-to-wire-lightning arrester" and of mutual induction of neighbouring circuits. At the same time, it allows to calculate accurately for the relay protection (10 % of error) the distribution of the zero-sequence currents (phase currents) of the lines having significant mutual induction between circuits in the load-mode, which is hard to achieve by the means of the typical programs used in Russia to calculate the short-circuit currents.

This paper has been prepared with the support of Ministry of the Education and Science of Russian Federation within the framework of the Federal Target Program "Research and development in priority areas of Russian scientific and technological complex for 2014-2020", the Agreement № 14.579.21.0083, PNIER theme "Development of technical solutions of distributed system of smart grid backup, proving increased reliability and survivability of electrical equipment" (a unique identifier RFMEFI57914X0083).

References

1. V. I. Nagay, G. N. Chmihalov, P. S. Kireev, I. V. Nagay, S. V. Sarry,Proceedings of the 8th International Scientific Symposium on Electrical Power Engineering, 8, 360 (2015)
2. A. B. Checherin, Korotkie zamykania pri nepolnofaznyh regymah v electricheskih sistemah - M.: Gosenergoizdat, (ЧерниnА.Б. Короткие замыкания при неполнофазных режимах в электрических системах. – М.:Госэнергоиздат, 1952. – 167) [in Rassia]
3. A. V. Ukraincev,Scientific and Technical Journal Russian Electromechanics, **3**, 119 (2014)
4. A. Brilinskyi, A. Sevastyanova, Sravnitelnyi analiz raboty program dlya rascheta nesimmetrichnyh korotkih zamykanyi v energosistemah, CADmaster, **5**, 35 (2014)
5. http://niieng.ru/production-obsego-primenenia/software/rtkz_2/

EXPERIENCE OF INTERNAL AND EXTERNAL EDUCATION QUALITY RATE: ESG AND THE LEGISLATION OF THE RUSSIAN FEDERATION

Ivan N. Nikanorov[1,*], and *Svetlana* V. Volkova [2]

[1]The Ministry of Education and Science Council's youth affairs' education quality commission of Russian Federation, Moscow
[2]Candidate of Sciences (Pedagogics) (PhD), The Ministry of Education and Science Council's youth affairs' education quality commission of Russian Federation, Moscow

Abstract. Overview of regulation of internal and external education quality rate has been presented. The development of regulation an independent assessment of higher education in Russia in the context of ESG has been considered. Authors noticed, that instruments of internal and external quality rate interact in Russian legislation insufficiently.

1 Introduction

Rapid changes in the world over the past 20-30 years have particularly transformed the global economy, radically changing the requirements for personnel training and, relatively, for the higher education system. The formation of "lifelong learning" conception has become a solemn challenge for the traditional institutions and the educational management principles.

Common problems and challenges inherent to a numerous systems of higher education, among which are countries of Western and Eastern Europe, as well as some countries in Asia and CIS, have determined the necessity of corporate solutions to form an integrated area that provides a mutual educational levels recognition, quality assurance along with the expansion of its availability.

The creation of European Higher Education Area (EHEA) has become the result of these mutual efforts after the Declaration at the conference in Budapest – Vienna in March 2010 was signed. It currently includes 48 participating countries.

One of the key documents organizing the establishment of unified principles of the higher education quality in the designated area has become "Standards and Guidelines for Quality Assurance in the European Higher Education Area" (ESG) [1]. This document has been adopted by the conference of Ministers, responsible for Higher Education of EHEA, in Yerevan on May 14–15, 2015.

2 Results

* Corresponding author: nikanorov@zaobrazovanie.ru

It has added opportunities for participation of educational institutions in international projects and expanded mutual cooperation. It has also introduced new principles of educational process for many educational systems, in particularly in Russia these are "student-centered learning" and "on-going monitoring and periodic review of programs". The European Students Union, which is a recognized member of the European education quality assurance system, actively involves students in the assessment mechanisms and the "internal quality insurance" of educational institutions. With the goal of developing student feedback tools and to improve quality of education, The European Students Union has become one of the key developers of "Standards and Guidelines for Quality Assurance in the European higher Education Area"(ESG).

The system of professional certification of qualifications and independent assessment of education quality based on the activities of professional communities is an essential guarantee for the independent education quality assessment of personnel training systems of the countries participating in the EHEA.

Thus, the European higher education area (EHEA) provides a complex self-developing system of the education quality assessment based on student-centered principles, based on independence and transparency.

In this regard, let us consider the development of Russian national legislation in the field of education that has formed the background for the devolvement of practices of involving students in the assessment of the education quality. The reforms of Russian higher education system within the last decade has been aimed at gradual transition to the competitive development of educational institutions. In this respect, each educational institution of higher education has received greater autonomy and, accordingly, greater responsibility within the Federal law from December 29, 2012 No. 273-FZ "On education in the Russian Federation", which came into force on 1 September 2013[2].

The development of the higher education system is impossible without creating multi-level tools of assessing the effectiveness and quality of education. Legal framework of formation of the system are defined by the above-mentioned Federal law as well as by the Federal law of July 21, 2014 № 256-FZ "On amendments to certain legislative acts of the Russian Federation on the point of independent evaluation of service delivery by the organizations in the sphere of culture, social services, health and education" (came into force on October 21, 2014) [3].

In this regard, the changes of article 95.1 "Independent evaluation of the quality of training of students" № 273-FL "On education in the Russian Federation" were adopted. It stated that "independent assessment of the quality of students training is held at the initiative of participants of relations in the sphere of education with the aim of preparing information about the level of mastering of the educational program or its parts, as well as providing the participants of relations in the sphere of education with information about the quality of training".

It is noteworthy that the coordination of activities and general methodological support of independent assessment of quality of services in the field of education is realized by the Federal Executive body, responsible for drafting and implementing state policy and legal and regulatory framework in the sphere of social protection for the public (Ministry of Labour and Social Development of the Russian Federation), in the order established by the Government of the Russian Federation. Earlier, in the context of absence of established mechanisms of legal regulation of this issue, the Federal executive authorities responsible for public policy in education (Ministry of Education and Science of the Russian Federation), carried out this activity independently on the basis of expediency.

Along with this, there is formed the system of professional-public accreditation, directly focused on the involvement of employers in assessing the quality of higher education. Its results can be considered by the founder of the higher education organization in making management decisions, including allocation of control figures of enrolment of students for training at the expense of budgetary appropriations of the Federal budget (CFE). Control figures of enrolment serve as an indicator of the demand for graduates of educational programs of higher education by employers in the eyes of potential applicants. In the emerging space of professional-public accreditation, the most notable associations are the Association of Russian Lawyers (AYUR), the Federation of Restaurateurs and Hoteliers, Opora of Russia, the Association of Managers of Russia and others. However, it should be noted that not all of the integrated groups of specialties have associations of employers that are interested in conducting qualified professional-public accreditation. The current practice of conducting professional-public accreditation allows to note that the criteria used by associations of employers are completely different criteria database and evaluation mechanisms, which makes it difficult to unify and correlate of the results of different accreditation procedures.

Thus, the legislator forms a system of education quality assessment oriented not primarily to the state education authorities, but to employers and their associations. Now that the assessment system of education quality by employers is in its infancy, it can be assumed that its regular functioning is a matter of long-term prospects.

Further development of legal regulation on this issue is expressed in the Federal law "On the independent assessment of qualification" from 03.07.2016 N 238-FZ [4] that has not come in force yet. The rules of this act assume full transfer of responsibility for certification of qualifications to employers and authorized centers of the qualifications completely excluding educational organizations and the academic community from the procedures. The implementation of this law is still a matter of debate among experts, as well as its correlation with the labor legislation of the Russian Federation.

3 Conclusion

In the framework of existing educational system there has formed a clear understanding of the necessity of assessment of the higher education system, not only external (public and expert) assessment, but also the internal one that expresses the opinion of the student community in respect of quality of educational services received. The analysis takes into consideration not only the work of the faculty, but also the satisfaction of the students with the conditions and results of education. Final results of the assessment are subsequently considered in the system of indicators of efficiency of higher educational institutions.

Only a system based on the maximum involvement of all the parties can solve social-cultural problems that Russian education faces. In this regard, it is most desirable to develop the system of shared government and public administration in the education system by developing the activities of the representative bodies of direct participants of the educational process, of students in particular.

All-Russian public organization "For quality education" in the framework of the international youth project "Involvement of students in evaluating and improving the quality of education" has formed a strong model for collaboration between student associations, the faculty and administration of educational institutions, aimed at improving the quality of education through the mechanisms of the internal assessment.

Thus, it should be noted that along with the development of the system of independent assessment of qualifications it is necessary to have tools for the development of internal education quality assessment, which will help to harmonize the interests of all stakeholders in the education system, as well as enhance the integration of the Russian Federation in the European higher education area (EHEA).

DOI: 10.1051/matecconf/20179101022

References

1. *Standards and Guidelines for Quality Assurance in the European Higher Education Area (ESG)* (Brussels, Belgium, 2015).
2. Federal Law 273-FZ, dated 29 December, "On education in the Russian Federation", 2012.
3. Federal Law 256-FZ, dated 21 July, "On amendments to certain legislative acts of the Russian Federation on the point of independent evaluation of service delivery by the organizations in the sphere of culture, social services, health and education", 2014.
4. Federal Law 238-FZ, dated 3 July, "On the independent assessment of qualification", 2016.

THERMAL CONVERSION OF THE PEAT TO COMBUSTIBLE GASES

Maxim E. Burnusov[1, *], *Alexander* V. Kazakov[1], *Konstantin* V. Buvakov[1], and *Maitree* Polsongkram[2]

[1]Tomsk Polytechnic University, 634050 Tomsk, Russia
[2]King Mongkuts University of Technology Thonburi, Department of Mechanical Engineering, Bangkok, Thailand

Abstract. Presented thermal characteristics of the peat and semi-coke derived on its basis. Presented the results of studies of the pyrolysis and catalytic conversion of peat and semi-coke at temperatures of 300-450 °C. According to the results of experiments to determine the composition and calorific value of combustible gases, obtained as the result of peat thermal conversion and semi-coke based on it. Presented the structure and analyzed the material balance and the energy distribution of derived products by the thermal conversion.

1 Introduction

Perspective directions of the Power Engineering development of Russia cause the relevance of centralized power system combinations with off-grid power plants using low-grade raw materials as a fuel.

Scientific direction of the research is to develop an approach to the deep conversion of fuels organic matter into combustible gases at low temperatures (maximum 500 °C). Advantages of this approach: low process temperatures that allows to refuse expensive constructive materials; minimum ballast in gas; the possibility to generate hydrogen-rich gas; the possibility of organizing process in autothermal mode. [1]. Obviously, that it is extremely difficult to carry out a deep conversion of carbon fuels into combustible gas at the low-temperature heating of solid organic material that is the main problem of research. Thus, disadvantages of the process is reduced intensity, additional energy costs of preparing the steam [1]. The solution of this problem will allow us to develop a new direction in the study and creation of technique and technology of resource-efficient production of fuels and energy.

The object of study is the process of low-temperature conversion of low-grade fuels, characterized by the following provisions: converting organic matter into combustible fuel gases flows under the influence of temperature in the range of 300-450 °C at atmospheric pressure and without supplying oxygen blast; into active reaction zone entered the catalytic

* Corresponding author: kazakov@tpu.ru

additive and steam with temperature corresponding to the level of the conversion process (300-450 °C) ; achieved maximum level of organic matter fuels conversion to hydrogen-rich combustible gas with the minimum ballast share [2].

2 The features of studying samples

The features of peat sample and the semi-coke obtained on its basis are given in the Table 1.

Table 1. The features of studied samples (IS – initial sample, SC – semi-coke from the initial sample).

Calculated mass / Features	Operating mass		Dry mass		Dry & ashless mass	
	IS	SC	IS	SC	IS	SC
VOC, %	–	–	–	–	75.40	7.08
Moisture, %	69.90	0.66	–	–	–	–
Ash levlel, %	4.49	24.20	14.92	24.36	–	–
Carbon, %	12.50	73.06	41.53	73.54	48.81	97.24
Hydrogen, %	4.11	0.98	13.65	0.99	16.05	1.30
Oxygen, %	8.50	1.00	28.24	1.01	33.19	1.33
Sulfur + Nitrogen, %	0.50	0.10	1.66	0.10	1.56	0.13
The net calorific value, MJ / kg	7.5	25.40	24.92	25.57	29.29	33.80

The peat has a high moisture content (69.9 %) at low ash level value (4.49 %) in the operating mass, that considerably determines the slight calorific value equal 7.5 MJ/kg. The emission of volatile organic compounds (VOC) is characterized by a high value (75.4 %) to fuel of the early stages of metamorphism that the indicating about high reactivity of peat capacity provided to reduce its moisture content.

Ash level of the dry mass is 14.92 %, that slightly higher than milled peat (Rostorf). In dry state, the peat has a quite high calorific value (24.92 MJ / kg) due to the high carbon (41.5 %) and hydrogen (13.7 %) content. High oxygen content is negatively affects on the calorific value of dry mass, however, this amount of oxygen is inherent in "young" coals. At the same time, high oxygen content in the fuel leads to the conclusion that during the thermal conversion of fuels will be observed exothermic heat effect, which will allow, either reduce the amount of heat depending on the process temperature, involved into the process, or to arrange it in autothermal mode [3, 4].

Semi-coke derived from the peat has a low VOC value (7.1 %) and almost no moisture (0.7 %), compared with the initial state of peat. In this case, the ash level has a quite high value (24.2 %). The organic part consists mainly of carbon (97.2 %) with low hydrogen (1.3%), oxygen (1.3 %) and nitrogen (0.1 %) content in fuel. Due to the small value of VOC and huge carbon content, it is difficult to organize the burning of peat's semi-coke. However, it has been known that obtaining heat from the semi-coke can be performed in a glow mode. At the same time due to the small content of VOC, the combustion gases are colorless, and the intensity of the process is maintained by highly porous fibrous structure of the semi-coke, which ensures the supply of atmospheric oxygen to the required extent.

Despite close to the limit of carbon content in the organic part of peat's semi-coke the net calorific value of operating mass is 25.4 MJ/kg (with the calorific value of 33.8 MJ/kg of a dry ashless mass), due to the high ash content. In general, we can say that the peat's semi-coke is a promising energy feedstock.

3 The features of obtained gases

According to the results of experiments to determine the composition and calorific value of combustible gases, obtained as the result of peat thermal conversion and semi-coke based on it (Fig. 1).

By increasing the temperature of treatment increases the content of hydrocarbon gases, first due to the increase of hydrogen and methane. The concentration of carbon monoxide, conversely, is reduced, which is caused by reactions in the early stages of the heat treatment. High oxygen content in the gas ballast caused by high surface area of the initial fuel with a large amount of adsorbed air.

According to the composition of combustible gases defined their net calorific value. Obviously, that the calorific value of the gases is determined primarily by the amount of methane. The increasing of gas calorific value from the peat throughout the entire studied temperature range provided an increase in the content of the gas methane and hydrogen with a slight fall of carbon monoxide. During the catalytic process, there is a noticeable increase in the concentration of combustible components first of all – hydrogen.

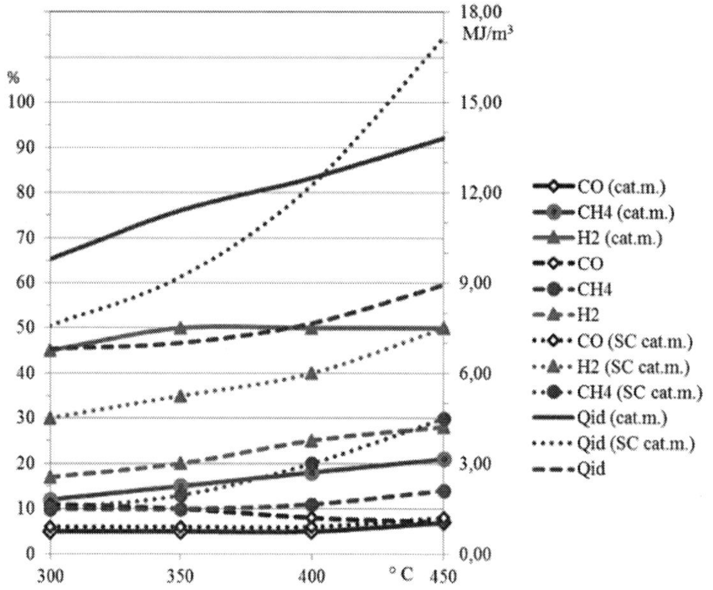

Fig.1. The content of combustible components in the gas from peat and semi-coke, % vol. The net calorific value of the gas from peat, MJ/m^3.

4 The material balances

During the thermal treatment of solid fuels the following products are formed: carbon residue (G_C); dry gas ($G_Г$); resin and paragenetically moisture of fuel (G_{CM}); the mineral part of fuel (G_M); the moisture of fuel, due to the operating moisture (G_W). Thus, the material balance is represented in the equation:

$$G_C + G_\Gamma + G_{CM} + G_M + G_W = 1 \tag{1}$$

According to the equation form of material balance, Fig. 2 provides information about products obtained by the process of semi-coking (SC) and the low-temperature catalytic conversion (LTCC).

The equation form of material balance for the LTCC has the fundamental difference between the balance of semi-coking, as in the LTCC take part are additional entered steam (G_n) and catalytic additives (G_κ). Noting the features of the LTCC, should be considered that the part of the input high-temperature steam is involved in the reactions of hydrogen generation and reduction of ferrum oxides, so due to the steam there is a change of the material balance components, as the volume of gas and undecomposed steam. In this case, when compiling the material balances of LTCC should be considered the amount of injected steam. As for the catalytic additive, the volume of direct transformation does not occur, and rule of additive balance are not apply to them, so there is no need to consider the presence of catalytic additives during compiling the material balances of LTCC [5]. Thus, the equation (1) takes the form:

$$G_C + G_\Gamma + G_{CM} + G_M + G_W = 1 + G_\Pi \tag{2}$$

Figure 2 shows the material balances based on dry mass of the low-temperature catalytic conversion of fuels, according to the equation (2).

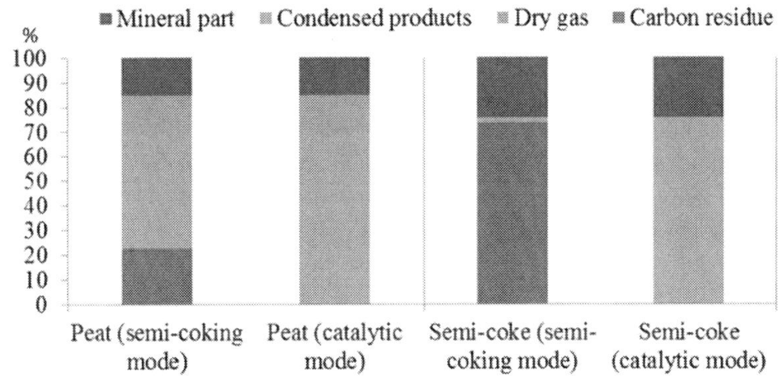

Fig. 2. The material balance of semi-coking and catalytic mode.

5 The heat balances

The heat balance is represented as the redistribution of the initial peat energy consumption by products of conversion. The determination of material balance is based on the quantitative data of material balance and the analysis of gas quality data. Figure 3 shows the results are based on dry ashless mass.

Obviously, that during the implementation of the semi-coking mode considerable part (about 30 %) of energy consumption accounted to the solid carbon residue. During the usage of peat as the initial feedstock, as in the semi-coking mode, and well in the catalytic mode, the part of the energy is distributed into a liquid fraction - resin. During the work with semi-coke in the semi-coking mode allocated only the VOC products and energy distribution between the gas and the carbonaceous residue makes the proportion of about 1:1. In the case of catalytic mode, all organic matter is converted to the gas and, respectively, the entire power consumption of the initial fuel converts to net calorific value of the gas.

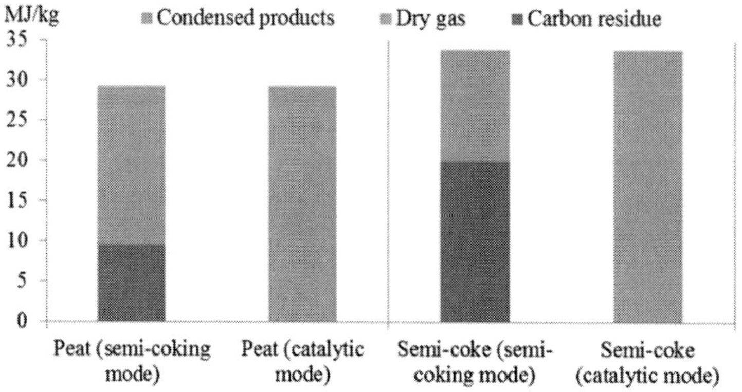

Fig. 3. The redistribution of power consumption initial sample by products of conversion.

6 Conclusion

Shown the results of studies of peat thermal conversion in the classic pyrolysis mode and the catalytic mode, characterized that in the process are injected iron oxide and steam with temperature of 300-450°C.

The intensity of the catalytic mode is low, which imposes a limitation for its industrial application. It seems appropriate to use this mode in power plants for small-scale power generation [6].

The embodiments of the conversion can be either manufacturing the hydrogen-rich gas, or combined gas and carbon product.

The reported study was partially supported by the Ministry of education and science of The Russian Federation, state order No. 13.948.2014/K

References

1. R. B. Tabakaev, A. V. Astafev, A. V. Kazakov, A. S. Zavorin, M. Polsongkram, IOP Conference Series: Materials Science and Engineering, **93**, 012017 (2015)
2. R. B. Tabakaev, A. V. Astafev, A. V. Kazakov, A. S. Zavorin, MATEC Web of Conf., **23**, 01039 (2015)
3. R. B. Tabakaev, A. V. Astafev, A. V. Kazakov, A. S. Zavorin, MATEC Web of Conf., **23**, 01040 (2015)
4. A. V. Kazakov, T. M. Plakhova, R. I. Popov, MATEC Web of Conf., **19**, 01019 (2014)
5. P. S. Gergelizhiu, S. A. Khaustov, R .B. Tabakaev, P. Y. Novoseltsev, A. V. Kazakov, A. S. Zavorin, Proc. of 2014 Int. Conf. on Mechanical Engineering, Automation and Control Systems, 6986901 (2015)
6. A. V. Gil, D. A. Baturin, MATEC Web of Conf., **72**, 01009 (2016)

TESTING THE DESIGN OF TECHNICAL SOLUTIONS FOR LIQUID HYDROCARBON WASTES FIRE DISPOSAL

Sergey V. Dolgov [1,*], *Ruslan* A. Visloguzov [2], *Sergey* A. Khaustov [2], *Roman* B. Tabakaev [2], and *Anton* N. Ermolaev [3]

[1] MUE «Teplosnabzhenie», 628606 Nizhnevartovsk, Russia
[2] Tomsk Polytechnic University, 634050 Tomsk, Russia
[3] Federal State budget institution of higher education "Tyumen industrial University», 625017 Tyumen, Russia

Abstract. The relevance of the work is caused by the need to develop the environmentally friendly heat-generating devices for disposal liquid technological wastes. The main aim of the research is to optimize the constructive implementation of technical solutions for disposal liquid hydrocarbon wastes. Using the software package «Ansys Multiphysics» the authors have simulated the combustion front at various flow conditions and measured reaction rate, temperature and components concentrations in the combustion zone. The results of the investigation have shown that the proposed technical solution has a number of features.

1 Introduction

The problem of the man-made liquid wastes disposal is one of the priority tasks to be solved by the international community. Liquid wastes are produced by oil refineries, chemical industry and transport, and may contain more than 200 dangerous chemicals causing irreparable harm to humans and the environment. This type of wastes contains large amounts of pollutants, harming to environment even when stored in specially engineered landfills. The magnitude of the problem is obvious, because in the most environmentally-advanced countries the proportion of the recycled wastes does not exceed 60 %, resulting in accumulation of wastes in large amounts.

Hundreds of millions of tons of sludge are accumulated, and 2-7.7 millions of tons of used lubricants are produced annually [1]. The possible ways of recycling can be divided into four main areas: dumping, recovery, recycling [2-4] and fire disposal [5, 6]. But only 15 percent of the total amount of wastes are recycled [1].

The most promising areas are fire disposal or recycling into a new kind of product. Fire disposal can withal solve the problem of power supply of remote regions [7, 8]. At present power supply in these areas is provided by expensive imported diesel fuel [7, 9].

Accordingly, development of the technical solutions for fire utilization of liquid hydrocarbon wastes and use of the produced energy is an important task of research.

* Corresponding author: tabakaevrb@tpu.ru

2 Fire disposal of liquid hydrocarbon waste in porous media

The main problems of the liquid wastes fire disposing are explosion and environmental safety of the combustion [10]. The possible option of solving these problems above is the combustion of liquid hydrocarbons wastes in the porous layer. The flameless combustion for radiative heat release makes possible the reduction of temperatures in the combustion zone to the values lower then thermal threshold of nitric oxide [11, 12].

Previously, with the participation of the authors the burner design modifications for fire disposal of liquid fuels [13] have been developed, using kerosene TS-1 and the mixture of waste flammable liquids. Physical research of the developed models was performed on experimental stands. Structures, actions and methods of these tests are described previously in [13].

A review of studies made using the mathematical models [14], and the analysis of experimental results suggest that all examined burners require a high excess air ratio (EAR = 2-6) because of the low mixing speed in the porous layer for complete fuel oxidation process. Reducing the excess air ratio results in excessive levels for unburnt fuel, causing burner to operate in "environmentally unfriendly" flare mode that uses the diffusion of atmospheric oxygen [5, 6]. Organization of the combustion with the high values of EAR and aerodynamic resistance of the workspace requires additional energy consumption for blasting, and therefore it has limitations on the capacity when firing low-calorie waste liquid hydrocarbon.

The objective of this study is to study the characteristics and regularities of the combustion process in the porous layer to be used in the development of the design of the burner device for effective fire disposal of liquid wastes.

3 The object and method of study

Using the application software "Ansys Multiphysics" the mathematical modeling of the combustion front in different conditions was produced and detailed contours of total and static pressure, temperature and concentration of the components in the combustion zone were obtained. The initial data for mathematical modeling were geometric and operational parameters of the chosen burner. The study proposed a modernized system for fire disposal of liquid hydrocarbon wastes in the form of a heat generating unit (Fig. 1), where:

- a combustion chamber (2) is a hollow cylinder, which prevents leakages of the air-fuel mixture through the chamber walls;
- an air injector (5) is formed with perforations along the entire length, ensuring the uniform air flow into the combustion chamber in both directions: from the center - to the periphery and upwards;
- fuel injectors (4) are evenly spaced around the air injector, ensuring the uniform mixing of the oxidant (air) and fuel.

Considered technical solution is based on the combustion chamber (2), designed as a gas-tight, hollow, filled with chemically inert porous filler cylinder with the diameter of 57 mm and the height of 200 mm. Front of the combustion is stabilized via portion of the porous filler (6) mounted on the fuel-flow path. An important initial condition for the mathematical modeling is the filler temperature higher than the fuel ignition temperature. During the study of variants with different values of this parameter it was determined that the convergence of the steady-state solutions of combustion and heat transfer equations can't be achieved with the initial temperature of the porous layer below the fuel ignition temperature. The autothermal combustion is concluded to be only possible in the case of preheating the porous medium above the fuel ignition temperature due to the thermal inertia of the porous filler in the investigated burner device. That glowing filler fraction is a source of ignition.

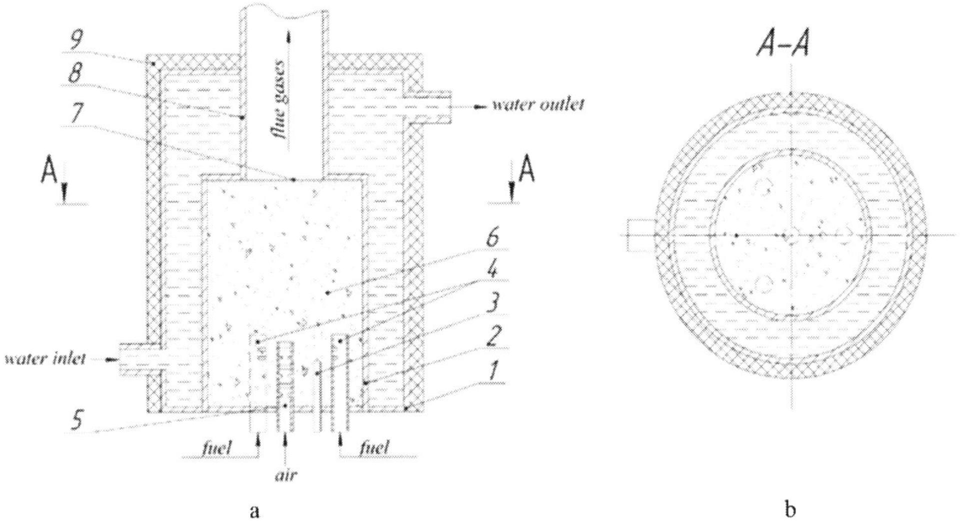

Fig.1. Longitudinal (a) and transverse (b) sections of a heat generating unit utilizing liquid hydrocarbon wastes: 1 is the casing; 2 is the combustion chamber; 3 is the ignition element; 4 are the fuel injectors; 5 is the air injector; 6 is the porous filler (carbon beads); 7 is the steel mesh nozzle; 8 is the channel for combustion products; 9 is the thermal insulation.

3 Results of the study

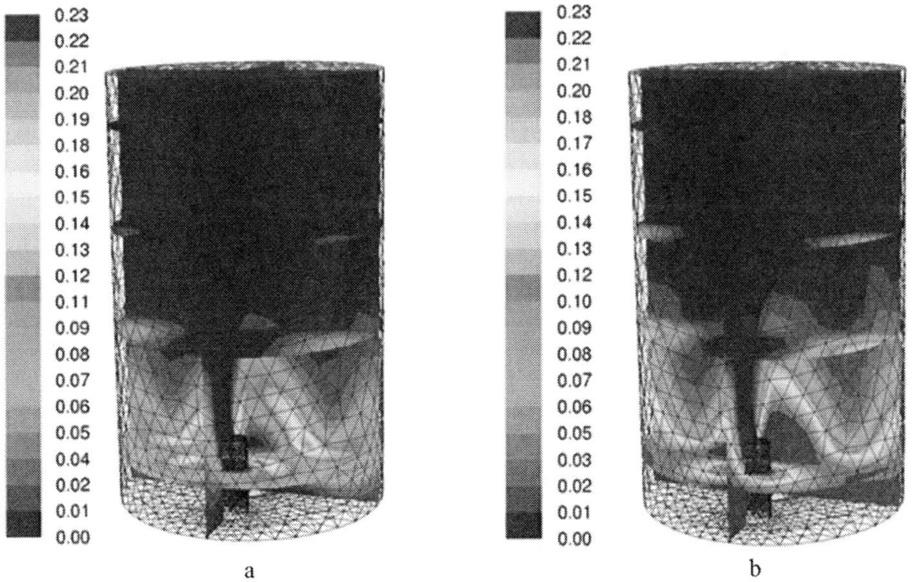

Fig. 2. Reaction rate contours, kgmol/(m3·s) (a) and the oxygen mass fraction, kg/kg (b) in a longitudinal section of the combustion chamber.

Contours of the chemical reaction rate and the oxidant mass fraction (Fig. 2) allow to estimate the geometric dimensions of the reaction zone, highlight the areas of the kinetic and diffusion combustion. Liquid fuel moves in the layer of heated porous filler and

DOI: 10.1051/matecconf/20179101024

vaporizes. The diffusion of the fuel jet into the air stream occurs on the external borders of the flame. The air-fuel mixture ignites forming the narrow angular combustion front, which burns less than 50 % of the fuel. The highest values of the chemical reaction rate - 200 mol/(m^3·s) are observed near the fuel injectors. Further combustion occurs in the diffusive mode as mixing. The resulted combustion front occupies one third of the porous filler (Fig. 2a). Estimated average reaction rate value is 90 mol/(m^3·s) in this region.

Despite the fact that the reaction zone is localized in the bottom of the combustion chamber, the temperature distribution on the combustion chamber volume is almost uniform due to the thermal conductivity in the porous layer (Fig. 3). It should be noted that the steady state values of the mean temperatures of the filler granules are equalized with the temperature of the surrounding flue gases.

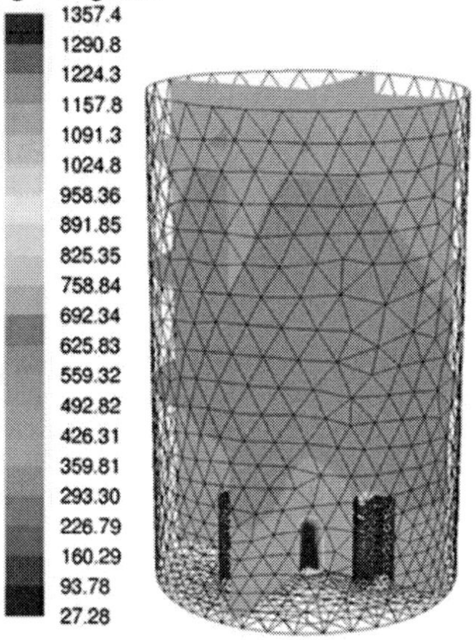

Fig. 3. Temperature contours (K) in the longitudinal section of the combustion chamber.

The maximum temperature in the combustion zone (1357 K) is less than 1473 K – the threshold for thermal NOx nitrogen oxides formation [17, 18]. In this regard, it can be argued that the combustion of liquid hydrocarbon wastes in this device will be accompanied by low emissions of harmful substances into the environment. away from the center the temperature is reduced down to 1036-1220 K in the wall region. Maximum of the emission spectrum (the infrared wavelength range with a length of 2377-2799 nm) is rated by Wien's displacement law at these temperatures. The radiation power calculated by the Stefan-Boltzmann law is 91.8 kW / m^2 with the mean average temperature of 1127K. Thus, the total area of the examined combustion chamber surface (0.04 m^2) radiates 3.7 kW (61.7% of the total thermal power of the device). The portion of the convective heat carried away by the flue gases (the remaining 38.3%) can be reduced by increasing the thickness of the porous layer complicating the regulation of the unit due to higher warm-up time. Therefore, water cooling [12] channel (fig. 1, pos. 8) is placed after the combustion chamber (pos. 1) in order to reduce heat loss with the exhaust gases.

4 Conclusion

The results of the mathematical modeling show that the proposed device for fire disposal of the liquid hydrocarbon wastes has a number of features.

– The high aerodynamic resistance of the porous filler causes a prevalence of the diffusion combustion.

DOI: 10.1051/matecconf/20179101024

– The thermal inertia of the burning core makes a power control difficult and takes time for warming up the device and entering the autothermal mode.

– The uniform heat dissipation in the combustion chamber enables a useful heat removal using the heating surface (water jacket).

– The low temperatures in the reactor are the reduction factor for the emissions of nitrogen oxides and other harmful substances during the utilization of liquid hydrocarbon wastes.

A low thermal conductivity of the filler particles forming the thermally inertial core causes an intensive heat exchange of a fuel-air mixture with the porous medium. Thus, the temperature distribution becomes uniform in volume, and the reaction zone fills a large part of the combustion chamber. However, the prevalence of the diffusive combustion requires a large air excess and long cocurrent trajectories of fuel and oxidant flows in order to eliminate the unburned carbon.

This work was carried out with the financial support from the grant of the Russian Scientific Fond (project No. 16-38-50126) and the grant of the Ministry of education and science of The Russian Federation (Government Order No. 13.948.2014/K).

References

1. S. Barişçi, M.S. Öncel, International Journal of Green Energy, **11**, 255 (2014).
2. A. V. Kazakov, R. B. Tabakaev, P. Y. Novoseltsev, A. V. Astafev, MATEC Web Conf., **19**, 1014 (2014).
3. A. S. Zavorin, R. B. Tabakaev, P. Y. Novoseltsev, A. V. Astafev, MATEC Web Conf., **19**, 1015 (2014).
4. R. B. Tabakaev, A. V. Kazakov, A. S. Zavorin, Solid Fuel Chem., **49**, 5, 267 (2015).
5. A. V. Kazakov, S. A. Khaustov, R. B. Tabakaev and Y. A. Belousova, IOP Conf. Series: Materials Science and Engineering, **124**, 012110 (2016).
6. S. A. Khaustov, A. S. Zavorin, K. V. Buvakov, L. D. Kudryashova, A. V. Tshelkunova, EPJ Web Conf., **82**, 1041 (2015).
7. S. A. Khaustov, A. V. Kazakov, G. A. Cherkashina, L. A. Sobinova, The European Physical Journal Conferences, **110**, 01028 (2016).
8. I. I. Shanenkov, A. A. Sivkov, A. Ya. Pak, Yu. L. Kolganova, Advanced Materials Research, **1040**, 813-818 (2014).
9. I. I. Shanenkov, A. Ya. Pak, A. A. Sivkov, Yu. L. Shanenkova, MATEC Web of Conf., **19**, 01030 (2014).
10. Z. Arsenijević, Z. Grbavčić, B. Grbić, N. Radić, R. Garić-Grulović, S. Miletić, G. Savčić, B. Dordević, Journal of the Serbian Chemical Society, **75**, 523–535 (2010).
11. K.A. Al-attab, J.C. Ho, Z.A. Zainal, Experimental Thermal and Fluid Science, **62**, 1–8 (2015).
12. X.-R. Chen, X.-S. Li, Z.-Y. Chen, Y. Zhang, K.-F. Yan, Q.-N. Lv, Energies, **8**, 1242–1255 (2015).
13. S. Dolgov, E. Savchenko, S. Khaustov, R. Tabakaev, A. Zavorin, EPJ Web of Conf., **110**, 01074 (2016).
14. S. A. Khaustov, A. S. Zavorin, K. V. Buvakov, and V. A. Sheikin, EPJ Web Conf., **82**, 1039 (2015).

AUTOMATED COMPLEX PROJECT ON THE ORGANIZATION CLAIM WORKS WITH FUEL

Vadim Kleshnya[1], *Elena* S. Vorontsova[1,*], and *Tatyana* S. Taylasheva[1]

[1] Tomsk Polytechnic University, 634050 Tomsk, Russia

Abstract. Consideration of problems with the quality of the account and the amount of fuel in the power plant. It is proposed to create an automated system that includes forms of documents, promptly filled energy facility staff, as well as related economic calculations. Consideration of the algorithm of reference works on fuel, laid in the automated system. Compiled by doing work scheme with the fuel.

1 Introduction

When the flow of fuel to the plant, an important point, that's will happened when working with him is the quantity and quality of the fuel. It is worth noting that there is now virtually no technical means for carrying out qualitative and representative input and control of fuel. In particular, the volume of mechanized input quality control of solid fuels to energy facilities is not more than 10 % [1]. In addition, there is no certified fuel laboratory with the possibility of using the results to conduct claim activities [2].

Another problem is that most businesses, there is no established system of acceptance of a service fuel and Legal Department of the company. And thus, the organization claims to be considered the work of high quality limited.

The efficiency claim work in greater influence compliance deadlines for filing claims (table 1), which in turn depend on the timing of the acceptance of the fuel quantity (10 days) and quality (20 days).

Table 1. Presentation of claims [4].

Type of claim	Respondent	Time
	Solid and liquid fuels	
Due to fuel shortage	Carrier	6 months
	Supplier	1 month
By supplying poor quality fuel	Supplier	1 month
Frozen fuel supply	Supplier	1 month
For delay in supply of fuel	Supplier	2 months
For delay in delivery of fuel	Carrier	45 days
By fuel, the lost in transit	Carrier	6 months
	Gas fuel	
Quantity	Supplier	By contract

* Corresponding author: ves@tpu.ru

DOI: 10.1051/matecconf/20179101025

Quality	Supplier	1 month
For mismatch pressure	supplier	1 month
For failure	supplier	By contract

In order to avoid problems arising from the need to observe the algorithm of reference works on fuel [3] that in the future will greatly facilitate their implementation and facilitate efficient and reliable operation of thermal power equipment.

2 Results

It is proposed to create an automated system that includes forms of documents, promptly filled energy facility staff, as well as related economic calculations. The diagram is shown (Fig. 1) reflects the algorithm of works on fuel, laid in the automated system.
The scheme is divided into two main blocks: technical and economic. Technical block includes an input control in the enterprise, which is divided into the control of fuel quantity and quality.

By controlling the number of fuel include documents [4]:

- Protocol mass evaluation;
- A statement of account;
- Transport documents (waybill, invoice, etc.);
- Calculation permissible weight differences.

By monitoring fuel quality [4]:

- Certificate of quality (quality certificate);
- Minutes of quality assessment indicators;
- Control sampling instrument;
- Transport documents (waybill, invoice, etc.).

If the fuel meets all the requirements for documents control of quantity and quality, it is drawn up acceptance certificate, and then the fuel is used in the power plant.

If the fuel has a deviation in the amount of shortage of fuel is drawn up, there is a calculation value of the missing fuel and put up claims as the carrier and the supplier.

The presence of deviations in quality is considered separately from the vendor and separate from the carrier. The incoming fuel poor quality shall be drawn on the actual fuel, there is a calculation of this cost of fuel and put up a claim to the supplier. Arriving at the energy company in the frozen state of the fuel also includes a procedure for preparation of the report of the arrival of the frozen fuel going assessment of the rolling stock and the cost of delay paths exhibiting claims to the carrier. The economic unit is a check violation of terms of delivery. Fuel which has no deviations shall be drawn as fuel acceptance and in the future it is used in power plants.

Depending on such deviations as “did not come the whole party”, “not delivered part of the party”, “lost in transit”, is drawn up violating the terms of the supply of fuel. Further, there is also a penalty for delay estimation and invoicing vendor claims. With regard to the refusal of recognition and claim after the procedure in two blocks in the event of failure by one of the parties claims documents for its consideration and the proceedings are sent to the Arbitration Court of the Russian Federation. In the event that claims are considered options such as the return of the fuel supplier, selling to a third party, the use of fuel in the power plant.

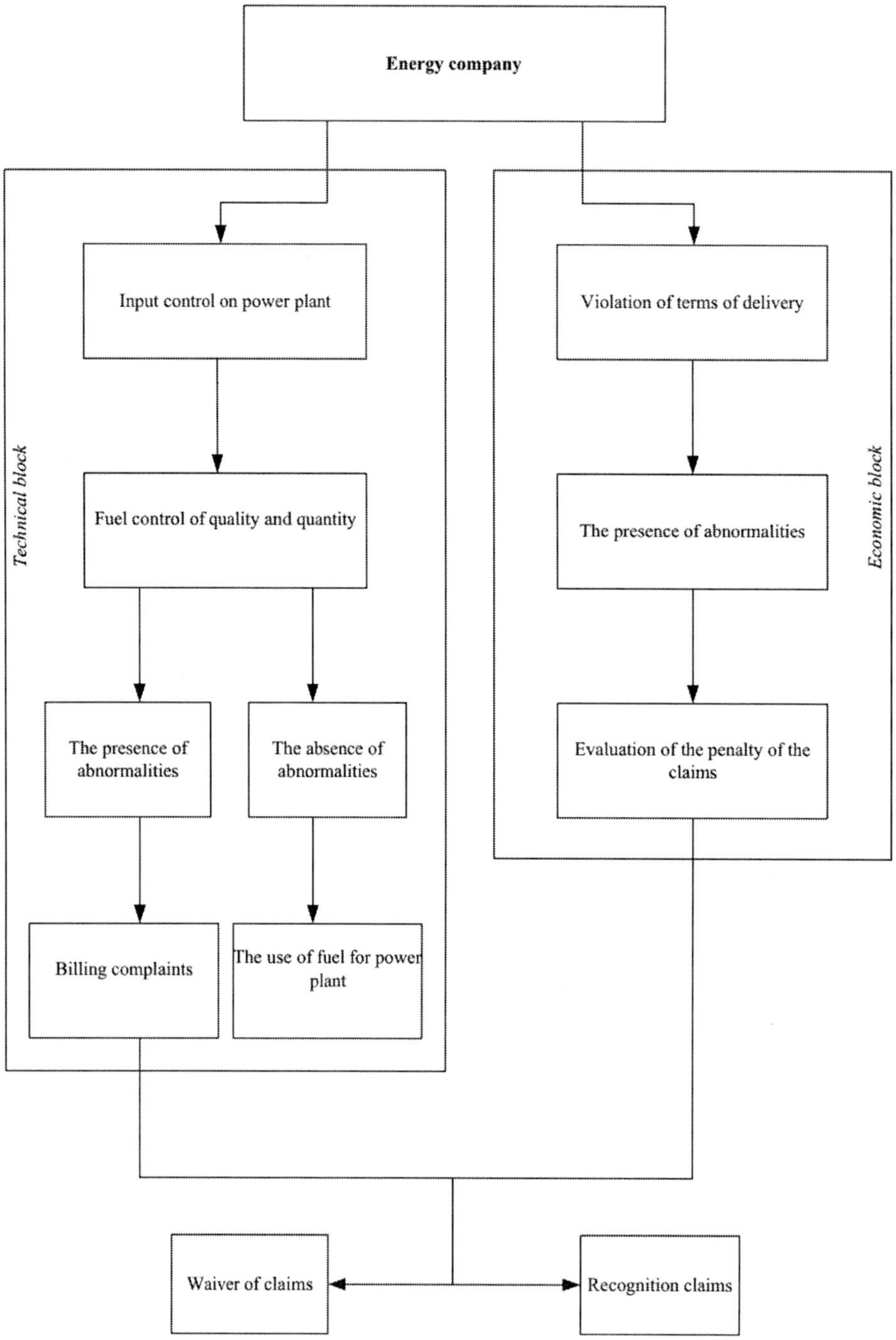

Fig. 1. Driving conduct work with the fuel in the power plant.

4 Conclusion

The possibilities of modern information technology allow you to create a new generation of software systems that have the flexibility and scalability that helps regular updating of the regulatory framework. At the moment the authors of the article set of documentation collected, necessary for the realization of the creation of the automated complex. The determining factor in the use of the automated complex in the energy company is to improve the quality of doing work on the fuel having the technical and legal in nature, according to the following criteria:

1. Reduction of the time and to perform certain operations on the collection and processing of data;
2. Improving staff efficiency;
3. Error-free and timely filling of documents;
4. Updating of normative and technical documentation on all sides;
5. Increasing the level of control and performance discipline;
6. Reducing the risk of deterioration or interruption of power equipment in the presence of any wrongdoing.

References

1. V. Kazakov, P. Mingalev, D. Linev, Input control of the solid fuel quality in the Russian power, **6**, 27 (2015)
2. S. Remes, V. Edelman, E. Govsievich, R. Alechinsky, Recommendations to improve the work with claims on fuel at thermal power plants of RAO "EEC of Russia", **10**, 4 (2006)
3. V. Kleshnya, V. Vorontsova, T. Taylasheva, The demand of the organization of reference works with a fuel in the power plant, **2**, 83 (2016)
4. CO 34.44.223. Recommendations for claims work on fuel [in Russia]

MECHANISM OF CONTACT LINE MOVEMENT OF A DROPLET SPREADING OVER A SOLID SURFACE

Dmitry V. Feoktistov[1, *], *Evgeniya* G. Orlova[2], and *Anastasia* G. Islamova[2]

[1] Kutateladze Institute of Thermophysics, 630090, Novosibirsk, Russia
[2] National Research Tomsk Polytechnic University, 634050, Tomsk, Russia

Abstract. The contact line movement, advancing and receding, of the distilled water droplet was studied on the polished and laser patterned aluminum surfaces. The obtained dependences of the dynamic contact angles on the contact line speed were fitted by the combined theory consisting of molecular-kinetic and hydrodynamic theories. Results obtained on laser patterned surface are found to fit better to the theory in comparison with polished surface. The assumption was made that this result is connected with the low contact angle hysteresis on the patterned surface compared to the polished one. It means that combined theory does not consider the contact angle hysteresis.

1 Introduction

Research [1-9] aimed at studying wetting, spreading, and phase transition of a droplet is currently very relevant. It is connected with the extensive application of the obtained results. Dynamic wetting phenomenon is widespread in the technology, for example, at droplet cooling of surfaces [10], application of various coatings [11], inkjet printing [12], oil extraction, lubrication, bonding and impregnating of various materials [13], growth of biological crystals [14]. Insufficient understanding of the three-phase contact line properties leads to deter the development of a number of technologies where rivulet flow [15], boiling [16], and droplet irrigation [17] are implemented.

A droplet placed on a solid surface spreads, tends to take an equilibrium state. The characteristics used to describe the spreading of droplets are dynamic contact angle, contact line speed and contact diameter. The most significant of these is the dynamic contact angle (DCA). There are several theories of spreading [18-20], in which DCA is mathematized as a function of the contact line speed.

In this study the possibility of applying the combined theory of spreading (as the most integrated, since it takes into account the adsorption and desorption of molecules, the forces of viscous friction in the liquid) to describe the wetting and dewetting of polished and micropatterned surfaces of aluminum substrates by liquid droplet was proved.

* Corresponding author: fdv@tpu.ru

This theory assumes that the equilibrium contact angle depends on the contact line speed [18]:

$$\theta_D^3 = \left[a\cos\left(\cos\theta_0 \pm \left(\frac{2k_B T}{\sigma\lambda^2} \right) a\sinh\left(\frac{U}{2k_0\lambda} \right) \right) \right]^3 \pm \left(\frac{9\mu U}{\sigma} \right) \ln\left(\frac{L}{L_m} \right), \qquad (1)$$

where θ_0 is the equilibrium contact angle, L is the characteristic capillary length, L_m is the slip length, k_B is the Boltzmann constant, T is the temperature, k_0 is the equilibrium frequency of the random molecular displacements occurring within the three-phase zone, and λ is the average distance of each displacement.

This theory includes hydrodynamic [19] and molecular-kinetic [20] theories. From the first theory it is known that the bulk viscous friction prevents the contact line motion. The second theory supposes that the dependence of the dynamic contact angle on the contact line speed is the result of molecular adsorption and desorption on the three-phase contact line.

2 Experimental apparatus

Research was conducted on the experimental setup with shadow method implementation. The principle of operation is described in detailed in [21-25]. To form advancing and receding contact line, liquid was pumped in/out through an opening in the surface (Fig.1)

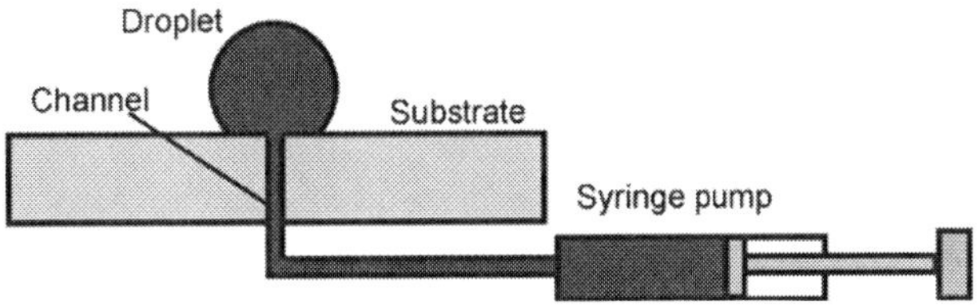

Fig. 1. Scheme of pumping in/out liquid for droplet formation.

Two substrates made of aluminum were used in the experiment. Surface of the first substrate was polished; the second was micro patterned by laser treatment. The energy in the laser pulse was 1.0 mJ, pulsed frequency was 20 kHz, traverse speed was 500 mm/s. The surfaces were studied on the profilometer "Micro Measure 3D station" (Fig. 2). The roughness parameter was obtained for each substrate, Ra=0.0558 µm and Ra=4.2096 µm, respectively. Microstructures are shown in Fig.2.

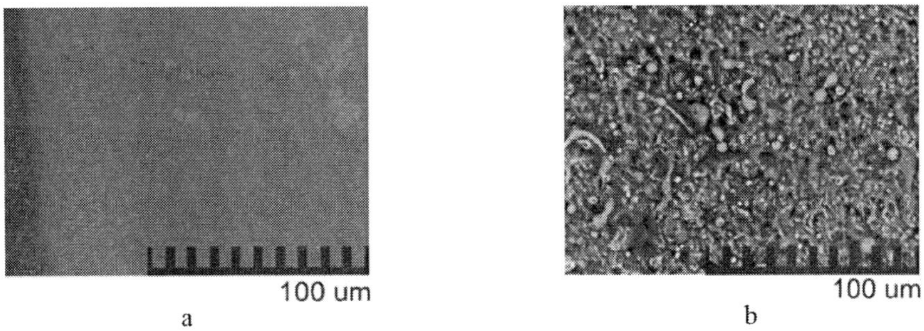

Fig. 2. Surfaces of substrates (×1000): (a) polished; (b) patterned.

 DOI: 10.1051/matecconf/20179101026

Liquid flow rate supplied by an electronic pump was varied in the following range: 0.05 mL/sec – 0.11 mL/sec. The random error was calculated from the results of experiments repeated under the identical conditions. Thereafter, the mean values of the contact angle and their standard deviations were calculated. The random measurement errors were mainly due to the small deviations of flow rate during dosing the liquid through a silicone tube.

3 Results and discussion

The experimental data processing for the following comparison with theory [18] was performed by least squares. To evaluate fitting of the experimental data to the theory the R-squared coefficient was obtained. Defined parameters of the combined theory for polished surface are listed in Table 1.

Parameters k^0 and λ were obtained by approximation. The combine theory is found to describe well the experimentally obtained values of θ_D on polished surface. It is explained by the fact that this theory takes into account both the contact line friction and viscous forces in the bulk of liquid.

Table 1. Parameters for polished surface.

Liquid flow rate, mL/sec	Measured equilibrium contact angle θ_0,°	Calculated parameters			
		θ_0,°	k^0,10^3 *Hz*	λ, nm	R^2
0.05	96.5	91.6	0.47	3.02	0.758
0.07	92.8	90.4	4.69	2.73	0.650
0.09	85.3	80.5	0.86	2.98	0.634
0.11	99.0	96.7	1.80	4.85	0.765

Fitting of the combined theory was conducted simultaneously to the advancing and receding lines both for polished and patterned surfaces. Dependences for polished surface are presented in Fig.3.

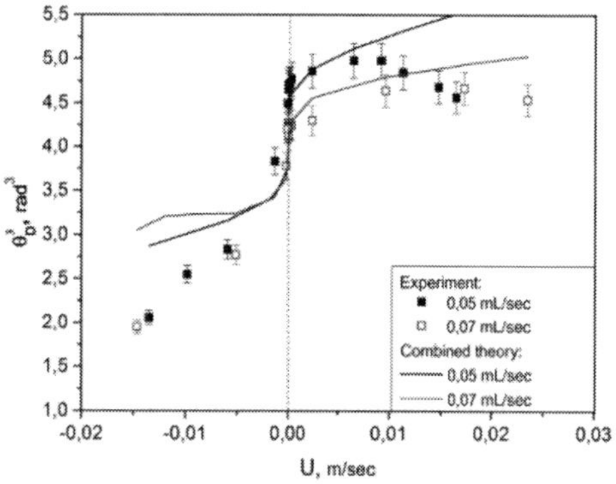

a

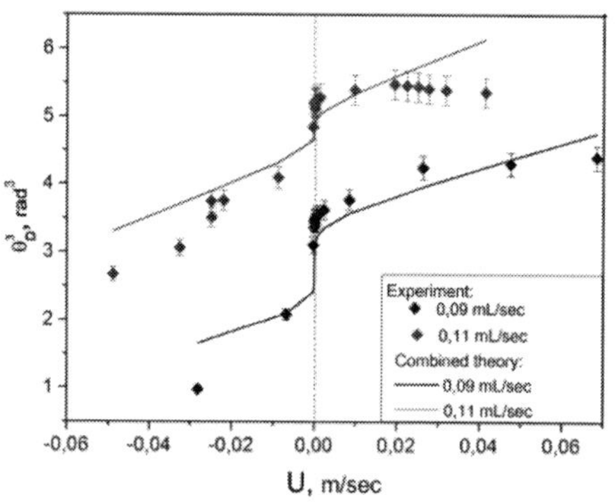

b

Fig. 3. DCA versus contact line speed on polished surface at different values of flow rate: (a) 0.05 mL/sec and 0.07 mL/sec; (b) 0.09 mL/sec and 0.11 mL/sec.

Resulting parameters of the combined theory for patterned surface are listed in Table 2. Obtained dependences of the dynamic contact angle on the contact line speed are presented in Fig.4.

DOI: 10.1051/matecconf/20179101026

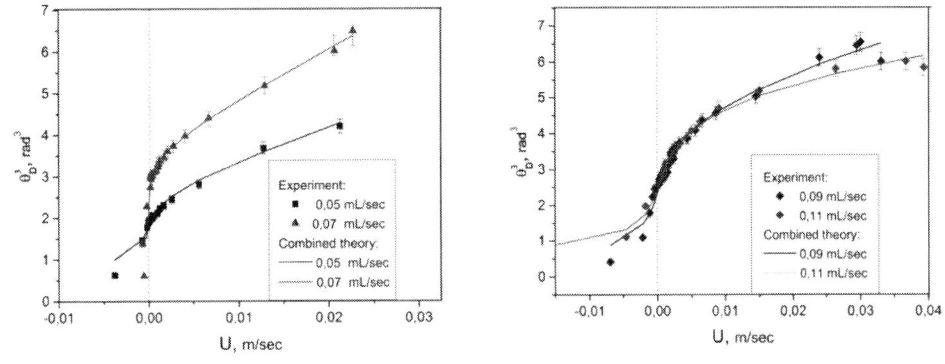

a b

Fig. 4. DCA versus contact line speed on patterned surface at different values of flow rate: (a) 0.05 mL/sec and 0.07 mL/sec; (b) 0.09 mL/sec and 0.11 mL/sec.

Table 2. Parameters for laser patterned surface.

Liquid flow rate, mL/sec	Measured equilibrium contact angle θ_0,°	Calculated parameters			
		θ_0,°	k^0,10^3 Hz	λ, nm	R^2
0.05	71.9	69.7	132.8	1.59	0.975
0.07	80.2	74.2	4.71	1.75	0.965
0.09	75.8	76.2	394.0	1.07	0.980
0.11	78.9	77.3	534.3	1.02	0.989

According to the results of fitting, dependences of the advancing and receding dynamic contact angle on the contact line speed obtained for patterned surface are found to fit better with the theoretical dependence compared to the results obtained on polished surface. It is connected due to the fact the dependence of combined theory (Eq. (1)) do not take into account the contact angle hysteresis. For this reason, the best fitting is observed for patterned surface with the lower hysteresis value (H<1.5°) compared to the polished surface (H=5°). Thus, when the liquid flow rate equaled 0.05 mL/sec R-squared value for polished surface was 0.758, for patterned – 0.975.

Acknowledgements

This work was carried out at the Kutateladze Institute of Thermo-physics SB RAS and financially supported by the Russian Science Foundation (project number 15-19-10025).

References

1. G. Kuznetsov, D. Feoktistov, E. Orlova, K. Batishcheva, Colloid J., **78,** 335 (2016)
2. D. Zaitsev, E. Chinnov, O. Kabov, I. Marchuk, Tech. Phys. Lett., **30**, 231 (2004)
3. V. Nakoryakov, S. Misyura, S. Elistratov, Int. J. Energy Technol., **4**, 1 (2012)
4. X. Li, L. Zhang, X. Ma, H. Zhang, Surf. Coat. Technol., **307**, 243 (2016)
5. Y. Lyulin, D. Feoktistov, I. Afanas'ev, E. Chachilo, O. Kabov, G. Kuznetsov, Tech. Phys. Lett., **41**, 665 (2015)
6. D. Glushkov, G. Kuznetsov, P. Strizhak, Math. Probl. Eng., **2014**, 920480 (2014)
7. G. Kuznetsov, A. Zakharevich, N. Bel'kov, Chem. Pet. Eng., **50**, 424 (2014)

8. V. Nakoryakov, S. Misyura, S. Elistratov, J. Eng. Thermophys., **22**, 1 (2013)
9. G. Kuznetsov, D. Feoktistov, E. Orlova, Thermophysics and Aeromechanics, **23**, 17 (2016)
10. D. Zaitsev, D. Rodionov, O. Kabov, Technical Physics letters, **35**, 680 (2009)
11. A. Sivkov, Y. Shanenkova, A. Saigash, I. Shanenkov, Surf. Coat. Technol., **292**, 63 (2016)
12. A. Oko, P. Claesson, P. Niga, A. Swerin, Nord. Pulp Pap. Res. J., **31**, 156 (2016)
13. S. Basu, S. Shravan, Pet. Sci. Technol., **26**, 1559 (2008)
14. A. Sivkov, A. Pak, D. Nikitin, I. Rakhmatullin, I. Shanenkov, Nanotechnol. Russ., **8**, 489 (2013)
15. Y. Lyulin, O.Kabov, Tech. Phys. Lett., **39**, 795 (2013)
16. V. Nakoryakov, S. Misyura, Dokl. Physics, **59**, 441 (2014)
17. V. Nakoryakov, S. Misyura, S. Elistratov, Dokl. Physics, **58**, 40 (2013)
18. P. Petrov, J. Petrov, Langmuir, **8**, 1762 (1992)
19. O. Voinov, J. Fluid Dyn., **11**, 714 (1976)
20. T. Blake, J. Haynes, J. Colloid Interface Sci., **30**, 421-423 (1969)
21. E. Orlova, E. Malyhina, D. Feoktistov, L. Zhidkova, MATEC Web of Conf., **72**, 01062 (2016)
22. K. Batischeva, E. Orlova, D. Feoktistov, MATEC Web of Conf., **19**, 01001 (2014)
23. E. Orlova, D. Feoktistov, G. Kuznetsov, EPJ Web of Conf., **82**, 01053 (2015)
24. K. Ponomarev, E. Orlova, D. Feoktistov, EPJ Web of Conf., **110**, 01060 (2016)
25. E. Orlova, G. Kuznetsov, D. Feoktistov, EPJ Web of Conf., **76**, 01039 (2014)

NUMERICAL ANALYSIS OF THREE-DIMENSIONAL NATURAL CONVECTION IN A CLOSED RECTANGULAR CAVITY UNDER CONDITIONS OF RADIANT HEATING AND CONJUGATE HEAT EXCHANGE

Alexander E. Nee[1,*]

[1] National Research Tomsk Polytechnic University, 634050, Tomsk, Russia

Abstract. The numerical simulation results of three-dimensional natural convection in a closed cavity were presented under conditions of the bottom horizontal solid-fluid interface radiant heating and conjugate heat exchange. Conservation equations of mass, momentum, and energy were formulated in terms of vorticity vector – vector potential – temperature dimensionless variables and solved by means of the finite difference method. It was found that the heat transfer process under study had a significant unsteady nature. According to the results of conjugate heat exchange integral analysis, it was shown that similar trends of mean Nusselt numbers versus dimensionless time were formed for both two and three dimensional problem formulations.

1 Introduction

Investigation of convective heat transfer in closed cavities plays an important role for science and engineering. Research of heat transfer regularities in technical systems and technological processes contributes to increasing of their energy performance. However, to conduct full-scale experiment in order to reveal the basic heat exchange characteristics is not always possible to perform under operating conditions of thermal engineering equipment. In many cases, numerical simulation methods of physical processes, based on the solution of differential equations in partial derivatives, are reasonable to implement.

Natural convection takes place in many technical systems. The solutions of various natural convection problems in nonconjugate [1-3] and conjugate [4-6] formulations were performed last years. The numerical analysis is conducted basically for the two-dimensional problems of heat transfer, which is associated with the significant complexity when implementing the difference schemes for time-dependent three-dimensional conservation equations of mass, momentum, and energy. Investigations of three-dimensional regimes of natural convection [7, 8] were mostly performed for nonconjugate problems. Moreover, the effect of radiation heat exchange on formation of differential and

* Corresponding author: nee_alexander@mail.ru

integral heat transfer characteristics was not analyzed in [7, 8]. Therefore, investigations of natural convection in a closed volume with radiant energy source in conjugate formulation are of great interest.

The main purpose of this study is numerical analysis of three-dimensional natural convection in a closed rectangular cavity under conditions of radiant heating and conjugate heat exchange.

2 Problem formulation and solution method

A typical infrared chamber were considered and presented in Fig. 1. Solution domain consisted of the air-filled cavity bounded by heat-conducting finite thickness walls. The radiant energy source was fixed on the top horizontal boundary. I assumed that the infrared emitter had time - independent surface temperature. Heat exchange with the environment was negligibly small at the external boundaries. The equalities of temperatures and heat fluxes were set at the solid – fluid interfaces. I assumed that the radiant energy coming from the infrared emitter was uniformly distributed along the bottom horizontal solid – fluid interface. Thermophysical properties of the system elements under study were considered as temperature independent. The flow regime was laminar.

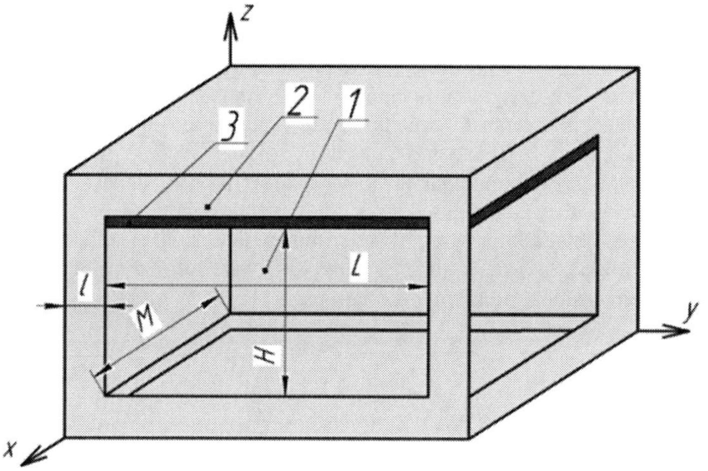

Fig. 1. Solution domain: 1 – air; 2 – walls; infrared emitter.

Heat transfer process under formulated physical model is described by unsteady three-dimensional conservation equations mass, momentum, and energy. In terms of vorticity vector – vector potential – temperature dimensionless variables for viscous heat-conducting fluid, satisfying the Boussinesq approximation, these equations are as follow [6, 9]:

$$\frac{\partial \Omega_x}{\partial \tau}+U\frac{\partial \Omega_x}{\partial X}+V\frac{\partial \Omega_x}{\partial Y}+W\frac{\partial \Omega_x}{\partial Z}-\Omega_x\frac{\partial U}{\partial X}-\Omega_y\frac{\partial U}{\partial Y}-\Omega_z\frac{\partial U}{\partial Z}=\sqrt{\frac{\mathrm{Pr}}{\mathrm{Ra}}}\cdot\Delta\Omega_x+\frac{\partial \Theta_1}{\partial Y}, \qquad (1)$$

$$\frac{\partial \Omega_y}{\partial \tau}+U\frac{\partial \Omega_y}{\partial X}+V\frac{\partial \Omega_y}{\partial Y}+W\frac{\partial \Omega_y}{\partial Z}-\Omega_x\frac{\partial V}{\partial X}-\Omega_y\frac{\partial V}{\partial Y}-\Omega_z\frac{\partial V}{\partial Z}=\sqrt{\frac{\mathrm{Pr}}{\mathrm{Ra}}}\cdot\Delta\Omega_y-\frac{\partial \Theta_1}{\partial X}, \qquad (2)$$

$$\frac{\partial \Omega_z}{\partial \tau}+U\frac{\partial \Omega_z}{\partial X}+V\frac{\partial \Omega_z}{\partial Y}+W\frac{\partial \Omega_z}{\partial Z}-\Omega_x\frac{\partial W}{\partial X}-\Omega_y\frac{\partial W}{\partial Y}-\Omega_z\frac{\partial W}{\partial Z}=\sqrt{\frac{\mathrm{Pr}}{\mathrm{Ra}}}\cdot\Delta\Omega_z, \qquad (3)$$

$$\Delta\Psi_x=-\Omega_x,\ \Delta\Psi_y=-\Omega_y,\ \Delta\Psi_z=-\Omega_z, \qquad (4)$$

$$\frac{\partial\Theta_1}{\partial\tau}+U\frac{\partial\Theta_1}{\partial X}-V\frac{\partial\Theta_1}{\partial Y}-W\frac{\partial\Theta_1}{\partial Z}=\frac{1}{\sqrt{\mathrm{Ra}\cdot\mathrm{Pr}}}\cdot\Delta\Theta_1, \quad (5)$$

$$\frac{1}{Fo_2}\cdot\frac{\partial\Theta_2}{\partial\tau}=\Delta\Theta_2,\ \frac{1}{Fo_3}\cdot\frac{\partial\Theta_3}{\partial\tau}=\Delta\Theta_3. \quad (6)$$

The initial and boundary conditions for the equations (1)-(6) were set analogously to [4].

The boundary value problem (1)-(6) was solved by means of the finite-difference method [10, 11] on a uniform grid. (71x71x71). Developed numerical code was tested on the benchmark problems of three-dimensional natural convection [7, 8]. Comparison of the obtained results (isotherms and streamlines) showed their good agreement.

2 Results and discussion

Numerical analysis of three-dimensional conduction and natural convection was carried out for the following dimensionless criteria corresponding to laminar flow regime: Rayleigh number $Ra=10^6$, Prandtl number Pr=0.71, Kirpichev number Ki=13, Fourier numbers $Fo_2=1.96\cdot10^{-8}$, $Fo_3=9.5\cdot10^{-7}$. The results of conjugate heat exchange integral analysis are presented in terms of the mean Nusselt numbers at the bottom horizontal ($Nu_1=\int_{0.1}^{1.1}\int_{0.1}^{1.1}\left|\frac{\partial\Theta}{\partial Z}\right|_{Z=0.1}dXdY$) and vertical ($Nu_2=\int_{0.1}^{1.1}\int_{0.1}^{1.1}\left|\frac{\partial\Theta}{\partial Y}\right|_{Y=0.1}dXdZ$) solid-fluid interfaces versus dimensionless time dependences and shown in Fig. 2.

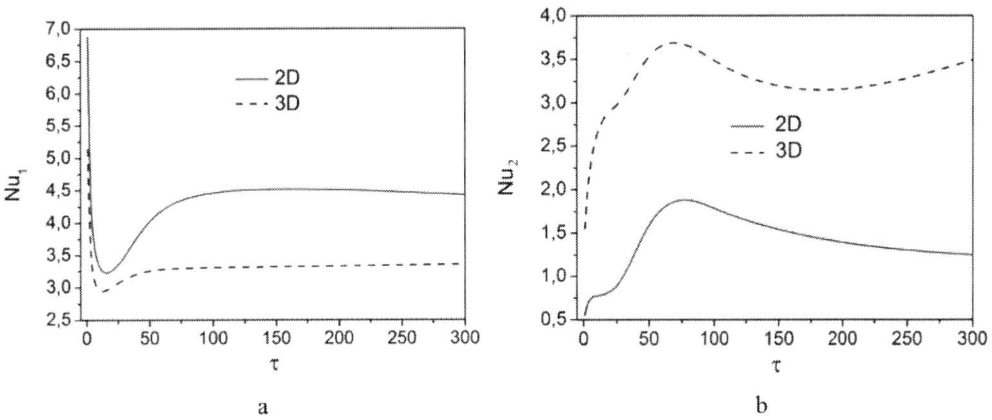

Fig. 2. Dependences of the mean Nusselt numbers versus dimensionless time at the: a) bottom horizontal solid-fluid interface; b) vertical solid-fluid interface.

It is clearly seen that the mean Nusselt numbers at the horizontal and vertical solid-fluid interfaces for both two and three dimensional problem formulations had similar trends. During the initial time period, Nu_1 was reduced before τ was equal to 12, which was associated with the formation of heated air layer at this interface. An increase in the mean Nusselt number at the bottom horizontal solid – fluid interface in a time range of 12 to 50 was due to the formation of circulating flows in the air cavity. Hereafter, Nu_1 changed slightly, which was associated with the constant value of temperature gradient near the bottom horizontal wall. Along with that, an increase in dimensionless time led to a rise in the mean Nusselt number at the vertical solid – fluid interface, which was due to the increasing of air velocity along this interface. Obviously, this factor led to the intensification of heat exchange process. As the result, the mean Nusselt number increased. Higher values of Nu_2 for the case of three-dimensional problem formulation was due to the more intensive heat loss from the heated air to the surfaces where vertical walls were connected. As a result, air velocity along the vertical solid-fluid interfaces was increased. This factor led to the Nu_2 growth.

3 Conclusion

The mathematical model of three-dimensional heat transfer in a closed volume taking into account the basic relevant factors (conduction, convection, radiation and conjugate heat exchange) are suggested. Obtained results showed the possibility of further use of the approach [6, 9] when solving more complex problems of conjugate convective – radiative heat transfer in closed areas under conditions of the radiant heating.

The reported research was supported by Russian Federation President Grant for state support of the Russian Federation leading scientific schools SS-7538.2016.

References

1. D. Han, B. Yu., J. Chen, Y. Wang and Y. Wang, Int. Commun. Heat Mass, **68** (2015)
2. K. Szewc, J. Pozorski and A. Tanire, Int. J. Heat Mass Transf., **54** (2011)
3. Y. Wei, H.-S. Dou, Z. Wang, Y. Qian and W. Yan, Comput. Fluids, **124** (2016)
4. G. V. Kuznetsov and M. A. Sheremet, Int. J. Heat Mass Transf., **52** (2009)
5. E. Bilgen, Appl. Therm. Eng., **29** (2009)
6. G. V. Kuznetsov and M. A. Sheremet, Int. J. Heat Mass Transf., **54** (2011)
7. D. C. Lo, Int. Commun. Heat Mass, **37** (2010)
8. Z. Li, M. Yang, and Y. Zhang, Int. J. Heat Mass Transf., **94** (2016)
9. G. V. Kuznetsov and M. A. Sheremet, Int. J. Therm. Sci., **50** (2011)
10. G. V. Kuznetsov and A. E Nee, Thermophys. Aeromech., **23** (2016)
11. A. Nee IOP Conf. Ser. Mater. Sci. Eng., **93** (2015)

EVAPORATION RATE OF A LIQUID LAYER STREAMLINED BY GAS FLOW IN MINICHANNEL

Evgeniya G. Orlova[1*], *Anastasia* G. Islamova[1], and *Oleg* A. Kabov[2]

[1] Tomsk Polytechnic University, 634050, Tomsk, Russia
[2] Kutateladze Institute of Thermophysics, 630090, Novosibirsk, Russia

Abstract. Paper presents the experimental results of mass evaporation rate from ethanol layer surface into a moving in parallel way air flow. Dependences of the mass evaporation rate on the gas velocity, temperature difference between gas and liquid layer, and liquid layer thickness were obtained. The assumption about the formation of the concentration boundary layer at the interface was made. It is found that with increasing the gas flow velocity the gradient of vapor concentration in the gas phase increases; as a result the evaporation rate rises. When changing the liquid layer thickness, the local maximum of evaporation rate was observed, but for 2 mm thickness layer two maxima were found. Probably, it is due the appearance of unstable structures on the liquid surface.

1 Introduction

Recently cooling systems with mini- and microchannels [1–3] are widely used in the microelectronic equipment and aerospace machines. It is connected with the fact that their application significantly reduces the coolant flow rate [4–9]. To intensify heat exchange in thermal control systems with motion of the coolant in minichannels, it is possible to use gas flow streamlining the liquid surface. In this case, the evaporation rate from the interface will depend not only on the velocity and thickness of the liquid, its temperature, but also on the velocity and temperature of gas flow.

Study of the interface formed at streamlining of the liquid layer by the gas flow is a complex problem. In this case, it is necessary to take into account simultaneously such transfer mechanisms as forced and natural convection, thermocapillary convection, diffusion [10–17].

According to analyses of previous studies devoted to this subject [10–21] a significant lack of experimental studies taking into account the evaporation is found out. The majority of works were devoted to the study of hydrodynamics and the emergence of various instabilities at the interface. The purpose of this work is to study experimentally the influence of gas velocity, temperature difference between gas and liquid layer, liquid layer thickness on the evaporation rate from the liquid/gas interface.

* Corresponding author: lafleur@tpu.ru

DOI: 10.1051/matecconf/20179101028

2 Experimental apparatus

The research was conducted on the experimental setup with stationary layer of ethanol streamlined by the air flow in the test cell (Fig. 1). The principle of operation is described in detail in [1]. To maintain a constant level of evaporating liquid layer in the evaporation zone, a peristaltic pump was used. The gas temperature was controlled with distilled water heated in a thermostatic bath and circulating through the channels of the test cell case. A thermoelectric module provided the necessary temperature of the liquid. The position of the interphase presented in Fig.1 was controlled by the shadow method through the optical window.

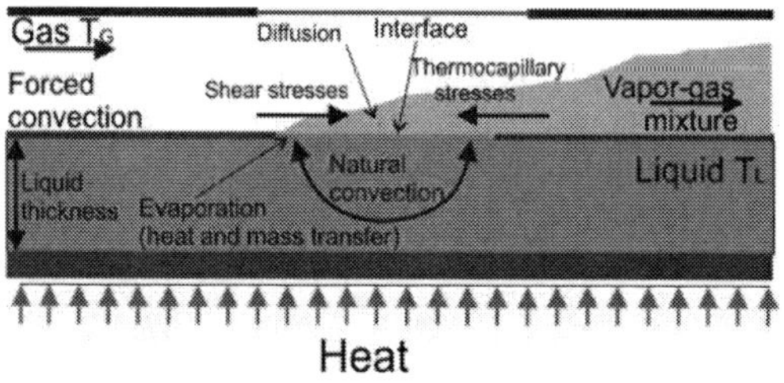

Fig. 1. Schematic representation of the experimental conception.

The test cell consists of the liquid chamber with 40 mm in width and variable height from 1 to 10 mm and rectangular gas channel 19 (3 mm in height, 40 mm in width, 150 mm in length). The upper wall of the liquid has a square opening where liquid and gas are in contact. The gas flow rate at the inlet of the test cell was varied from 0.0139 m/s to 0.1389 m/s. The gas temperature was constant (25 °C), and the values of liquid temperature were 15, 20, 25, 30, and 35 °C. The interface area was constant (100 mm^2). The values of liquid layer thickness were 1, 2, 3, 4, 5, and 8 mm. The evaporation rate from the liquid layer surface was defined by measuring the difference between mass flow rate of gas (air) at the inlet of the test cell and air-coolant vapours mixture at the outlet.

3 Results and discussion

According to results of conducted experiments the mass evaporation rate is found to increase at increase in the temperature difference between liquid layer and moving in parallel way the gas flow $\Delta T = T_L - T_G$ (Fig.2).

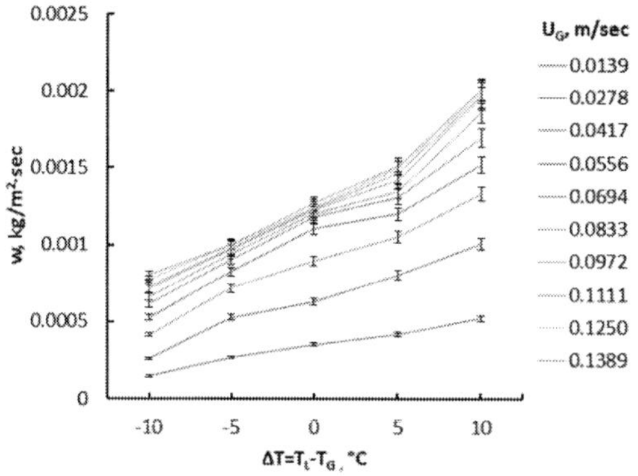

Fig. 2. Mass evaporation rate versus temperature difference $\Delta T = T_L - T_G$. Liquid layer thickness is h=1 mm.

The difference between the temperature of the liquid and gas does not influence the power form of the growth curves of the evaporation rate. The mass evaporation rate increases monotonically for all thicknesses of the layer. It can be explained by the fact that the vapor concentration near the liquid/gas interface corresponds to the saturated vapor pressure at the interface temperature and increases with increasing the liquid temperature. Also with an increase in the liquid temperature the average kinetic energy of the liquid molecules increases. Thus, the numbers of molecules with sufficiently large kinetic energy to pass to the evaporation zone increase.

Dependences of the mass evaporation rate on the liquid layer thickness are shown in Fig. 3.

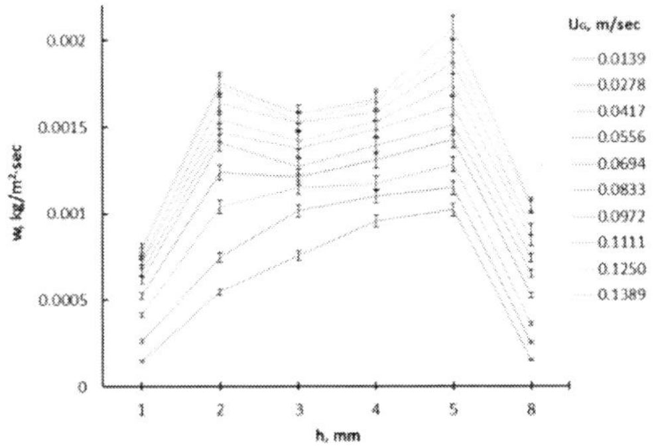

Fig. 3. Mass evaporation rate versus liquid layer thickness. Temperature difference is $\Delta T = -10°C$.

With increasing layer thickness of the evaporating liquid the local maximum of the evaporation rate, which depends on the gas flow rate, was observed. It should be noted that the maximum of the evaporation rate from 5 mm thickness liquid layer was observed for all dependences (Fig. 3). In addition, the evaporation rate from 5 mm thickness liquid layer increases by 35% compared to the 4 mm thickness. However, when the gas flow rates range from 0.0556 to 0.1389 m/sec, another maximum can be observed at a layer thickness of 2 mm. It can be connected with a liquid layer instability, which in turn causes an increase in the evaporation rate.

The mass evaporation rate is found to increase with an increase in the gas flow velocity (Fig. 4.).

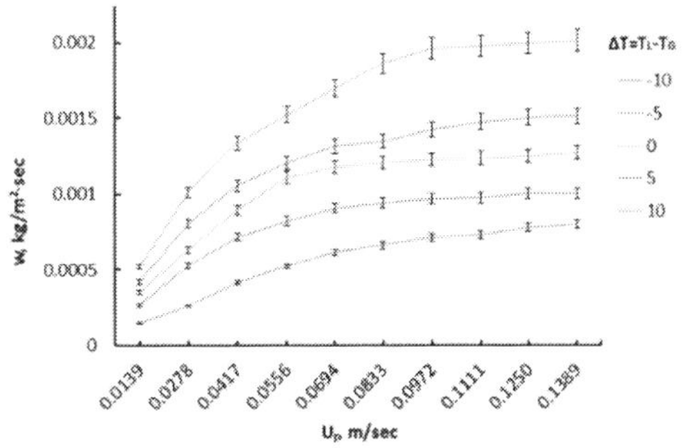

Fig. 4. Mass evaporation rate versus gas flow velocity. Liquid layer thickness is h=1 mm.

It can be explained by the fact that the vapor forming above the interfacial surface is removed in the channel by the gas flow. Alcohol vapors are moved from the liquid/gas interface due to forced convection and diffusion. Thus, the evaporation rate increases with increasing gas flow velocity due to an increase in the concentration gradient of the gas phase.

The range of the gas flow velocity from 0.0139 to 0.0694 m/sec should be pointed out because the mass evaporation rate increases dramatically in this range. It is due to the relatively low velocities of alcohol vapors removal in the gas phase. With further increase in the gas velocity the removal of vapors becomes more intensive; as a result the evaporation rate growth reduces.

It is also found that the mass evaporation rate increases sharply by 25% for the entire range of the liquid layer thickness when the temperature difference is 10°C. When the liquid temperature increases, there is an imbalance of forces during evaporation process. The influence of convection in the liquid layer amplifies, liquid is mixed more intensive, and also with an increase in the liquid temperature the numbers of molecules capable of leaving the free path area and passing to the evaporation zone increase.

The reported study was supported by RFBR, research project No. 16-38-50220 mol_nr. The Schlieren optical system was elaborated under the support of Russian Science Foundation (Agreement No. 14-19-01755).

References

1. Y. Lyulin, D. Feoktistov, I. Afanas'ev, E. Chachilo, O. Kabov, G. Kuznetsov, Tech. Phys. Lett., **41**, 665 (2015)
2. D. Zaitsev, D. Rodionov, O. Kabov, Tech. Phys. Let., **35**, 680 (2009)
3. O. Kabov, D. Zaitsev, Multiphase Sci. Technol., **21**, 249 (2009)
4. A. Sivkov, A. Saygash, J. Kolganova, I. Shanenkov, IOP Conf. Ser.: Mater. Sci. Eng., **66**, 012001 (2014)
5. A. Sivkov, Y. Shanenkova, A. Saigash, I. Shanenkov, Surf. Coat. Technol., **292**, 63 (2016)
6. D. Glushkov, J. Legros, P. Strizhak, A. Zakharevich, Fuel, **175**, 105 (2016)
7. D. Glushkov, G. Kuznetsov, P. Strizhak, R. Volkov, Therm. Sci., **20**, 131 (2016)
8. D. Glushkov, G. Kuznetsov, P. Strizhak, Therm. Sci., **19**, 1541 (2015)
9. G. Kuznetsov, A. Zakharevich, N. Bel'kov, Chem. Pet. Eng., **50**, 424 (2014)
10. V. Maksimov, T. Nagornova, EPJ Web of Conf., **76**, 01006 (2014)
11. K. Batischeva, E. Orlova, D. Feoktistov, MATEC Web of Conf., **19**, 01001 (2014)
12. E. Orlova, D. Feoktistov, G. Kuznetsov, EPJ Web of Conf., **82**, 01053 (2015)
13. V. Nakoryakov, S. Misyura, S. Elistratov, A. Manakov, A. Sizikov, J. Eng. Thermophys., **22**, 169 (2013)
14. V. Lebedev, V. Lemanov, S. Misyura, V. Terekhov, Fluid Dyn., **28**, 624 (1993)
15. V. Nakoryakov, S. Misyura, S. Elistratov, R. Dekhtyar, J. Eng. Thermophys., **23**, 257 (2014)
16. D. Feoktistov, E. Orlova, A. Islamova, EPJ Web of Conf., **110**, 01018 (2016)
17. K. Ponomarev, E. Orlova, D. Feoktistov, EPJ Web of Conf., **110**, 01060 (2016)
18. R. Volkov, G. Kuznetsov, P. Strizhak, Thermophysics and Aeromechanics, **21**, 255 (2014)
19. O. Vysokomornaya, G. Kuznetsov, P. Strizhak, J. Eng. Phys. Thermophys, **86**, 62 (2013)
20. G. Kuznetsov, M. Sheremet, Fluid Dyn., **41**, 881 (2006)
21. E. Orlova, G. Kuznetsov, D. Feoktistov, EPJ Web of Conf., **76**, 01039 (2014)
22. G. Kuznetsov, P. Kuybin, P. Strizhak, High Temp., **53**, 254 (2015)

EVAPORATION RATE OF AQUEOUS SALT SOLUTIONS DROPLETS

Konstantin O. Ponomarev[1], *Dmitry* V. Feoktistov[1,*], and *Igor* V. Marchuk[2]

[1] Tomsk Polytechnic University, 634050, Tomsk, Russia
[2] Kutateladze Institute of Thermophysics, 630090, Novosibirsk, Russia

Abstract. The evaporation of aqueous salt solution droplets from metal surfaces has been studied experimentally. The volumetric evaporation rate is found to decrease in time for any initial droplet volume due to an increase in salt concentration and the efforts of system to take a state of thermodynamic equilibrium. Crystalline hydrate film was being formed during desorption of $CaCl_2$ 10%, LiCl 10%, LiBr 30% salts. When NaCl 10% salt solution evaporated, there was no film. The average evaporation rate of NaCl salt is higher than for other salts. The lowest values of the average evaporation rate were found for LiBr 30% salt solution.

1 Introduction

Nowadays evaporation of thin liquid films, rivulets or droplets on a solid is very urgent problem for all scientific community. Spreading of a droplet over a heated surface is a complex problem, and to solve it, it is necessary to take into account many factors. Theoretical and experimental results about the mechanisms describing such processes [1-21] are not enough. For this reason, there are some difficulties of modelling the spreading and evaporation of droplets on the surfaces of heat exchange systems, which are used in all industrial processes of petrochemical processing.

The agreement about modernization of the Russian refining complex up to 2020 was signed between eleven large oil and gas companies and Russian Government in July 2011. As a result, all fuels produced on factories must comply with the Euro-5 standard, and oil conversation ratio must be increased from 72 to 85 % or more. These plans can be effectively implemented through using of absorption lithium bromide refrigerating machines and heat pumps in various technological processes.

In addition to water, different aqueous solutions, for example, lithium bromide (LiBr), are used as working liquids (absorbent) in heat pumps, thermotransformers and absorption refrigerating machines. Thin films, jets, and droplets of salt solutions, are formed on the walls of the heat exchange equipment. To date, there is not enough experimental data on the evaporation of salt solution droplets of LiBr, $CaCl_2$, LiCl и NaCl under the crystalline formation to simulate effective absorption heat pumps. So the purpose of this work is to determine experimentally the effect of the composition and the volume of aqueous salt solution droplets on their evaporation rate.

* Corresponding author: fdv@tpu.ru

2 Experimental technique

Research was conducted using the experimental setup including shadow system. For implementation of this system, the high speed video camera and source of parallel light were used. The principle of operation of the setup is described in detail in [16].

A disk-shaped substrate (54 mm in diameter, 4 mm in thick) used in the experiments was made of aluminium. It is heated up by the Peltier element. Temperature under the substrate was kept constant 80°C. Geometric characteristics, such as, contact angle, height and diameter of a droplet were obtained after processing the shadow images in Drop Shape Analysis software.

The experimental setup was isolated from external influences (convection, radiation) by transparent box 11 made of 3 mm thick polymer glass. It ensures constancy of heat and mass transfer conditions.

We used the following working liquids: 10 % NaCl, 10 % CaCl2, 10% LiCl, 30 % LiBr. The liquid volume was varied in the following range: 10, 20, 30, 40, 50, 60 μL.

The volumetric evaporation rate was calculated as follows:

$$Es = \Delta V/\Delta t,\ \mu L/sec, \quad (1)$$

where $\Delta V = (V_{i+1} - V_i)$ is the change of volume of aqueous solution, μL; $\Delta t = (t_{i+1} - t_i)$ is the period of time when the volume changed, sec.

3 Results and discussion

Figure 1 presents time dependences of the evaporation rate of aqueous salt solution droplets.

The initial values (Fig. 1) correspond to the time when a droplet was placed on the substrate, final values – to the crystallization when the shape of a droplet was distorted by crystalline hydrates. It did not allow to use optical methods of processing.

During evaporation of aqueous salt solution droplets it is found that the evaporation rate decreases with time for any initial droplet volume. There is one molecule of water for one molecule of salt in solutions of LiBr, LiCl at 80 °C temperature and atmospheric pressure. But in $CaCl_2$ solution there are two molecules of water for one molecule of salt. It causes the intensive decrease in the evaporation rate of $CaCl_2$ with time compared to the lithium salts. In the case of NaCl salt at 80 °C crystals of pure salt are formed (without crystalline or crystalline hydrates are unstable and transit quickly in pure crystals of NaCl). In contrast to the above-mentioned salts, for aqueous salt solution of NaCl the evaporation rate in the initial period of time decreases, then the curve has a quasi-linear nature. It is connected with the early crystallization.

The evaporation rate decreases in time because of an increase in salt concentration and due to the efforts of system to take a state of thermodynamic equilibrium, i.e. the difference between the equilibrium concentration at the droplet interface and the average volumetric concentration decreases with time. Changing of the droplet volume occurs only due to desorption of water from the solution. In this case, mass of salt in the droplet remains constant as there is no boiling crisis, which can lead to droplet detachment with periodic discharge of salt with water.

In comparison with single-component liquids, in solutions during desorption the third phase (solid state) forms. It is known that with increasing initial concentration of salt solution there is a significant decrease in the intensity of desorption. This law is determined by decrease in the saturated water vapour pressure above the surface at the solution/gas

DOI: 10.1051/matecconf/20179101029

interface. Probably, it is caused by intensive change in the evaporation rate of LiBr 30 % droplets and the lowest values of their rates in comparison with other salts.

Evaporation along the edges of the droplet is more intensive so compensation capillary flow occurs inside the droplet, which transfers a dispersed phase from central part of the droplet to the three-phase contact line. As a result, crystal growth does not start from the centre, but from the periphery of the droplet. In this case, the movement of crystallization is always directed from the periphery to the centre. It is connected to the fact that the thickness of the solution along droplet perimeter is minimal and, consequently, the local heat flow from the heated surface to the droplet along perimeter is larger than in the centre. This leads to a more intense evaporation of the solution, its super saturation and solid phase extraction. Regions located in the droplet periphery stretch central part causing radial flow, with that the ratio of components in inside part of the droplet varies. The lack of solution near the droplet top is compensated by an upward flow in the centre. What is more, the temperature of the flow is higher than at the surface. The surface tension continues to decrease, and circulation enhances in this manner. It continues to decrease the surface tension and enhances circulation. Concentration heterogeneity of the aqueous solution leads to non-uniform evaporation of droplet; the central part of the droplet flattens and becomes thinner.

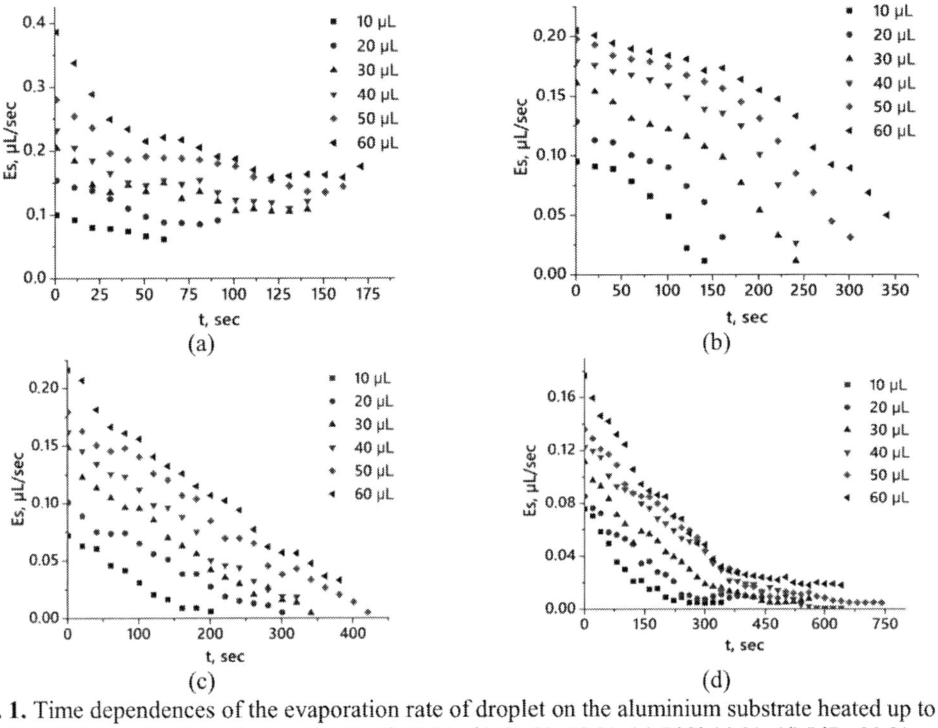

Fig. 1. Time dependences of the evaporation rate of droplet on the aluminium substrate heated up to 80 °C at different droplet volumes: (a) NaCl 10 %; (b) $CaCl_2$ 10 %; (c) LiCl 10 %; (d) LiBr 30 %.

Crystals formed at the edges of the droplet create a film, which moving to the centre, completely covers its entire surface. After forming a solid surface of the crystal or crystalline hydrates film, desorption process continues. Water diffuses, or penetrates the crystalline film, and then evaporates, but this process is very slow. The dimensions of formed crystals depend on the conditions of crystallization process, which can be divided into two stages: formation of crystal nuclei and their growth due to a substance in solution. For the formation of finely crystalline sediments it is required that the first stage (formation of crystal nuclei) occurred quickly. Many nuclei of crystallization form small crystals because a little substance is deposited on their surfaces. The transition from slow to rapid crystallization is accompanied by changes in the morphology of structures. The emergence of new well-ordered crystalline structures is advantageous from an energy point of view, since it allows to increase vapour desorption flow and significantly reduce total time of droplet evaporation.

4 Conclusion

After conducting the experimental studies of aqueous salt solution droplets evaporation on metal surface, it is found that the volumetric evaporation rate decreases in time for any initial droplet volume due to an increase in salt concentration and the efforts of system to take a state of thermodynamic equilibrium. During desorption of $CaCl_2$ 10%, LiCl 10 %, LiBr 30% salts thin film was being formed, and after that it moved from the periphery to the droplet centre and covered its entire surface. During evaporation of NaCl 10 % salt solution, there was no film.

The reported study was supported by RFBR, research project No. 16-38-50221 mol_nr.

References

1. V. Nakoryakov, S. Misyura, S. Elistratov, Int. J. Heat Mass Transfer, **55**, 6609 (2012)
2. E. Gatapova, A. Semenov, D. Zaitsev, O. Kabov, Colloids and Surf., A, **441**, 776 (2008)
3. A. Sivkov, Y. Shanenkova, A. Saigash, I. Shanenkov, Surf. Coat. Technol., **292**, 63 (2016)
4. V. Nakoryakov, S. Misyura, J. Eng. Thermophys., **25**, 24 (2016)
5. V. Nakoryakov, S. Misyura, S. Elistratov, Thermal Sci., **16**,997 (2012)
6. D. Zaitsev, D. Rodionov, O. Kabov, Tech. Phys. Lett., **35**, 680 (2009)
7. D. Zaitsev, D. Kirichenko, O. Kabov, Tech. Phys. Lett., **41**, 551 (2015)
8. G. Kuznetsov, A. Zakharevich, N. Bel'kov, Chem. Petr. Eng., **50**, 424 (2014)
9. V. Maksimov, T. Nagornova, I. Shestakov, EPJ Web of Conf., **82**, 01048 (2015)
10. D. Glushkov, J. Legros, P. Strizhak, A. Zakharevich, Fuel, **175**, 105 (2016)
11. D. Glushkov, G. Kuznetsov, P. Strizhak, R. Volkov, Therm. Sci., **20**, 131 (2016)
12. O. Vysokomornaya, G. Kuznetsov, P. Strizhak, Russian Journal of Physical Chemistry B, **5**, 668 (2011)
13. E. Orlova, G. Kuznetsov, D. Feoktistov, EPJ Web of Conf., **76**, 01039 (2014)
14. R. Volkov, G. Kuznetsov, P. Strizhak, Int. J. Therm. Sci., **88**, 193 (2015)
15. G. Kuznetsov, G. Mamontov, G. Taratushkina, Combust., Explos. Shock Waves, **40**, 70 (2004)
16. K. Batischeva, E. Orlova, D. Feoktistov, MATEC Web of Conf., **19**, 01001 (2014)
17. E. Orlova, D. Feoktistov, G. Kuznetsov, EPJ Web of Conf., **82**, 01053 (2015)
18. G. Kuznetsov, D. Feoktistov, E. Orlova, J. Eng. Phys. Thermophys., **89**, 317 (2016)
19. D. Feoktistov, E. Orlova, A. Islamova, MATEC Web of Conf., **23**, 01054 (2015)
20. K. Ponomarev, E. Orlova, D. Feoktistov, EPJ Web of Conf., **110**, 01060 (2016)
21. G. Kuznetsov, D. Feoktistov, E. Orlova, K. Batishcheva, Colloid J., **78,** 335 (2016)

THE PROBLEM OF EXPERIMENTAL EVALUATION FOR CORONA RESISTANCE OF VFD CABLES

Andrey P. Leonov[1], and *Dmitry* I. Charkov[1,*]

[1] National Research Tomsk Polytechnic University, 634050, Tomsk, Russia

Abstract. This article describes an approach for evaluating experimental cable insulation resistance electro-thermal aging, typical for the drive frequency being adjusted. It estimates the average time until the insulation breakdown when exposed to a sample of high electrical loads. The possibility for an adequate choice of insulating materials, which are used in the insulation of VFD cable and shown ways to improve testing methods.

1 Introduction

When using a VFD on the basis of pulse-width modulation having electro-thermal aging. This reduces the life of the insulation of supply cables and reduce the reliability of the entire system. [1]

The nature and magnitude of operating voltages in the VFD depends on the length of the supply cable [2], the method of modulation voltage [3], the stator winding structure, composition and parameters of elements in the VFD. With that said insulated power cables for variable frequency drive to withstand higher operating load for the entire lifetime. First of all it is determined by the ability of insulation to withstand the action of electric discharges [4].

2 Experimental part

The work carried out to determine the average time of the breakdown products of cable samples with an extruded polymeric insulation.

The objects of study are the samples were insulated from the block copolymer, ethylene-propylene, polyurethane and polyethylene. Samples for testing are to a twists wire with a working area of 125 mm.

The samples were placed in an oven heated to a temperature class Insulation, kept up to breakdown when exposed to a sample of the AC voltage. It has been tested several batches

* Corresponding author: dic1@tpu.ru

of twist at different temperatures, under the influence of an elevated voltage and power frequency.

3 Experimental results

The results showed: when the frequency of the applied field surface discharges reduce the mean time to breakdown (Figure 1). This is due to the fact that increased heterogeneity of the field and the intensity of any electrical discharges.

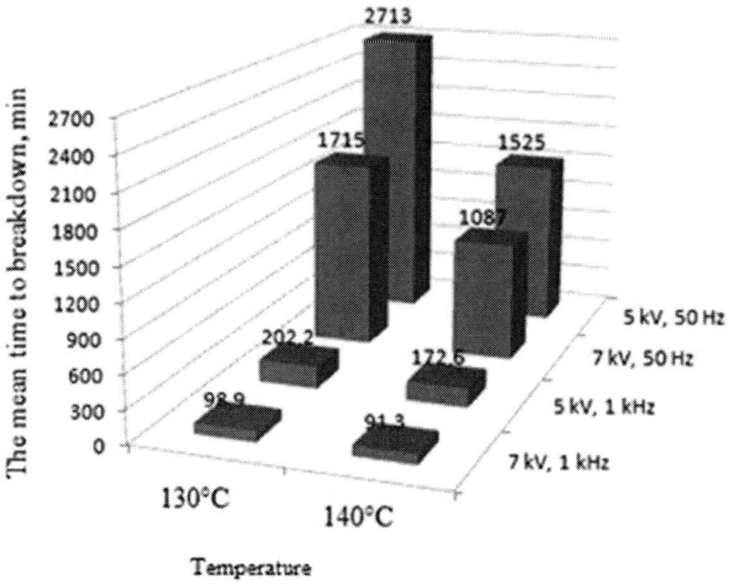

Fig. 1. The results of the mean time to breakdown of the insulation twists (block copolymer insulation).

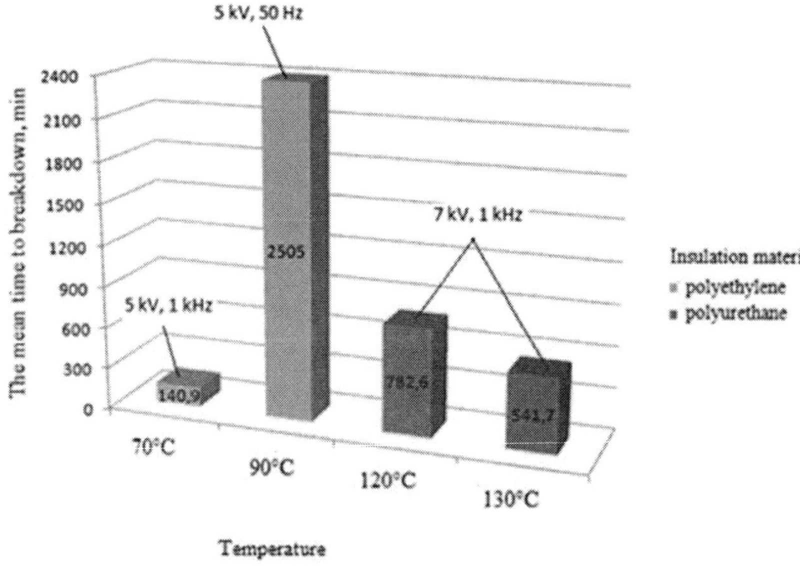

Fig. 2. The results of the mean time to breakdown of the insulation prototypes.

DOI: 10.1051/matecconf/20179101030

For polymeric insulation noted that the increase in frequency, intensity increases abruptly discharges and reduces the mean time to breakdown, leading to insulation breakdown.

4 Conclusion

Analyzing the results obtained are the following:

1. Hanging frequency of the applied field dramatically reduces mean time to breakdown, due to the fact that the increased heterogeneity of the field, there are discharges, which accelerates the process of electro-thermal aging of insulation.
2. Increasing the level of exposure to electric voltages greater impact on reducing the mean time to breakdown of the insulation than the temperature rise. The intensity of the electric aging prevails over the heat, thereby accelerates the process of destruction of the insulation.
3. Such tests allow to adequately assess the insulation resistance of cable products to the electro-thermal load and select materials for the insulation of VFD cables.

In future completion of the experimental setup and the definition of insulation resistance to high-frequency modulated load, both during the operation of the variable frequency drive.

References

1. A. Garganeev, A. Leonov, V. Merkulov, D. Charkov, IOP Conf. Ser.: Mater. Sci. Eng., **135,** 012013 (2016)
2. C. Rodriguez Valdez, R. M. Tallam, R. J. Kerkman, Electro/Information Technology, 2008. EIT 2008. IEEE International Conference on, 151 (2008)
3. Leon M. Tolbert, Fang Zheng Peng, Thomas G. Habetler. IEEE Trans. Ind. Appl. **35**, 36 (1999)
4. Jaroslaw Luszcz. Industrial Electronics (ISIE), 2011 IEEE International Symposium on. 538 (2011)

DOI: 10.1051/matecconf/20179101031

IMPROVING THE METHOD OF SOLAR RADIATION DURABILITY DETERMINATION OF CABLE PRODUCTS

Vladimir Rybakov[1], and *Elvira* Sokolova[1,*]

[1] Tomsk Polytechnic University, 634050, Tomsk, Russia

Abstract. The authors suggest an extra physic-mechanical procedure to determine durability of cable goods to solar radiation exposure. The test check was carried out using the standard and the proposed procedures. The results of two tests were compared and analyzed. The use of the proposed technique allows improving the validity and reliability of the laboratory experiment conducted in a real production environment and exclude the low-quality cable items. The only disadvantage of the proposed technique is its labor intensiveness.

1 Introduction

One of the key factors negatively affecting the exploitative characteristics is solar radiation. It leads to early aging and wear of a cable jacket and requires the replacement of cables, which is uneconomical and leads to some additional financial expenses. The existing technique intended to assess sunlight durability of cables is imperfect, due to the fact only visual observation is carried out during the test. Thus, defective goods may be allowed for their use in industry.

The goal of the research is to improve the existing traditional technique used for durability assessment of cable goods exposed to solar radiation.

Research objectives are as follows:

1) To carry out a standard test procedure to determine the deterioration process in cable items.
2) To carry out the proposed procedure with physic-mechanical test, which is considered to be crucial in this research due to its ability to improve accuracy and efficiency of durability assessment.
3) To compare the results of two tests aiming to confirm the accuracy of the proposed testing procedure.

2 Materials and Methods

[*] Corresponding author: weiberd@tpu.ru

To improve the technique concerning the assessment of solar radiation impact various cables were chosen as research objects. Selection of certain brands is determined by the necessity of their quality verification.

The tested cables are used as power cables for outdoor applications and explosion/fire hazardous structures. Various polymeric materials are used to produce insulation and jackets for these cables, providing they have different mechanical and chemical properties. It is to be mentioned that the composition of polymeric mixtures are kept in secret by the producers.

One of the most important criteria relative to operation properties is cable durability to sunlight exposure. Therefore, an external cable jacket is made to provide the protection against external various factors, among which solar radiation is. Thus, the outer jacket has been chosen as the study object.

The resistance to sunlight is determined as the ability of a cable jacket made of polymers to keep the desired mechanical and chemical properties over a long period. Solar radiation is one of the factors having significant impact on degradation, which influences polymer properties negatively. The most destructive is high-energy radiation having small wavelength [3]. Ionizing radiation as well as ultraviolet and visible parts of light spectrum of wavelengths belongs to these negative effects. The vulnerability of polymers to sunlight is defined by the presence of chromophore chemical groups capable to absorb light.

3 Standard procedure for deterioration of jackets assessment caused by solar radiation

A standard method of deterioration assessment of electrical products is in a visual observation and testing signs of destruction after exposure to solar radiation in a climatic chamber. The exposure lasts from 5 to 10 days depending on the purpose. In a Heat, Moisture and Solar Radiation Chamber (HMSRC) specific conditions are kept.

The inference about durability or vulnerability to sunlight exposure is based on visual detecting of cracks, color change or intermediate deterioration reactions products emissions.

4 Physical-mechanical test

Physical-mechanical tests contribute to the increase of accuracy and effectiveness of the method defining cable durability to sunlight exposure since they define this characteristic both qualitatively and quantitatively. The main indices of durability to sunlight in this work are unit elongation and tensile strength. The comparison of these two values before and after being exposed to the sunlight allows to assess the deterioration and the aging rate.

As test samples shovel-shaped pieces (according to State Standard IEC 60811-1-1-2011) were chosen (fig. 1).

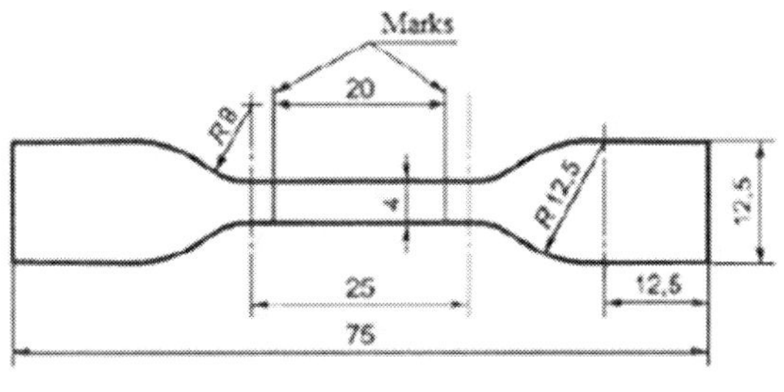

Fig. 1. Shovel-shaped sample.

 DOI: 10.1051/matecconf/20179101031

5 Calculations

Tensile strength and elongation values are calculated using the following expressions: $\sigma_p = \frac{P}{S}$, where P is the force applied to a sample, [kilogram-force]; S is the area of transverse section of the sample, [mm^2].

$$\sigma_p = \frac{15.8}{7.52} = 2.101 \text{ kgs/mm}^2$$

$\theta = \frac{(l_i - l_0)}{l} \cdot 100\%$, where l_0, l_i are the lengths of measuring section before the test and at rupture.

$$\theta = \frac{63 - 20}{20} \cdot 100\% = 215\%$$

Arithmetic mean value is accepted as the calculation result.

6 Result analysis

The standard test according to State Standard 20.57.406-81 did not reveal any defects in the samples, like cracks, color change or depolymerization products emission. The external appearance of the tested sample has not changed. Thus, all types of cables have successfully passed the test.

Tension test results are shown in Fig.2. (bar chart). One can see, tensile strength values of aged samples are below the same results for the samples, which were not exposed to radiation. It can be assumed that there is an impact of solar radiation on mechanical strength of the jackets.

For the sample of HF(AF) jacket it is notable, tensile strength has dropped by 18 %. Elongation at rupture values of aged samples have risen by 34% after light exposure.

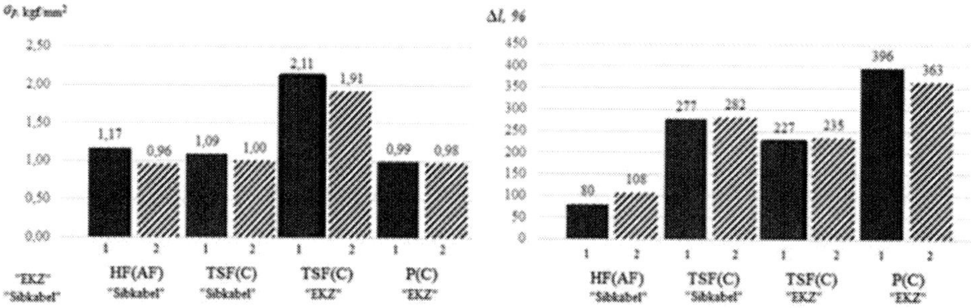

Fig. 2. Tensile strength changing and elongation at rupture changing of samples after solar radiation exposure: 1 – initial state values, 2 – values after aging in a climatic chamber.

For polymeric insulation noted that the increase in frequency, intensity increases abruptly discharges and reduces the mean time to breakdown, leading to insulation breakdown.

For the samples of TSF(C) cable jackets made by two different manufacturers there is 8 – 9 % drop of tensile strength after the light exposure.

Elongation values are 2 – 3 % higher for aged samples. It should be noted that the tensile strength absolute values of initial samples made by "EKZ" are 95 % higher, than the values of those made by "Sibkabel". Wherein, elongation values of initial "EKZ" samples are 18 % lower, than of those made by "Sibkabel".

The last jacket sample unlike the previous ones is made from thermosetting mixture, (rubber) and thus is subject of concern to our research. Experiment results state 1 % change of tensile strength. Average elongation value of aged samples is 8 % less than of the initial ones.

Upon the results of the test the cable jacket №1 Hold Cab EPR LV HF(AF) 4x2,5ок(N)-1 is admitted to be insufficiently durable to solar radiation impact on physic-mechanical properties of a make. The degradation of tensile is about 18%, which is the same value as the analogous one for thermal aging of this cable and is equal to 40% according to technical documents [6]. The change of elongation value for thermal aging is also 40%.

Cable products like Hold Cab EPR LV TSF(C), made by different companies have shown the same durability to solar radiation, and the impact on physic-mechanical properties is comparable to thermal aging. Thermal deterioration causes 25% decline of mechanical properties [6].

A blend compound of rubber №4 Hold Cab EPR LV P(C) may contain vulcanizing agents and activators leftovers, which are able to maintain cross-linkage at 50° C, which is the temperature being kept during the whole aging process. However, the impact of these components is lesser and is comparable to vulcanization irregularity within a sample

7 Conclusion

The obtained results concerning break durability and elongation at break allows stating the necessity of the standard technique improvement. The improved technique will increase the precision and efficiency of the resistance assessment of cables to solar radiation.

We can conclude that the proposed technique aimed at the research of the cable jackets resistance taking into account their physical-mechanical properties is perspective in terms of scientific and practical significance. In accordance with the standard test technique [5] all the produced items were recognized as resistant to solar radiation. The research with applying the proposed technique, however, showed the produced items being exposed to solar radiation have some defects like photochemical destruction and procure. Photochemical destruction of cable products leads to their precocious aging and the necessity to replace them and procure causes the flexibility decrease. Thus, the use of the proposed technique allows improving the validity and reliability of the laboratory experiment conducted in a real production environment and exclude the low-quality cable items. Thereby, the only disadvantage of the proposed technique is its labor intensiveness, since the majority of tests should be carried out manually.

References

1. A. Garganeev, A. Leonov, V. Merkulov, D. Charkov, IOP Conf. Ser.: Mater. Sci. Eng., A. A. Gerasimenko, **1** (1987)
2. A. A. Gerasimenko, *Corrosion, aging and biological protection of machines, equipment and installations V. 2* (Mashinostroenie, Moscow, 1987)
3. L. S. Pinchuk, A.S. Neverov *Polymer film containing corrosion inhibitors* (Chemistry, Moscow, 1993)
4. State Standard IEC 60811-1-1-2011, *The general testing methods of insulation materials and jackets of electrical and optical cables.* (Publishing House of standards, Moscow, 2013)
5. State Standard 20.57.406-81, *Electronic engineering product, quantum electronics products and electromechanical products. Test methods* (Publishing House of standards, Moscow,1982)
6. Specification 16.K73.130, (2015)

 DOI: 10.1051/matecconf/20179101032

ANALYSIS HYSTERESIS CLUTCH TECHNICAL CHARACTERISTICS OF THE SHUT-OFF VALVES ELECTRICAL DRIVE SYSTEM

Maria V. Saveleva[1], and *Nadejda* Y. Sipaylova[1,*]

[1] Tomsk Polytechnic University, 634050, Tomsk, Russia[1]

Abstract. The results of calculations of the hysteresis clutches of various designs for use in electric shut-off valves of the pipeline. The possibility of using the clutch with an axial air gap for a wide range of electric power.

1 Introduction

The shut-off valves are used to shut off the pipeline sections. In particular, the shut-off valves are installed in the main oil pipelines every 15–20 km. Often, the shut-off valve is equipped with the electric drive, which are essential components of an asynchronous motor and gearbox (worm or wave) [1].

Power shut-off valves should provide gate breakdown from rest and sealing the end of the course, be able to remote control, comply with explosion-proof and climate performance. As often as the shut-off valve is used wedge gate [2], the tightness of the system in the first place depends on the wear of the wedge. To increase its service life it is necessary to force the end of the course provides a reliable contact, but do not over-tighten the wedge.

2 Formulation of problem

The problem of reducing the seals of the wedge in the shut-off valve with the passage of time can be solved with the help of:

- Using a frequency converter as a part of electric drive, which will allow to precisely control the force developed by the motor at all stages of valve operation. However, this option is quite expensive for widespread implementation;
- Replacement of the induction motor synchronous hysteresis motors, which control will be performed using a PWM [3], which is also an expensive option;
- The introduction to the electric drive hysteresis clutch.

Hysteresis clutch belongs to a class of synchronous magnetic coupling and is used for non-contact transmission of mechanical torque from the drive shaft to the output. The advantages of the hysteresis coupling, important for the operation as part of the drive of valves are:

* Corresponding author: sny@tpu.ru

- Having its own starting point, which will allow for the failure of the gate from the rest;
- Set the maximum transmitted torque, eliminates pinched wedge at the end of the course;
- The clutch can be introduced into the electric drive, slighty complicating the structure thereof;
- Compliance with the terms of explosion and low sensitivity to temperature changes [4].

The disadvantages which can be isolated from the hysteresis coupling include the use in the manufacture of complex materials hard-magnetic and functioning own characteristics different constructions described below.

The basic design of the hysteresis clutches are shown in Figure 1.

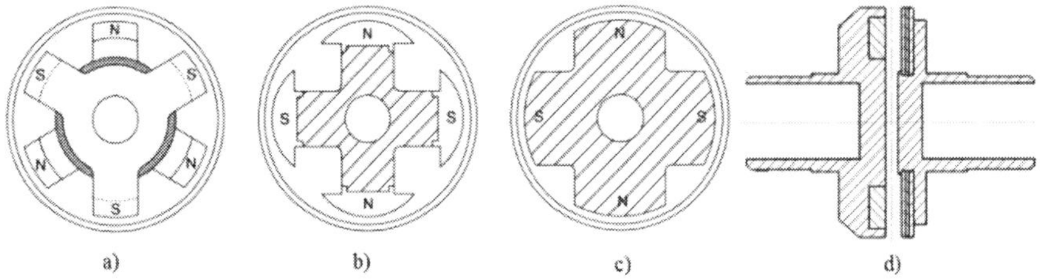

Fig. 1. The hysteresis clutch a) clawlike inductor; b) radially spaced from the magnets and pole piece; c) with the inductor of the "asterisk"; g) butt magnetization.

Clawlike inductor (Figure 1, a) consists of a cylindrical permanent magnet having on the ends of the soft magnetic steel washers with clawlike projections. The main advantage of the inductor is that the permanent magnet is protected from the external elements of the soft magnetic field, and its initial magnetization is performed in the assembled external uniform field. Therefore, the degree of utilization of the magnet is high and labor induction is 0.6-0.7 Tesla. Furthermore, the magnet has a simple shape and is situated near the center of the inductor, which allows to realize the inductor circumferential speed of 80-100 m / s, because the outer soft magnetic elements have sufficient mechanical strength. Therefore, power of machines with clawlike inductor can reach 10–20 kV · A [5].

Inductor of hysteresis clutch with radially arranged magnets and pole pieces (Figure 1, b) has an average value of the induction of labor among the presented designs, but has a worse fill factor magnet and a more complex structure compared with the inductor of the "asterisk".

The design of inductor type "asterisk" (figure 1c) comprises a six-pointed molded permanent magnet which is fastened to the shaft by casting a nonmagnetic alloy (based on aluminum or zinc). Inductor Advantages - simplicity and high filling rate of a magnet. Disadvantages include low mechanical strength because of the fragility hard-magnetic alloy and residual stress in the casting. Maximum allowable inductor circumferential speed is 40-50 m / s, and the workers do not exceed 0.2-0.4 induce T, thereby asterisk inducer is usually used at a relatively low power machine (10 kVA).

Hysteresis clutch with an axial air gap (Figure 1, d) and the butt is the magnetization inductor with alternating poles clawlike type on the drive shaft. The driven coupling part comprises a hysteresis layer which holds a laminated or cast and enclosed in the armature of soft magnetic or non-magnetic material and may also be carried out without any reinforcement.

Energy losses in the magnetic hysteresis clutches in synchronous transmission mode is small, because they are determined by the friction losses in the bearings, air friction losses and eddy currents in the electrically conductive nonmagnetic wall. Thus the efficiency of the magnetic of hysteresis clutch is extremely high, approaching unity.

 DOI: 10.1051/matecconf/20179101032

Estimation of optimal design of magnetic hysteresis clutches in terms of the energy characteristics of secondary importance, as the loss of energy in it is small.

In couplings with a radial air gap misalignment due to manufacturing errors technology leads to uneven air gap, which adversely affects the characteristics of the coupling and creates radial forces on the bearings due to violations of magnetic symmetry. The clutches with axial air gap misalignment less impact on non-uniformity of the gap and the coupling characteristics, so manufacturing techniques clutches with radial air gap requires more precision than for the coupling with the axial air gap.

The clutches with radial air gap, it is possible to develop a series of couplings with different hysteresis torque value based on one of the external diameter with the same geometry of the active parts. The clutches with axial air gap, these possibilities are limited, since their optimal axial length is operably linked to the outer diameter.

For large values of the desired hysteresis torque coupling with a radial air gap becomes more rational than the clutch with an axial air gap since the first allows you to get the required amount of hysteresis torque by increasing the axial length, without increasing the outer diameter. With a uniform air gap in the sleeve with radial clearance no radial forces of magnetic attraction and unilateral axial forces. In couplings with axial air gap having an axial magnetic attraction force acting on the bearing, inversely proportional to the magnitude of the air gap [6].

3 Conducted calculations

Based on the range of engines used in the composition and on the basis of the electric drive YUNDK35T5AA materials and Fe-Cr-Co sizes were calculated hysteresis clutches required for transmitting a predetermined torque value. Tables 1 and 2 were built on the information received.

Table 1. Dimensional specifications of hysteresis clutch.

Motor specifications			Clutch Specifications							
					Radial gap			Axial gap		
P	T	$D_{in.m.}$	$D_{out.h.}$	V_h	D_{inh}	L_h	λ	$D_{in.h}$	Δ_h	Δ_h*
power	torque	motor inner diameter	outer diameter of the hysteresis layer	volume of hysteresis layer	inner diameter of the hysteresis layer	length of the inductor	constr. factor		height of hysteresis layer	relative height of hysteresis layer
kW	N · m	cm	cm	cm^3	cm	cm		cm	cm	
0.55	0.4	20	18	23.26	17.5	5.46	0.27	6.30	0.109	0.01
1.5	1	20	18	58.15	17.5	13.65	0.68	6.30	0.271	0.02
2.5	1.7	25	22.5	98.85	22	18.51	0.74	7.88	0.295	0.01
4	2.7	25	22.5	157.00	22	29.40	1.18	7.88	0.469	0.02
7.5	5	35	31.5	290.74	31	38.77	1.11	11.03	0.443	0.01
11	7.3	35	31.5	424.48	31	56.60	1.62	11.03	0.647	0.02
18.5	12.3	35	31.5	715.22	31	95.36	2.72	11.03	1.09	0.03
30	20	40	36	1162.96	35.5	135.54	3.39	12.60	1.357	0.04
45	30	45	40.5	1744.44	40	180.58	4.01	14.18	1.608	0.04

Table 2. Torque values of different hysteresis clutches.

Axial gap clutch					Radial gap clutch								
					"Star" type inductor			Pole piece inductor			Clawlike inductor		
D	λ	$\sigma_{mg.m.}$	M_h	D	λ	$\sigma_{m\,g.m.}$	Mh	λ	$\sigma_{m\,g.m.}$	M_h	λ	$\sigma_{mg.m.}$	M_h
outer clutch diameter	constr. factor	active materials using factor	hysteresis torque										
cm		10^{-3}	N · m	cm		10^{-3}	N · m		10^{-3}	N · m		10^{-3}	N · m
18	0.5	1.34	3.91	17.5	–	1.5		–	1		–	0.63	
18	0.5	1.34	3.91	17.5	0.76	1.5	6.1	0.76	1	4.1	0.76	0.63	2.56
22.5	0.5	1.34	7.63	22.0	0.82	1.5	13.1	0.82	1	8.7	0.82	0.63	5.51
22.5	0.5	1.34	7.63	22.0	1.31	1.5	20.8	1.31	1	13.9	–		
31.5	0.5	1.34	20.9	31.0	1.23	1.5	54.9	1.23	1	36.6	–		
31.5	0.5	1.34	20.9	31.0	1.80	1.5	80.2	1.80	1	53.5	–		
31.5	0.5	1.34	20.9	31.0	–	1.5		–	1		–		
36	0.5	1.34	31.3	35.5	–	1.5		–	1		–		
40.5	0.5	1.34	44.5	40.0	–	1.5		–	1		–		

4 Conclusion

The preliminary assessment of the hysteresis clutch with an axial air gap are more promising for use in the electric drive of shut-off valves, as they meet all the specified requirements to the coupling as a whole, are able to work in all aspects ranging and significantly affect on the drive dimensions, unlike couplings with radial magnetization. In general, due to its simplicity and relatively low cost of the materials used can be to talk about the possible use of the hysteresis clutches and in other industries where currently used asynchronous motors with direct start.

References

1. G. Sevastyanihin, Science and Construction, **5**, (2006)
2. A. Garganeev, A. Karakulov, S. Landgraf, *Power shut-off valves: monograph* (Publishing House of Tomsk Polytechnic University, Tomsk, 2012) [in Russian]
3. A. Garganeev, D. Padalko, A. Chervatyuk, Reports TUSUR, **2**, 32 (2014) [in Russian]
4. *Operating Instructions Power torque limiting explosion-proof two-way clutch for valves DN 500-1200b PN 8,0-15,0* (TOMZEL, Tomsk, 2013) [in Russian]
5. V. Booth, *Contactless electrical machines* (Vysshaya Shkola, Moscow, 1990) [in Russian]
6. S. Mizyurin, M. Ermilov, *Design magnetic hysteresis clutches* (Publishing House of MAI, Moscow, 1966) [in Russia]

MATEC Web of Conferences **91**, 01033 (2017) DOI: 10.1051/matecconf/20179101033
Smart Grids 2016

THE USE OF A SPECIAL SOFTWARE FOR INDUCTION MOTOR DIAGNOSTICS IN THE DIAMOND INDUSTRY

Vladislav A. Shevchuk[1], *Oleg* P. Muravlev[1], *Olga* O. Stolyarova[1,*], and Vladimir P. Shevchuk[2]

[1] Tomsk Polytechnic University, 634050, Tomsk, Russia
[2]DC "ALROSA", Yakutia, Mirny

Abstract. The field of induction motor application in the diamond industry has considerably increased due to the transition to the underground mining method. This article is devoted to the research the induction motor diagnostic technique in the diamond industry using special software. The practice shows that unreliability of induction motors causes some serious economic losses related to the expenses on premature maintenance and idle time. This article is devoted to the research of diagnostic technique of induction motors in the diamond industry using a special software. The article presents some data obtained in the course of the research conducted at an industrial site of the diamond company "ALROSA" in Mirny using the vibration-based diagnostics. Thus, based on the analysis of the findings some recommendations concerning the improvement of maintenance and servicing system have been given. The use of these recommendations allows assessing the operational reliability of induction motors in the diamond industry, minimizing the risk of sudden failures and reducing the maintenance and repair expenses.

1 Introduction

The field of induction motor application in the diamond industry has considerably increased due to the transition to underground mining methods. However, the induction motor unreliability leads to serious economic losses associated with the premature maintenance and idle time expenses. Therefore, there is an urgent problem to choose an adequate induction motor diagnostics technique without its stopping.

The main sources of induction motor failures are: bearing failures – 49 %; stator winding failures – 35 %; rotor winding failures – 10 %; other failures – 6 % [1].

The relevance of this work is in the necessity to improve the maintenance and service system of a.c. machines through the selection of diagnostics technique under the existing conditions in the diamond industry.

The aim of this work is to research the induction motor diagnostic technique in the diamond industry using special software. The vibration-based diagnostic technique on a

* Corresponding author: muravlevaoo@tpu.ru

root-mean-square value of vibration velocity is widely used in the diamond industry. This technique is based on the analysis of the induction motor vibration. Electric machines have some mechanical parts which cause vibrations. When tear and wear or destruction of mechanical part of the motor is observed, its frequency spectrum also undergoes some changes. Each fault (defect) causes extra vibrations of the induction motor with their distinctive features which can be measured, compared to the reference values and then the failures can be diagnosed [2]. Vibration is often measured by several sensors installed in different parts of the machine. The main idea is to analyze the root-mean-square value of vibration velocity (seldom – vibration acceleration) in different parts of the electric motor [3, 4, 5].

The main advantages of this technique are as follows [3]:

- It does not require full stopping of the electric motor;
- Simplicity in implementation;
- Low cost, since very simple apparatuses are used to measure this vibration.

2 Experimental procedure

The diagnostics is carried out using the vibration signal analyzer "Quartz" with its further processing in software "Diamond-2".

The device "Quartz" is an autonomous microprocessor-based vibration analyzer. The device allows defining natural frequency and the waveforms of machine components, analyzing the vibration signal shape, measuring acceleration rate (characteristics) and aggregate rundown and performing diagnostics of the aggregate condition using spectral analysis. Data processing is performed in special software "Diamond-2". This special software allows considerably simplifying the problem of vibration-based diagnostics through automation of routine inspection processes as well as personal computer data input in the database and their analysis [6, 7, 8].

At various plants of the DC "ALROSA" all operational information and technical reports on vibration-based diagnostics of electrical equipment has been thoroughly studied

Further we will consider the illustrations of this diagnostics. The diagnostics of the pump of the brand CPS 180-900 in the mine environment "International" is presented in Figures 1 and 2.

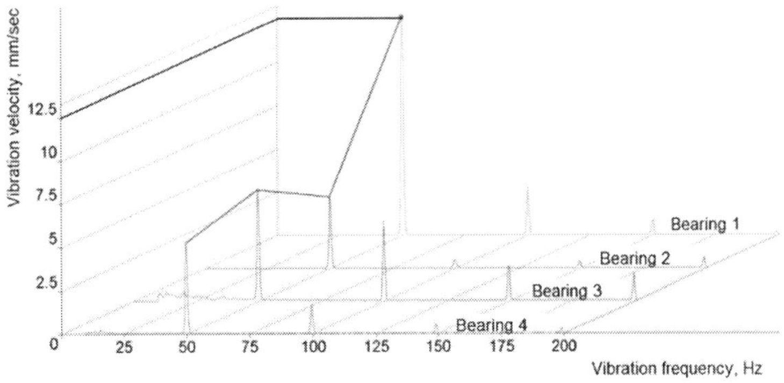

Fig. 1. Spectra cascade of vibration velocity, vertical direction.

Maximal level – on bearing 1 (electric motor). Rotational frequency 50 Hz and amplitude – 12.5 mm/sec. are designated by cursor.

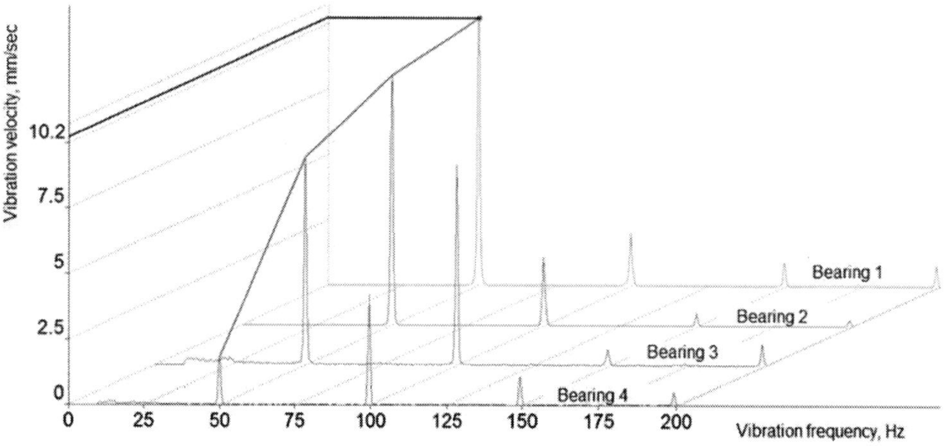

Fig. 2. Spectra cascade of vibration velocity, crosswise direction Maximal level (rate) – on bearing 1 (electric motor). Rotational frequency 50 Hz and amplitude 10.2 mm/sec. are designated by a cursor.

The maximal vibration level of the machine becomes localized on electric motor bearings. The data of root-mean-square value of vibration velocity are shown in Table 1 below.

Table 1. Data of vibration velocity of root-mean-square value.

Direction / point	Vertical direction , mm/sec	Crosswise direction mm/sec
Bearing 1 (electric motor) – rolling bearing	15.1	12.6
Bearing 2 (electric motor) – rolling bearing	4.7	11.08
Bearing 3 (pump) – sliding bearing	10.3	14.9
Bearing 4 (pump) – sliding bearing	7.1	5.3

According to the nameplate of the pump of the brand CPS 180-900 permissible (acceptable) level of the vibration velocity is 7.1 mm/sec.

3 Results

According to the diagnostic features (indicators) the major cause is likely high vibration:

- rotor out-of-balance of the electric motor (including half-coupling and blower of a stator frame);
- increased sliding bearing clearance;
- increased retaining clearance in the bearing bracket №3 of a pump.
- The analysis of results:
- Due to large balancing mass required the attempt to balance the rotor in its bearings on-site without disassembling is impossible;
- Rolling bearings №3 and №4 of a pump are without any evident defects, rotor and pump impellers out-of-balance condition is not diagnosed;
- the increase of vibration level of bearing housing №3 of the pump is likely associated with high vibration of the rotor of electric motor and the presence of clearings in the node.
- Measures and recommendations:
- It is necessary to balance the electric motor rotor in a set with a clutch on a balancer;

- To inspect the rotor located on the open storage in order to reduce the repair and maintenance time. It is necessary to test the quality of sliding bearings to be sure that they are of high quality;
- After rotor inspection and complete assembly of the induction motor with a pump it is reasonable to carry out its idle run.

4 Conclusion

To sum up, we came to the following conclusions:

- vibration-based diagnostics on the mean-square value of vibration velocity using the vibration analyzer "Quartz" and software "Diamond- 2" allows conducting diagnostic study during the induction motor operation. It makes this technique one of the most perspective in the diamond industry.
- This technique allows recording mechanical defects in the induction motor which make up more than 40 % of all faults.
- The findings enable to plan some required measures and recommendations for further maintenance and service which results in the reduction of the equipment idle time when there are some failures and also shows positive economic effect.

References

1. M. V. Vedyashkin, O. P. Muravlev, Russian Electrical Engineering, **84**, 397 (2013)
2. O. P. Muravlev, V. P. Shevchuk, V. V. Gusev, Proceedings of institutions of higher education. Power engineering problems, 183 (2008)
3. V. V. Gusev, O. P. Muravlev, V. P. Shevchuk, Bulletin of Tomsk Polytechnic University, **314**, 74 (2009)
4. S. I. Zakharov, Russian Engineering Research, **29**, 602 (2009)
5. A. N. Novozhilov, E. V. Kryukova, T. A. Novozhilov, Russian Electrical Engineering, **85**, 457 (2014)
6. K. A. Kuznetsov, Journal of Machinery Manufacture and Reliability, **45**, 32 (2016)
7. M. P. Kozochkin, F. S. Sabirov, A. N. Bogan, K. V. Myslivtsev, Russian Engineering Research, **33**, 486 (2013)

COMPARISON OF ENERGY EFFICIENCY DETERMINATION METHODS FOR THE INDUCTION MOTORS

Bator Tsybikov[1], *Evgeniy* Beyerleyn[1,*], and *Polina* Tyuteva[1]

[1] Tomsk Polytechnic University, 634050, Tomsk, Russia

Abstract. In the paper the effects on the induction motor efficiency determination due to the introduction of the EU, USA, Canada, Japan and Russian Standards have been analyzed and discussed. Industrial induction motor has been tested and a comparison between the efficiencies determined by various National Standards has been presented. In addition, the motor efficiencies have been compared with the efficiency classes defined by the EU Standardization.

1 Introduction

Currently, induction motors are played the most important part in the alternating electric drivers due to their robust construction, easy operation, self-starting and rather high reliability in comparison with direct current or synchronous motors that have many considerable drawbacks. Mostly such machines present small or average power motors with traction mode, induction generators are used quite rarely. As induction motors represent one of the most important power equipment, so the power consumption of them accounts for a large proportion [1, 2]. Induction motor efficiency states how much electrical power is converted to useful work (mechanical power); also what part of the electrical power is lost. Factors which influence efficiency of the induction motors used in industrial applications vary and these must be taken into account to minimize energy losses. The ways of power consumption reduction are discussed in considerable details in variety of articles. But not only the question of minimization of losses in induction motors should be under close attention, but also the approaches of accurate efficiency determination of the induction machines [3]. The purpose of our article is to provide comparison between methods of efficiency determination for the induction motors according to various National Standards.

2 Losses and efficiency in induction motors

According to the standards there are two major ways of efficiency determination of the induction motors are implemented: direct and indirect. The usage of the direct method means that the motor efficiency is determined from the direct measurements of the input electrical power P_{in} and output mechanical power P_{out}. The direct method is generally

* Corresponding author: beierleinev@tpu.ru

thought as less accurate as it applies the direct measurement of input and output values of power:

$$\eta = \frac{P_{out}}{P_{in}}. \quad (1)$$

In the indirect method, the input power and output power may also be measured and used to determine the constituent components of the losses in the motor this process is known as segregation or summation of the losses. The efficiency is determined from the total losses, P_{loss}, by equation:

$$\eta = \frac{P_{in} - P_{loss}}{P_{out}}. \quad (2)$$

The efficiency determination methods that are based on loss segregation are known to be more accurate. There are five categories of losses in the motor: P_{core} is the core losses, P_{fw} is the friction and windage losses, P_{stator} is the stator copper losses, P_{rotor} is the rotor copper losses, and P_{stray} is the stray load losses, the distribution of the losses is given on Fig. 1 (P_{ag} – air-gap power, P_m –mechanical power).

The core losses and the friction and windage losses are known as constant losses because they are considered independent of the value of the load. Core losses represent the amount of energy required to the core magnetization process. The friction and windage losses appear due to bearing friction and air resistance, which is primarily caused by the cooling fan.

The other losses are considered load dependent and therefore variable with load. The stator and the rotor copper losses occur due to current flow through the resistance of the stator and the rotor windings. The fixed losses could be determined from no-load tests at variable voltage and the conventional load losses are determined from load tests as described in various literature.

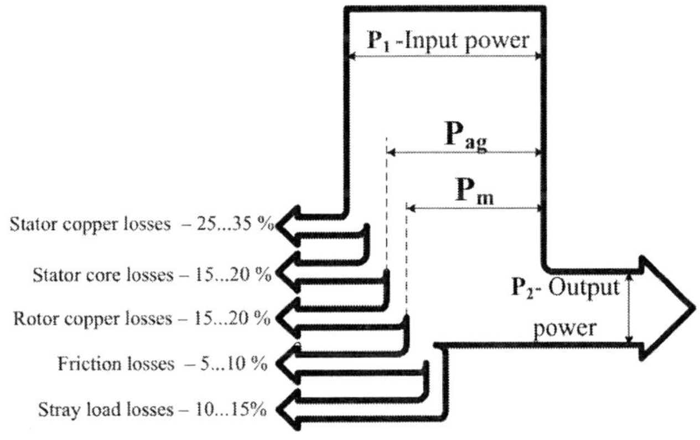

Fig. 1. Distribution of the losses in the induction motors.

Both methods have direct connection with all losses in the motor. Engineers are trying to minimize the percentage of losses, cause each percent is very important for electrical energy industry. There are several effects that could influence and minimize the losses that occur in the induction motors:

1. Stator core losses – could be reduced by using of improved permeability steel, lengthening core, using thinner laminations in the core.
2. Stator and rotor copper losses - could be reduced by increasing the volume of copper wire in the stator, through improved stator slot designs, and by using thinner insulation.

In a case of squirrel cage rotor the copper losses could be reduced by increasing the size of rotor conductive bars and end rings to reduce resistance.

3. Friction and windage losses – using the lower friction bearings, improved fan design and air flow could effect on the losses reduction.
4. Stray losses – as this type of the losses appear mostly due to leakage fluxes induced by load currents thus various design and manufacturing details may influence on them.

Energy losses are the determining factor in motor efficiency. The losses, as it shown earlier, could be divided into five main classes, Fig.1. In the next parts of this paper we will give an overview of various National Standards for efficiency determination and discuss the main differences between them.

3 National standards for induction motors testing

Since the induction motor consumes about 90 % of all the electrical energy used by all motors combined, its efficiency reporting on the nameplate has become the industry standard. However, various international standards for testing and reporting the induction motor efficiency on the nameplate can be significantly different. National and International energy efficiency test standards have been developed to reduce the errors when motor efficiency is measured and make the procedure standardized. There are exist a number of international standards that are used to determined real efficiency according to experimental data. The major and the most important Standards for poly-phase induction motors efficiency measurement are [3]: IEEE Std. 112 – USA Standard; IEC 60034-2-1 – EU Standard; CSA C390 – Canadian Standard; JEC 37 – Japanese Standard; IEC 60034-2 – Russian Standard; GOST 25941-83 – Russian Standard [4-9].

Most of these National Standards have more than one method for the efficiency determination, for example the IEEE Std. 112 has several methods identified as Methods A, B, B1, C, E, E1 F, F1, C/F, E/F, E1/F1. The IEEE 112 standard, method B, is commonly used for induction motor efficiency determination. The method is based on the summation of losses of the motor into conventional and additional (stray load) losses. The CSA 390 Standard uses an identical procedure to the IEEE Std. 112 method B for the stray load losses while the JEC 37 Standard ignores those losses [7]. The European standard for measuring motor efficiency, IEC 60034-2-1 has for a long time arbitrarily assigned 0.5 % input power as stray load loss, the same stray load losses determination we have in IEC 60034-2-1 and GOST 25941-83. The major differences between these standards is in the treatment of stray load losses, see Table 1 below.

Table 1. The comparison between major test Standards

Standard	Stray losses	Temperature correction
IEEE Std. 112	Loss segregation method	Yes
IEC 60034-2-1	Loss segregation method	Yes
CSA C390	Loss segregation method	Yes
JEC 37	Ignores SLL	Yes
GOST 25941-83	Assigns 0.5% of rated power of the motor	No

The loss segregation method (used in IEEE, IEC, GOST and CSA) determines efficiency through the direct method or separation of loss method. This method requires an open-circuit and load test, under open-circuit the core and windage and friction loss are segregated and the stator and rotor copper losses recieved under the load test. The stray load loss is determined by subtracting all the major losses from the total measured loss. As stray load losses are the smallest one, they are sensitive to errors in measurement. The IEEE

Std. 112 requires a correlation factor of 0.90 while the Canadian standard CSA C390 requires factor of 0.95. The GOST 25941-83 method allocates stray load losses to be 0.5 % of rated load and the Japanese one JEC 37 ignores stray load losses completely.

4 Efficiency levels

These standards also play an important role in induction motor classification according to efficiency levels. In the European Union (CEMEP) and the U.S (EPAC) laws have been set to define the efficiency class of induction motors [2]. Table 2 shows the requirement limits for the efficiency according to the CEMEP classification. These limits are determined through testing according to a particular standard (IEC in this case). Previously in Europe, low voltage three-phase motors have been graded and marketed in three efficiency classes – EFF3, EFF2 and EFF1 – based on a voluntary agreement between motor manufacturers and the European Commission. Unfortunately, other countries have also developed their own national systems, which are very different from the European system. That was the reason for the European motor manufacturers in CEMEP, to develop an energy efficiency standard for the International Electrotechnical Commission (IEC). The new international standard defines efficiency classes IE1, IE2 and IE3 for three-phase motors. The efficiency classes for induction motors according to the new classification is shown in Table 2.

Table 2. Efficiency classes of CEMEP classification for 4-pole, 50 Hz induction motor.

Rated power, kW	IE1 (EFF2) Standard efficiency	IE2 (EFF1) High efficiency	IE3 Premium efficiency
0.75	72.1	79.6	82.5
1.1	75	81.4	84.1
1.5	77.2	82.8	85.3
2.2	79.7	84.3	86.7

The differences in treatment of stray losses have been seen to produce different efficiencies on the same tested motor. A 1.1 kW induction motor was tested according to various National Standards. Due to the similarity in testing standards, only the IEEE Std. 112, IEC 60034-2-1 – direct method, JEC 37 and GOST 25941-83 loss segregation method were used.

5 Experimental study

For experimental studies of the induction motor energy performance it is required to select the type of installation unit. Selecting of circuit for testing of electrical motors is determined by the parameters of the equipment installed on testing platform2 that can be characterized by quantitative and qualitative parameters [10]. In general, the installation unit for experimental data determination includes: tested motor, loading machine and measuring equipment. In the function of measuring equipment we use current and voltage sensors that are connected to the phases of stator winding. CIRCUTOR AR.5 (portable power analyzer) is used as processing system for current and voltage sensors data processing.

The following experiments were provided for determining the efficiency of the tested induction motor: the measurement of phase resistance of the stator winding; idle running mode performance; load mode performance. Load mode performance is produced by the different connected to the motor shaft in the range from not less than 10 % in excess of the rated power to the idle running mode, the load changing is carried out by lowering the load.

Before the experiment performance the pilot running should be done to warm induction motor in the case stator and rotor windings temperature should achieve the steady value.

The following values are measured in the course of the experiments: voltages in each phase and the average value; currents in each phase and the average value; main supply frequency; power factor in each phase; rotor speed. The measurement results are undergoing the following treatment: the actual line voltage applied to the stator windings were takenc as the arithmetic average of the measured values, the actual line current were taken the arithmetic average of the measured values, the rotor speed was measured by direct current tachogenerator mounted on the load of the generator shaft.

The experimental test results are shows the difference in efficiency values using different standards, Fig. 2.

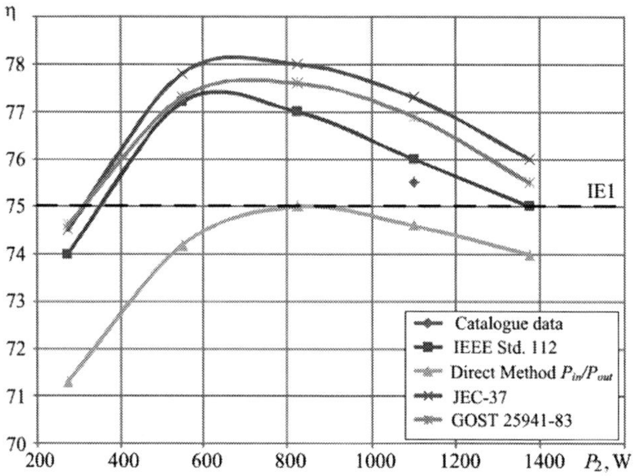

Fig. 2. Efficiency of 1.1 kW induction motor according various standards.

Fig. 2 compares the efficiency of a 1.1 kW induction motor derived from four international standards, namely, the IEEE Std. 112, IEC 60034-2-1 (direct method), JEC 37 and the GOST 25941-83. The direct method IEC 60034-2-1 and IEEE Std. 112 result in the lowest efficiency. The difference comes from the different treatments of the stray load losses. The IEEE Std. 112 derives it indirectly from tests, and the GOST 25941-83 assumes to be fixed 0.5% of the rated power, whereas the JEC-37 ignores it altogether, resulting in the highest efficiency. The difference between founded levels of efficiency results in that the classification according CEMEP could have this motor as IE1 Standard efficiency or in a case of direct method the value of efficiency is lower than desired level and the difference would be more significant in the case of higher rated power motors. The main point is that a standard efficiency motor made in the one country cannot be compared with a motor made in another one for energy-saving considerations.

Not only the assumed National Standardization, but also some other various aspects could have impact on the efficiency level:

1. Quality of power supply, according to the standards the power supply should not have the voltage unbalance exceed 0.5 % and frequency 0.1 % of rated values. The fluctuations of voltage magnitude and frequency, any unbalance has an impact on the accuracy of the experimental data.
2. Equipment of test platform, instrumentation plays a large role in determining efficiency with a high accuracy, as well as the process of measurement equipment calibration. Calibration of measurement instruments resulted in the confidence in the experimental results.

DOI: 10.1051/matecconf/20179101034

6 Conclusion

This paper highlighted the main factors that influence induction motors efficiency in industry. The value of induction motor efficiency can be affected by various aspects one of them is National Standardization. The paper considers the influence of various standards on the value of induction motor efficiency and the efficiency level in accordance with CEMEP is evaluated. As efficiency level determined in accordance with National standards depends on the stray load losses accuracy determination in the future it is planned to work on the development of the experimental procedure for stray load losses determination which takes into the account the impact from power supply characteristics and the features of test equipment.

References

1. H. Mzungu, P. Barendse, A. Khan, M. Manyage, Industrial and Commercial Use of Energy, 1 (2008)
2. E. Agamloh, Electric Machines and Drives Conference, 36 (2009)
3. O. Dobzhanskyi, R. Gouws, Industrial and Commercial Use of Energy (2014)
4. IEC 60034-2-1, Rotating Electric Machines Part 2-1: Standard methods for determining losses and efficiency, 53 (2007)
5. IEEE Std. 112: Test Procedure for Polyphase Induction Motors and Generators, IEEE, 79 (2004)
6. CSA C390: Energy Efficiency Test Methods for Three-Phase Induction Motors, Canadian Standards Association, 79 (2004)
7. JEC 37: Induction Machine, Standard of Japanese Electrotechnical Committee, 44 (1979)
8. GOST 7217-87. Rotating electrical machines. Asynchronous motors. Test methods, 42 (2003)
9. GOST 25941-83. Rotating electrical machines. Methods for determining losses and efficiency,31 (2003)
10. E. V. Beyerleyn, P. V. Tyuteva, Micro/Nanotechnologies and Electron Devices (2014)

MATEC Web of Conferences **91**, 01035 (2017) DOI: 10.1051/matecconf/20179101035
Smart Grids 2016

CURRENT ENERGY BUDGET OF AN ENTERPRISE

Galina Klimova[1], *Valeriy* Litvak[2,*], *Valeriya* Leonova[1], and *Anastasia* Chernaya[1]

[1]Tomsk State University, 634050, Tomsk, Russia
[2]Tomsk Polytechnic University, 634050, Tomsk, Russia

Abstract. Energy budget shows production volumes and energy consumption for the certain period of time as well as fuel storage, auxiliary power consumption, losses in utilities, energy sources consumption in structural units and depending on areas of operations [1, 2].

1 Introduction

Thus, it is of vital importance to develop approaches towards energy budget calculation for short period of time. This issue has a great value in case of development potential, plotting graphs of product transportation and fuel supply, assessing possibilities of energy resources demand coverage for severe conditions e.g. winter peak and summer low load, repair and maintenance schedule, specific climatic conditions (waterflood, hurricane, fallout and etc.), anthropogenic and natural disasters [5-9].

It is evident that energy budgets of any period should be based on the same methodological framework and database. The structure and content of balance items, collection, and calculation and correction techniques of energy budget base elements should be arranged.

2 Calculation of current energy budget

Calculation of current energy budget is based on reinforced concrete plant energy budget.

Figures 1 and 2 show pie charts of energy resources consumption structure in the plant's subdivisions (fractions of the whole).

* Corresponding author: litvak@tpu.ru

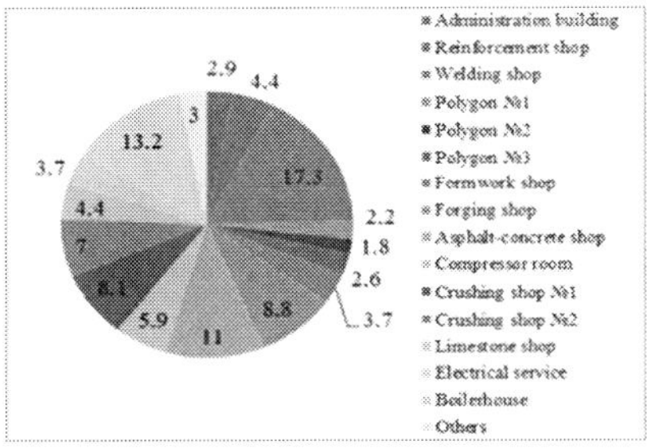

Fig. 1. Electrical energy consumption, %.

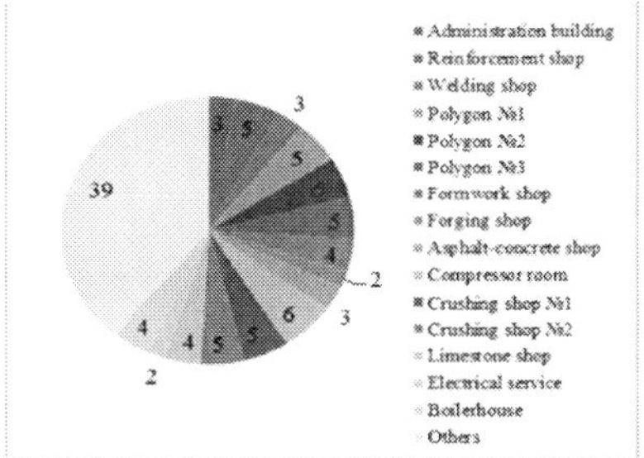

Fig. 1. Water consumption structure, %.

Current energy budget for short period of time supports financial, commercial, logistic, organizational and other branches' basis of controlling [3].

Average calculations of assigned year energy consumption are not difficult if we do not consider short periodical energy budgets structure perturbations:

$$A_t = \int_0^t P_t \cdot dt \cdot \frac{A_{\text{year}}}{N_t} \cdot \Delta t, \tag{1}$$

where P_t – current power, kW, A_{year} – annular consumption, kilowatt hour, T – basis time duration, hour, Δt – current period duration, hour, N – number of equal current time intervals per one year [4]. In case of known plant energy consumption per one year (40000 Ths. kilowatt hour) average short periodical consumption volumes will be calculated as follows.

There is actual difference between energy resources use for short period of time and average for one year. Thus it is necessary to adjust temperature, social, process and other factors. Element value adjustment should be done in the following way:

$$A_k = A_{av} \cdot (1 + k_1 \cdot k_2 \cdot k_3 \cdot k_4), \tag{2}$$

where A_k – current element value, A_{av} – average element value, k_1 – correcting factor considering energy resources volumes of operating days and holidays, k_2 – correcting factor considering demand's increment between start and end of the year; k_3 – correcting factor considering energy resources consumption depending on air temperature, k_4 – correcting factor considering energy resources consumption between heating and non-heating seasons.

Energy consumption ratio of holidays and operating days with correcting factor k_1 approaches 0.5:1. This ratio between holidays and operating days for one week reinforced concrete plant operation equaled 0.55:1.18. In relation to year average plant measurements level showed following k_1 values (Table 1).

Table 1. Correction factor k_1.

Correction factor k_1, relative unit	Electrical energy	Thermal energy	Gas	Water
Operating days	1.15	1.1	1.07	1.12
Weekends and public holiday	0.55	0.75	0.825	0.7

Correction factor k_2 considering demand's increment between start and end of the year is defined as:

$$k_2 = 1 + k_{\text{in}} \cdot \frac{i}{365}, \tag{3}$$

where $k_{in} = 0.02 \div 0.04$ – average incremental growth of demand equal 2 – 4% per year, I – number of the date starting from 1st of January.

Air temperature correction factor k_3 describes energy resources incremental growth with air temperature lower than $+10^0$C in relation to average temperature for same period of many years. In some cases, temperature may be close to zero degrees. Temperature reference scale origin is shifted to the point of – 100 ^{0}C to avoid division by zero. Then:

$$k_3 = \frac{100 - \Theta_d}{100 - \Theta_{av}}, \tag{4}$$

where Θ_d – desired air temperature of design day, ^{0}C, Θ_{av} – average long-term temperature of this day, ^{0}C.

Correction factor k_4 characterizes energy resources consumption between heating and non-heating seasons. It is important to include this factor into calculations because there is a great difference in heat and fuel consumption for both seasons.

Centralized annual heat and fuel consumption comprises the following:

$$Q = Q_{heat} + Q_{\text{hws}} + Q_{pr}, \tag{5}$$

where Q_{heat} – heat consumption for heating, Gcal, Q_{hws} – hot water supply, Gcal, Q_{pr} – heat consumption for process requirements, Gcal.

Thermal energy for heating Q_{heat} is supplied only during heating season, hot water supply Q_{hws} – all year round, and Q_{pr} according to operational program. In calculations hot water supply system volume in comparison with heating accounts for 10 – 12 %. Throwing away Q_{pr} due to its uncertainty, we may assume that:

$$Q = Q_{heat} + Q_{\text{hws}} = Q_{heat} + 0.12Q_{heat} = 1.12Q_{heat},$$

$$Q_{heat} = \frac{Q_{heat} + Q_{\text{hws}}}{1.12} = 0.89 \cdot (Q_{heat} + Q_{\text{hws}}),$$

$$Q_{heat} = \frac{Q_{heat} + Q_{\text{hws}}}{1.12} = 0.89Q.$$

Daily average heat transmission for heating volume during heating season is:

$$q_{\text{av}}^{heat} = \frac{0.89Q}{235} = 3.79 \cdot 10^{-3} Q,$$

and hot water supply all year round:

$$q_{\text{av}}^{hws} = (1 - 0.89) \cdot \frac{Q}{365} = 0.29 \cdot 10^{-3} Q.$$

Daily heat supply during heating season:

$$q_{day} = 3.79 \cdot 10^{-3} Q + 0.29 \cdot 10^{-3} Q = 4.08 \cdot 10^{-3} Q,$$

and during non-heating season:

$$q_{day} = 0.29 \cdot 10^{-3} Q.$$

Correspondingly, the correcting factor k_4 during heating and non-heating seasons is equal to:

$$k_4 = 4.08 \cdot 10^{-3} \cdot 235 = 0.9588,$$

$$k_4 = 0.29 \cdot 10^{-3} \cdot 130 = 0.0377.$$

This correcting factor for electrical energy and water supply can be taken as 1.

3 Experimental results

Correction factors for other periods of time are defined the same way. Table 2 shows correcting factors' values for accepted design conditions.

Table 2. Correcting factors' values.

Current period	k_1, k_o – operating days, k_h – holidays				k_2	k_3	k_4 heat and fuel	
	Electrical energy	Thermal energy	Gas	Water			Heating season	Non-heating season
Hour k_o k_h	 1.15 0.55	 1.1 0.75	 1.07 0.875	 1.12 0.7	$1+k_{in}\cdot\frac{i}{365}$	$\frac{100-\Theta_d}{100-\Theta_{av}}$	0.9588	0.0377
Day k_o k_h	 1.15 0.55	 1.1 0.75	 1.07 0.875	 1.12 0.7	$1+k_{in}\cdot\frac{i}{365}$	$\frac{100-\Theta_d}{100-\Theta_{av}}$	0.9588	0.0377
Week	$\frac{k_o \cdot n_o}{7}+\frac{k_h \cdot (7-n_o)}{7}$				$1+k_{in}\cdot\frac{i}{52}$	$\frac{100-\Theta_d}{100-\Theta_{av}}$	0.9588	0.0377
Month	$\frac{k_o \cdot n_o}{n_a}+\frac{k_h \cdot (n_a-n_o)}{n_a}$				$1+k_{in}\frac{j}{12}$	$\frac{100-\Theta_d}{100-\Theta_{av}}$	0.9588	0.0377
Heating season	$\frac{k_o \cdot n_o}{n_a}+\frac{k_h \cdot (n_a-n_o)}{n_a}$				$1+k_{in}\frac{n_a}{365}$	$\frac{100-\Theta_d}{100-\Theta_{av}}$	0.9588	0.0377
Non-heating season	$\frac{k_o \cdot n_o}{n_a}+\frac{k_h \cdot (n_a-n_o)}{n_a}$				$1+k_{in}\frac{n_a}{365}$	$\frac{100-\Theta_d}{100-\Theta_{av}}$	0.9588	0.0377

*Foot-note: k_o – operating load factor, k_h – weekend and public holidays load factor, $k_п$ – increment load factor of the whole year, Θ_d – desired temperature of period, Θ_{av} – average temperature of period, i – day number of the year, n_o – operating days number of period, j – week number starting from the beginning of the year, n_a – day quantity of the period, e – month number starting from the beginning of the year.

4 Conclusion

Current energy budget calculation represents simple computational task. Experience of current energy budget use for energy consumption and energy saving control allows to plan energy sources volume for different short periods of time more accurately, to calculate faults, to determine energy performance factors in terms of the year for fixed points based upon development prospect (or other side – production curtailment) and etc.

References

1. GOST 27322-87. Power plant balance. Main principles [in Russian]
2. GOST P 51379-99. Energy performance certificate of fuel and energy resources power consumer [in Russian]
3. International Energy Agency, Mind the Gap - Quantifying Principal-Agent Problems in Energy Efficiency (2007)
4. G. N. Klimova, V. V. Litvak, *Seven problems and seven keys of energy saving* (Red banner, Tomsk, 2013) [in Russian]
5. A. S. Nekrasov, Yu. V. Sinyak, *Power plant engineering control* (Energiya, Moscow, 1979) [in Russian]
6. Yu. N. Savenko, E. O. Steinhaus, *Energy budget* (Energiya, Moscow, 1971) [in Russian]
7. S. M. Romanov, G. F. Alekseev, Mountainous research and information bulletin, **1**, 143 (2010)
8. K.-H. Chai and C. Yeo, Energy Policy, **46**, 460 (2012)
9. P. Blomkvist and J. Larsson, Int. J. Commons, **7**, 113 (2013)

MATEC Web of Conferences **91**, 01036 (2017) DOI: 10.1051/matecconf/20179101036
Smart Grids 2016

DEVELOPMENT OF THE STRUCTURAL AND FUNCTIONAL DESIGN OF THE LABORATORY BENCH FOR EXPERIMENTAL RESEARCH DIESEL GENERATOR SETS ON VARIABLE SPEED

Sergei Obuhov[1, *], *Igor* Plotnikov[1], *Michael* Surkov[1], *Lyudmila* Sumarokova[1], and *Kirill* Savkin[1]

[1] Tomsk Polytechnic University, 634050, Tomsk, Russia

Abstract. A diesel generator set working at variable speed to save fuel is studied. A description is provided of a laboratory bench for conducting experimental studies of a variable speed diesel generator set. Its component parts are described, and its technical characteristics are given.

1 Introduction

Russia has decentralized power supply zones with the greatest area in the world. They encompass about 70 % of the country's territory, and house, according to different estimates, up to 20 million people. Most of these territories are located in areas with severe weather conditions - Siberia, the Far East, and the Far North. Traditionally, small-scale power generation sets have been used to supply consumers in decentralized zones with power – small power plants powering an autonomous electrical grid in one or several nearby communities. Most small-scale power plants working in various regions of Russia today have a host of similar problems, mostly due to significant wear of primary power plant equipment, a sharp lack of investment resources, difficult and expensive conditions of freight transportation, isolation from supply centers and maintenance services, etc. As the source for autonomous power supply systems, they have some clear advantages and well-known disadvantages, namely: high consumption of fossil fuels for the production of one kWh of electric power, environmental pollution, etc. However, no adequate replacement is available.

One of the most promising directions in improving energy efficiency of local power supply systems is using renewable energy sources in the regional energy balance and optimizing the operating modes of primary energy equipment. Weather and geographic characteristics of the decentralized zones in question make it practically possible to use only wind and solar energy out of all types of renewable energy sources. Since electric power consumers need an uninterruptible power supply, hybrid power plants on the basis of diesel generator sets [1–3] are the most promising way to build autonomous power supply

* Corresponding author: serob@tpu.ru

systems. One of the technical solutions to increasing the efficiency of autonomous power supply systems is changing the DGS operating mode from constant speed to variable speed, optimized according to the size of the electric load [4].

2 Diesel generator sets operating at variable speed

As stated above, the energy efficiency of a DGS can be increased significantly by changing the operating mode of the diesel engine from constant to variable speed and by using specialized regulators for controlling the plant [5]. A structural diagram of such a diesel generator set (DGS) is shown in Fig.1.

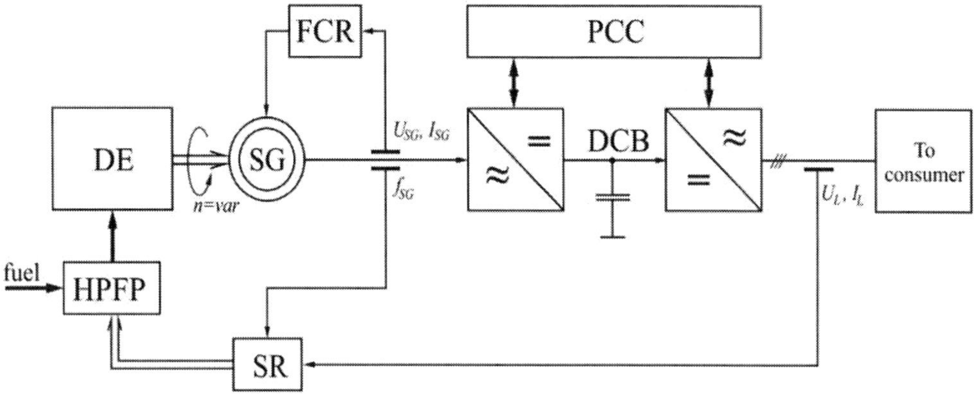

Fig. 1. Variable speed DGS structural diagram: DE – diesel engine; PCC – power converter controller; SR – speed regulator; FCR – field current regulator; SG - synchronous generator; HPFP – high-pressure fuel pump; DCB – direct current bus.

3 Laboratory bench

A laboratory bench was developed and manufactured in order to conduct experimental research on variable speed DGSs. A key development requirement was providing high level of automation for the control flow of bench components, and for control and recording of measured variables.

Development of the structural and functional design of the laboratory bench was conducted in view of the following technical requirements: the possibility of modeling static and dynamic DGS modes of operation for all main parameters of the operating mode; use of the modular construction principle, which allows for simple attachment/detachment of separate components from the power plant; satisfying all mandatory fire, environmental, and electrical safety regulations; convenience in use and maintenance; the possibility of further modernization of the bench.

The structural design of the laboratory bench for studying the operating modes of variable speed DGSs is presented in Fig. 2, and its appearance is in Fig. 3.

The laboratory bench is based on the Kipor (China) KDE12EA3 diesel power plant [6]. The following changes were made in order to conduct experimental studies of the KDE12EA3 diesel power station: the fuel supply system to the diesel engine high-pressure pump and the air-intake system were modified; the drive component of the fuel pump final element was developed; a new synchronous generator excitation system was installed; an intake system for moving exhaust fumes outside of the laboratory space was fitted.

A volumetric flow meter, for which the DARCONT OM 004A precision pressurized microrheometer with oval rotors was selected, was installed into the fuel line in order to receive a data signal on the DGS fuel consumption.

The BOSCH 028-0218 037 calorimetric film mass airflow sensor is used as the intake flowmeter. The G ¾ Water Flow Sensor TEM01071B rotor flowmeter is used for measuring the volumetric flow rate of cooling liquid. Semiconductor temperature probes and various types of thermal converters are used to control water temperature in the inlet manifold, cooling liquid, and diesel engine exhaust fumes, depending on the range of measured temperature values. The BOSCH 0261210113 variable inductance transducer, installed on the engine flywheel case, is used to measure the diesel engine shaft speed.

The synchronous generator excitation system developed within the framework of this project ensures that the regulation range for field current is from 0...5 A at a field voltage of up to 150 V.

In order to measure immediate values of electrical variables, currents and voltages, the relevant sensors are installed in the bench. Current sensors were built on the basis of Honeywell sensor CSNB-131.

The in-line simulation stand (ILSS) of an autonomous electrical supply unit makes it possible to simulate three-phase symmetrical and asymmetrical loads on the unit within the range of 0 – 15 kW in increments of 0.25 kW in manual control mode, as well as to measure the load according to set parameters.

The data measuring system, which includes the data collection system (DCS) and personal computer (PC) with necessary software, plays an important role in the laboratory bench. The data collection system is based on modular devices with an Agilent Technology USB interface.

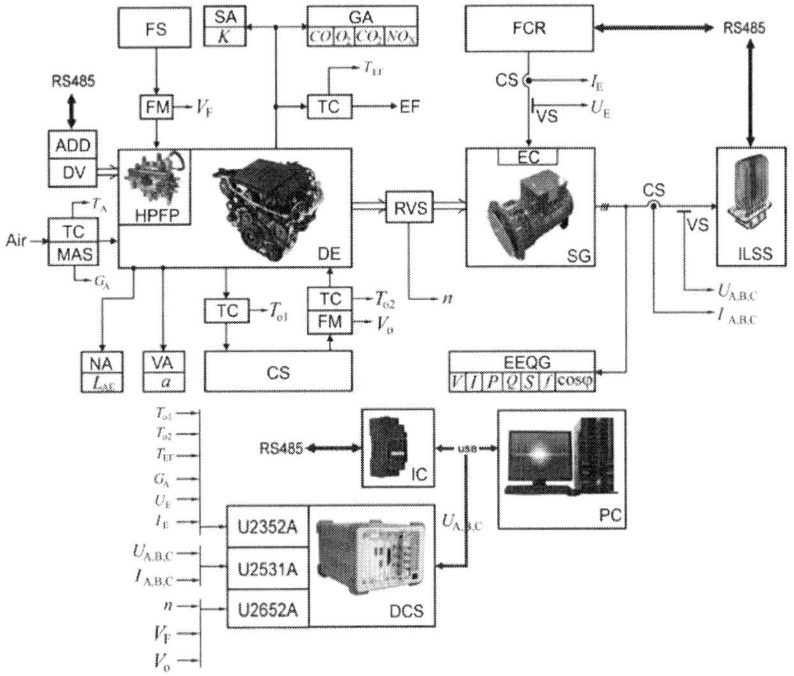

Fig. 2. The structural design of the laboratory bench: EEQG – Electrical energy quality gauge; GA – gas analyzer; DE – diesel engine; SA – smoke analyzer; MAS – mass airflow sensor; RVS – rotation velocity sensor; DV – drive controller; EC – exciting coil; EF – exhaust fumes; IC – interface converter; PC – personal computer; ADD – fuel pump actuating device drive; FM – flow meter; FCR – generator field current regulator; SG – synchronous generator; CS – cooling system; ILSS – in-line simulation stand of an autonomous consumer load; DCS – data collection system; HPFP – high

pressure fuel pump; TC – thermal converter (temperature sensor); FS – fuel system; CS – current sensor; VS – voltage sensor.

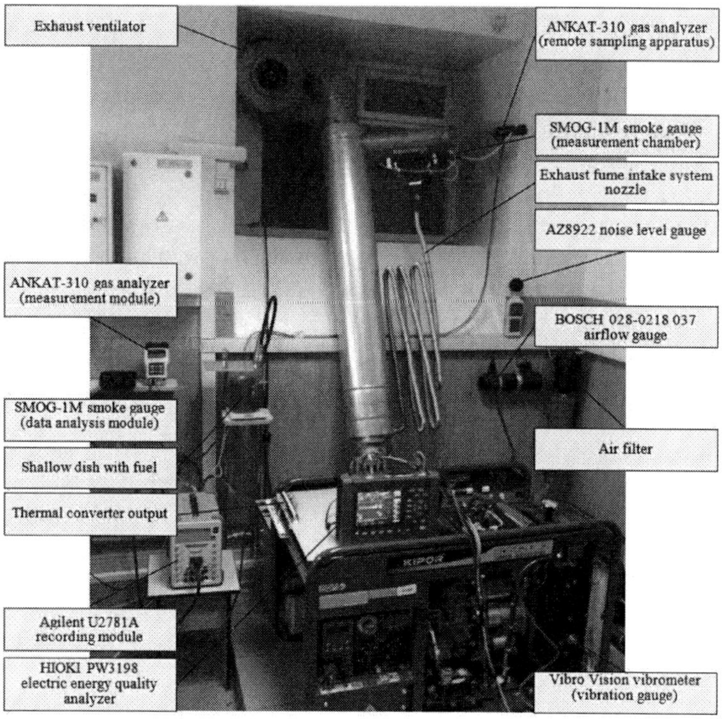

Fig. 3. Appearance of the laboratory stand.

4 Conclusion

The developed laboratory bench made it possible to conduct all of the necessary experimental research on proving the efficiency of changing the DGS operating mode from constant to variable speed and determining the areas for regulating the electrical unit through its operating modes.

References

1. Y. Sarsikeev, B. Lukutin, D. Lyapunov, M. Surkov, S. Obuhov, Adv. Mater. Res. **953-954**, 529 (2014)
2. B. Lukutin, E. Shandarova, D. Matukhin, A. Makarova, I. Fuks, IOP Conf. Ser.: Mater. Sci. Eng.. **124** (2016)
3. B. Lukutin, E. Shandarova, A. Makarova, I. Shvartsman, IOP Conf. Ser.: Mater. Sci. Eng. **127**, 012005 (2016)
4. D. Wang, C. Nayar, C. Wang, 2nd IEEE International Symposium on Power Electronics for Distributed Generation Systems, 1 (2010)
5. Y. Zhang, S. Li, G. Lu, Y. Zhou, International Conference on Systems and Informatics, 619 (2012)
6. Kipor Power Co Power plants Available on: http://kipor-power.ru/catalog/ [accessed 25.11.2016]

THE CONTROLLED ELECTRIC DRIVE OF THE AUTOMATIC COOLING SYSTEM OF THE ENGINE ROOM ON A VESSEL

Dennis V. Bevz[1,*], *Lubov* A. Payuk[1], and *Nataliya* A. Voronina[1]

[1] Tomsk Polytechnic University, 634050 Tomsk, Russia

Abstract. Development of an automatic cooling system control by the Zelio Logic controller. Development of the simulation model of the open-loop system "frequency inverter – asynchronous motor with IR-compensation" in Matlab Simulink. Analysis of the transient characteristics $\omega = f(t)$, $M = f(t)$ at the start of the asynchronous motor.

1 Introduction

The Engine room it is a room on board designed to accommodate machines and mechanisms for its moving [1].

In this room there are:

- the central control post, which derived all automation systems, control and alarm systems
- main engine, which rotates the rowing screw;
- electric motors, intended to generate electricity for lighting, heating, operation of various devices and mechanisms;
- pumps for supplying oil, liquid fuel, cooling water;
- separators intended for clarification fuel and oil from impurity;
- compressors, which compress air supplied for engines starting;
- steam boiler intended for heating, heating of liquid cargo, for technological needs.

The engine room is presented in figure 1.

The ventilation system of the engine room has two main objectives:

- to provide such an environment that will allow machines and mechanisms work reliably;
- to provide comfortable working conditions for staff.

Heat that is emitted from the engine and other mechanisms of the machine room, absorbed by the surface of the engine room. Part of the heat transferred to the atmosphere or water through the hull. The remaining heat has to be removed by the cooling system. Place the input of outside air into the engine room should be located as far away from heat sources and as low as possible. Since heat causes the air to rise upwards, the air outlet should be done at the highest point of the engine room.

* Corresponding author: d.v.bevz@mail.ru

DOI: 10.1051/matecconf/20179101037

Fig. 1. The engine room of a vessel.

2 Development of automatic control of the cooling system

A functional diagram of automatic ventilation system is presented in figure 2 [2].

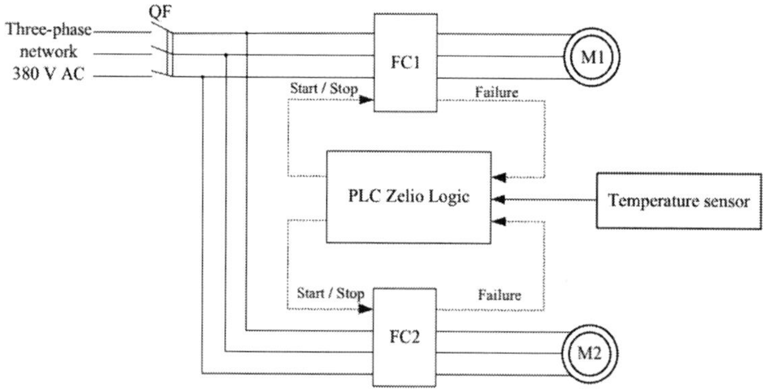

Fig. 2. Functional diagram of automatic cooling system: QF – the automatic circuit breaker, FC1 – the main frequency inverter, FC2 – the reserve frequency inverter, M1 – the main asynchronous motor, M2 – the reserve asynchronous motor PLC – the programmable logic controller Zelio Logic of Schneider Electric.

This system has two asynchronous motors. The first motor is main and the second motor is reserve. The temperature sensor has analog output in the range 0...10V which corresponds to a temperature range from 0 to 100 degrees Celsius. Analog output of temperature sensor connects to analog input of PLC. When the temperature rises exceeds 30 degrees Celsius PLC generates a start signal which comes from the digital output of the controller to the digital input of the first frequency inverter, then the main asynchronous motor is started and ventilation process is begins. Ventilation continues until the temperature of 20 degrees Celsius and then the PCL generates stop signal to the first inverter. In case of a malfunction of the main frequency inverter a fault signal is entered to the PLC digital input. The controller sends a stop signal to the first frequency inverter and a start signal to the second frequency inverter. If the temperature during the operation of the main asynchronous motor exceeds the preset critical value equals 40 degrees Celsius the controller will generate start signal to the reserve frequency inverter and then the reserve asynchronous motor is started. The standby motor is operated together with the main motor for a predetermined time interval. If after that time interval the temperature does not decrease alarm is triggered. The developed algorithm [3] that explains the operations of automatic cooling system is presented in figure 3.

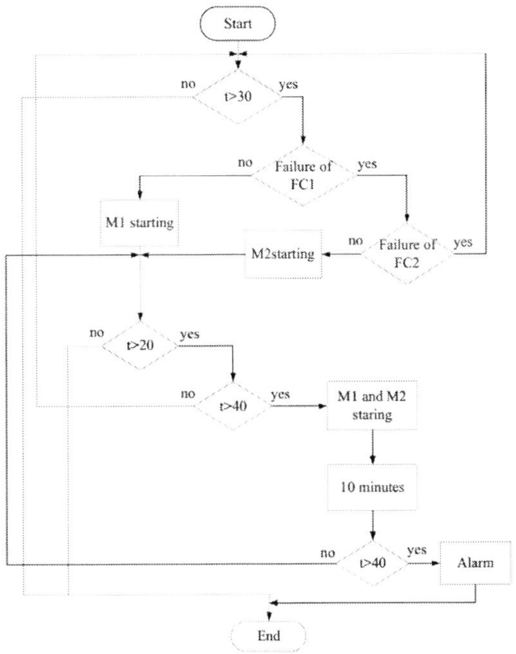

Fig. 3. The algorithm of the operation of the automatic cooling system.

The algorithm is written for the controller in the functional block diagram language. The exterior of the software window of the controller with algorithm [4] is shown in Fig. 4.

The developed algorithm provides work of the system of cooling in 3 modes:

1. start the main motor when the temperature exceeds the set value;
2. start the backup motor if there is an accident at the first frequency inverter and the temperature exceeds the set value;
3. operation of both motors for a predetermined time interval.

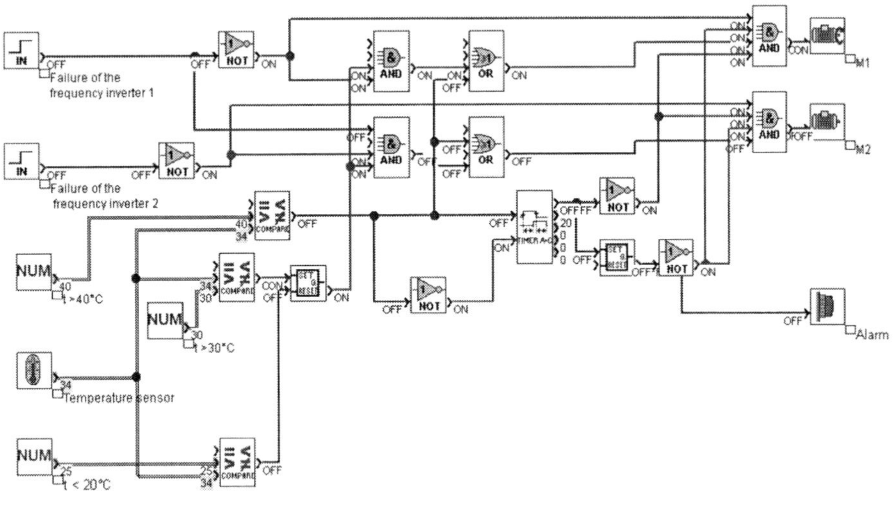

Fig. 4. The exterior of the software window of the PLC.

3 Development of a simulation model of the open-loop system "frequency inverter – asynchronous motor with IR-compensation"

To control the speed of the electric drive is used scalar control law – *U / f = const*. The simulation model of the open-loop system "frequency inverter – asynchronous motor with IR-compensation" is created in Matlab Simulink 2012 and presented in figure 5 [5].

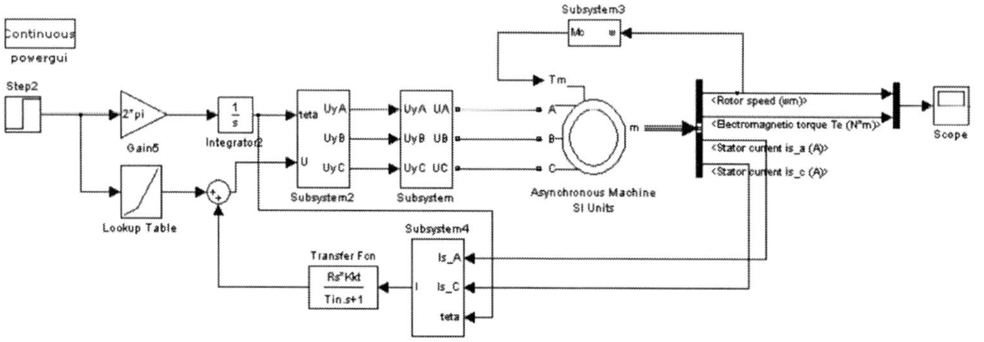

Fig. 5. The simulation model of the open-loop system "frequency inverter – asynchronous motor with IR-compensation".

Block "Lookup Table" forms the required relationship of the scalar control between the frequency and voltage [6, 7]. It sets the received frequency regulation law. The control signal is entered to the direct coordinate converter (block subsystem 2). Three sine wave control voltages U1yA, U1yB, U1yC which are shifted relative to each other at an angle ±2*pi/3 with amplitudes proportional to the control voltage are formed at the output of the converter. The signals U1yA, U1yB, U1yC form phase output voltage of the autonomous voltage (block subsystem). The transient characteristics of speed and torque are presented in figure 6.

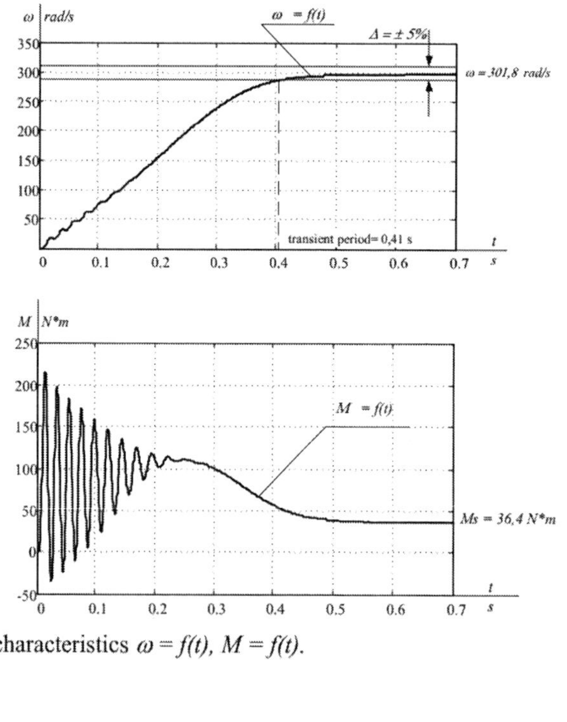

Fig. 6. The transient characteristics $\omega = f(t)$, $M = f(t)$.

4 Conclusion

The designed automatic cooling system of the engine room on a vessel has a high reliability, is simple in service and economic:
1. the system has two asynchronous motors one of them is main motor and another one is backup motor. The control of the asynchronous motors is provided by the programmable logic controller Zelio Logic depending on the signal from the tempearture sensor and the faulire signal from frequency inverters;
2. start and stop of the asynchronous machines are relized by the frequency inverters, which allow to relize smooth start of the motors. It leads to reduction of starting currents and therefore increases life cycle of the asynchronous motors. Also, use of the frequency inverter leads to saving of electricity.

References

1. N. F. Emelyanov, *Design, construction and elements of the theory of vessel* (Dalrybvtuz, Vladivostok, 2002) [in Russia]
2. M.P. Belov, A.D. Novikov, L.N. Rassudov, *Automatic electric drive of industrial machinery and technological complexes* (Academa, Moscow, 2007) [in Russia]
3. E.P. Ugrumov, *Digital circuitry,* (BHV-Petersburg, St. Petersburg, 2007) [in Russia]
4. R. Coris, X. Schmidt-Walter, *Reference circuit design engineer* (Technosphere, Moscow, 2008) [in Russia]
5. L.S. Udut, O.P. Maltseva, N.V. Koyain, *Asynchronous frequency controlled electric drive* (Publisher of Tomsk Polytechnic University, Tomsk, 2009) [in Russia]
6. I.V. Chernyh., Simulation of electrical devices in MATLAB, SimPowerSystems and Simulink (Piter, St. Petersburg, 2007) [in Russia]
7. L. Payuk, N. Voronina, S. Korepanov, O. Galtseva, N. Natalinova, MATEC Web of Conf. **79**, 01060 (2016)

MATEC Web of Conferences 91, 01038 (2017) DOI: 10.1051/matecconf/20179101038
Smart Grids 2016

FORMER OF TURN TRAJECTORY OF SLIDING VALVE SHAFT OF GAS LINE

Vladislav.Yu. Dementyev[1], *Nikolay*.V. Gusev[2,*], *Nikolay*.V. Koyain[2], and *Sergey*.Yu. Shiryaev[2]

[1]The 1st category expert of production-technical management, OOO "Gazprom transgaz Tomsk", 634000, Tomsk, Russia
[1]Tomsk Polytechnic University, 634050, Tomsk, Russia

Abstract. Former of turn trajectory of sliding valve shaft of gas line, that allows to provide desired motion trajectory of sliding valve and its full closing, is considered in that paper. Imitation model of that former, research results, which allow to detect influence of gain factor and time constant of posi-tion controller on value of speed error, that has impact on delay of output coordinate from setting, and that results to delay of sliding valve motion process to setting position point, are shown.

1 Introduction

Gases flow controlling is maintained by changing of production of gas-compressor unit (GCU) [1]. Gas turbine plants (GTPs) or electrical machines (EMs) are used as drive of GCU. For new designed gas-compressor station (GCS), problem of GCU type choice should be solved with obligatory consideration of factors such as: high efficiency, reliability and high durability, maintenance simplicity and even ability of creation of non-maintainable electric-drive units, ecological cleanness, wide regulation range. Using of regulated electric drive is economically right especially at objects, that have irregular load traffic, GTPs truly can't work with loads less critical. For rotation frequency regulated EMs, "surging" limits are reduced. Due to this, it is necessary to consider that implementation of regulated electric drives to GCS is the 1st step on the way to automation of all gas-transport system.

2 Concept headings

However, choice of application of frequency-regulated electric drive (FRED) was made, because FRED is the most economically effective with using of main gas-line, that work in variable technological modes of gas transit. The aim of this paper is development and researching of former of steering trajectory of sliding valve shaft of gas line. Electric drive of sliding valve is difficult multiple-loop structure of automatic regulation, that closes on position. The scheme incudes frequency-regulated electric drive, that based on KEB ST90

* Corresponding author: gusev@tpu.ru

MB type induction motor, that has nominal rotating speed; position sensor with gain factor, that installed on mechanism, and position controller with transfer function [2, 3]. To provide soft trajectory of sliding valve motion, it is neces-sary to limit not only value of maximal steering speed, and accelerate. It can be provided by application of preferred trajectory former of motion, by way of application of power-up sensor with S-characteristic is preferred [4, 5, 9, 10]. Characteristic and imitation model scheme of that former are shown on figure 1.

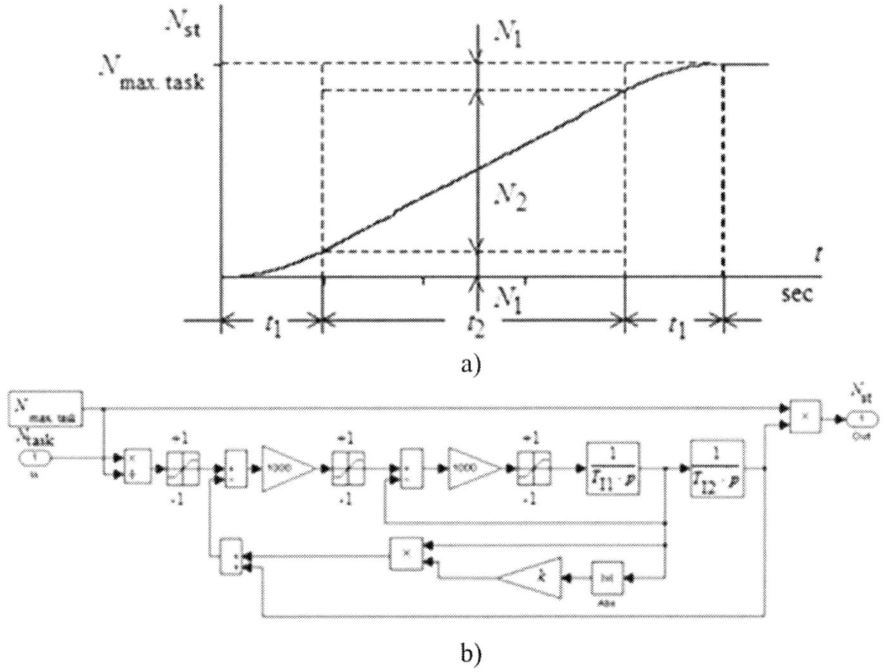

Fig. 1. Power-up sensor with S-characteristic: a – transient function; b – imitation model.

Preferences of optimal steering trajectory of sliding valve are defined from following equations:

$$t_1 = t_3 = \frac{\omega_{max}}{\left(\dfrac{d\omega}{dt}\right)_{max}}, \ sec, \qquad (1)$$

Where $\omega_{max} \leq 209.4\ rad/sec$ – used value of maximal rotation speed of motor with refining of given turn angle of sliding valve, rad/sec;

$\left(\dfrac{d\omega}{dt}\right)_{max}$ – used maximal value angular acceleration of motor with refining of given turn angle of sliding valve, rad^2/sec;

$$t_2 = \frac{N_{max}}{k \cdot k \cdot \omega_{max}} - t_1 = \frac{\varphi_{max}}{k \cdot \omega_{max}} - t_1, \ sec, 1 \qquad (2)$$

Where $\varphi_{max} \leq \dfrac{21600}{4} = 5400\ ang.min.$

For setting of trajectory former, following quantity values are used:

$N_{max} = 10000\ discrete;\ \varphi_{max} = 5400\ ang.\ min.$

$\omega_{max} = 52.35\ rad/sec$; $t_1 = t_3 = 0.2\ sec.$, that is considered of maximal value of angular acceleration

$$\left(\frac{d\omega}{dt}\right)_{max} = \frac{\omega_{max}}{t_2} = \frac{52.35}{0.2} = 261.75\frac{rad^2}{sec}.$$

Then additional preferences of former setting are:

$$t_2 = \frac{N_{max}}{k \cdot k \cdot \omega_{max}} - t_1 = \frac{10000}{1.3222 \cdot 1.85185 \cdot 52.35} - 0.2 = 41.9283\ sec;$$

$$k = 0.5 \cdot \frac{t_1}{t_1 + t_2} = 0.5 \cdot \frac{0.2}{41.9283 + 0.2} = 0.0023737$$

Researches were organized with task refining N_t=200 discrete, that is considered to problem

$$\varphi_t = \frac{N_t}{k} = \frac{200}{1.85185} = 108\ ang.\ min.$$

Transients in servo drive with P-position controller with task refining N_t=200 discrete (ϕ_t=108 angular minutes) are shown on figures 2 a, b.

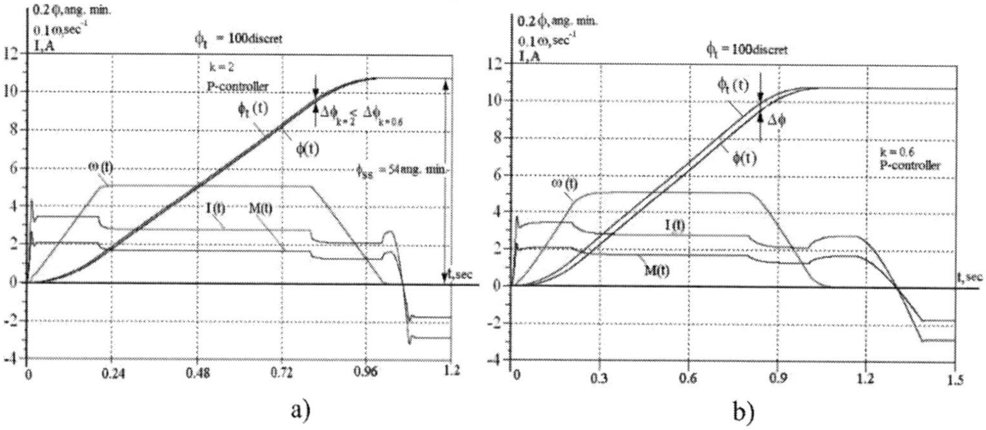

Fig. 2. The process of task refining a) N_t=200 discrete with P-position controller (k_{pc}=0.6); b) N_t=200 discrete with P-position controller (k_{pc}=2).

3 Experimental results

It will be seen from figure 2a, that velocity compensation Δφ exists with refining of given motion trajectory of sliding valve shaft in system with P-controller. The value of velocity compensation depends on chosen values of gain factor of P-position controller (figure 2b) and on maximal steering speed. Gain factor of controller is chosen on provide single sided approach of sliding valve to last spot of position, that is very important for its full closed. For this reason, gain factor of position controller can not be high and so velocity compensation can be serious with choice of maximal speed. That compensation leads to delay of output values to task and can be not preferred [6, 7, 8, 9]. Full elimination of velocity compensation is provided by application of PI-position controller. Curves of refining process of given motion trajectory of sliding valve shaft with PI-position controller are shown in figures 3 a and b.

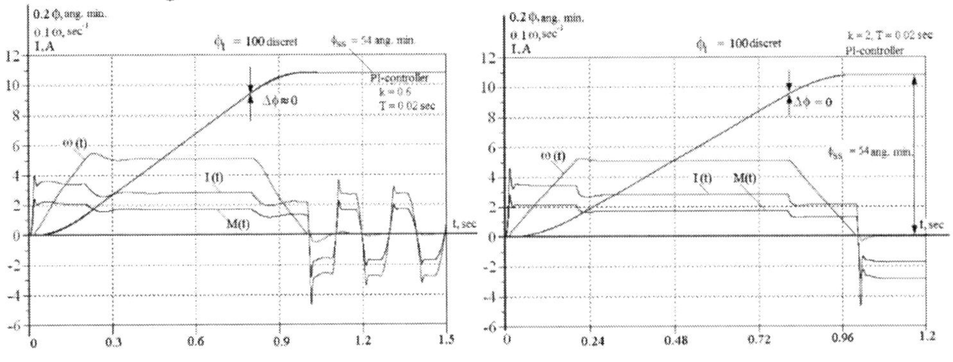

a) b)

Fig. 3. Process of task refining: a) N_t=200 discrete with PI-position controller (k_{pc}=2, T_{pc}=0.02 sec); b) N_t=200 discrete with PI-position controller (k_{pc}=0.6, T_{pc}=0.02 sec).

4 Conclusion

As it appears from figures 3 a and b, required character of electric drive motion of sliding valve and its way to position point are achieved by choice of parameters values of PI-controller settings: Gain factor and time constant.

References

1. F. M. Mustafin, *Pipeline rebar. Studying edition for university* (Ufa, 2007) [in Russian]
2. Yu. N. Dementyev, Fundamentalynie issledovaniya, **12**, 931 (2014) [in Russian]
3. Yu. N. Dementyev, A. D. Bragin, N. V. Koyain, L. S. Udut, 2015 International Siberian Conference on Control and Communications, 1, (2015)
4. Yu. N. Dementyev, K. N. Negodin, N. V. Koyain, L. S. Udut, Elektromehanika **4,** 49 (2016) [in Russian]
5. A. Yu. Chernishev, Yu. N. Dementyev, I. A. Chernishev, *AC electric drive* (Tomsk Polytechnic University, 2015) [in Russian]
6. I. G. Odnokopylov, Y. N. Dementyev, I. V. Usachyov, D. Y. Lyapunov, A. S. Petrusyov, 2015 International Siberian Conference on Control and Communications (SIBCON): proceedings, 1 (2015)
7. L. S. Udut, O. P. Malyceva, N. V. Koyain, *Design and research of automatic electric drives* **8** (Tomsk Polytechnic University, Tomsk, Russia, 2014) [in Russian]
8. E. V. Bolovin, A. S. Glazyrin, V. N. Brendakov, 2015 International Siberian Conference on Control and Communications (SIBCON): proceedings, 1, (2015)
9. R. Yu. Tkachuk, A. S. Glazyrin, V. I. Polichshuk, 7th International Forum on Strategic Technology, **2**, 586 (2012)
10. V. I. Polishchuk, V. L. Sergeev, Journal of Siberian Federal University. Mathematics & Phisics, **8 (2)**, 201 (2015)

AN ASYNCHRONOUS ELECTRIC DRIVE WITH THE INDIRECT CONTROL OF THE OUTPUT VARIABLES

Alexander V. Glazachev[1], *Yuri* N. Dementyev[1*], *Ivan* V. Rakov[1], and *Anara* D. Umurzakova[1,2]

[1]Tomsk Polytechnic University, 634050, Tomsk, Russia
[2]Innovative University of Eurasia, 637000, Pavlodar, Kazakhstan

Abstract. The article gives a mathematical description of an asynchronous motor with a device of indirect control of the electromagnetic torque and the angular velocity of the asynchronous motor in the electric drive that allows to prove on its basis the method of monitoring the output variables of the asynchronous electric motor. The simulation model of an asynchronous motor with a device of indirect control of the output variables of the asynchronous motor and the main results of the study have been given.

1 Introduction

A modern asynchronous electric drive for industrial mechanisms occupies a leading position in all industries. When using modern electric drives, such pump units in the liquid transmission systems [1], there is a constant need to control the output variables of the asynchronous motor (AM), which allows to have the information about the technological output parameters of pump units, to monitor the engines' workload, and maintain the speed within predetermined limits. With the development of semiconductor technology there is a real possibility of regulating the rotation frequency of the AM to provide the required technological parameters of the liquid transmission systems. The use of static frequency converters allows to control the AM of the pumping station in accordance with the specified rate of the technological process of the liquid transmission systems. In addition, knowing the current parameter values and conditions of asynchronous motors enables to provide the monitoring of the technological process of the liquid transmission systems and the behavior of the AM, to control its technical condition, to carry out diagnostics in identifying at the early stages the emerging defects in time to fix them. However, not in all technological processes of the liquid transmission systems it is possible to use pump units regulated by the electric drive with frequency converters because of the high cost and complexity of the applied equipment, performed at full capacity of the AM. Therefore, the most appropriate for the liquid transmission systems is the use of the functional diagram shown in Fig. 1, in which to control the output variables of AM of pump units in normal conditions or in case of failure of the main frequency converter, the indirect control device of the output variables of AM (DIC) is used.

* Corresponding author: dementev@tpu.ru

In the functional scheme of the liquid transmission systems n is logically connected to centrifugal pump units (CP_i), where H_i – pressure of i-th pump unit ($i = 1, 2, \ldots, n$), Q_i – feed and flow rate of i-th pumping unit ($i = 1, 2, \ldots, n$), L – distance.

The most common devices for measuring and controlling the speed are not pulse and tachogenerator sensors, but torque [3, 4] balanced and transmission dynamometers, torsion devices and converter installation. With the help of these devices one may control the output variables of the electric motor in the electric drive, but they have a complex design and require careful installation in conjunction with their rotating parts of the electric drive, which results in low reliability and significantly degrades the performance of the electric drive. In addition, their use may not be possible under the terms of the functioning of the electric drive. In this regard, industrial enterprises need the devices, circuit design and methods that allow to monitor the technological output parameters of the operating units through the output variables of the asynchronous motor, to maintain the speed within predetermined limits in the absence of speed and torque sensors when the necessary information is controlled by indirect methods, for example, the indirect control device.

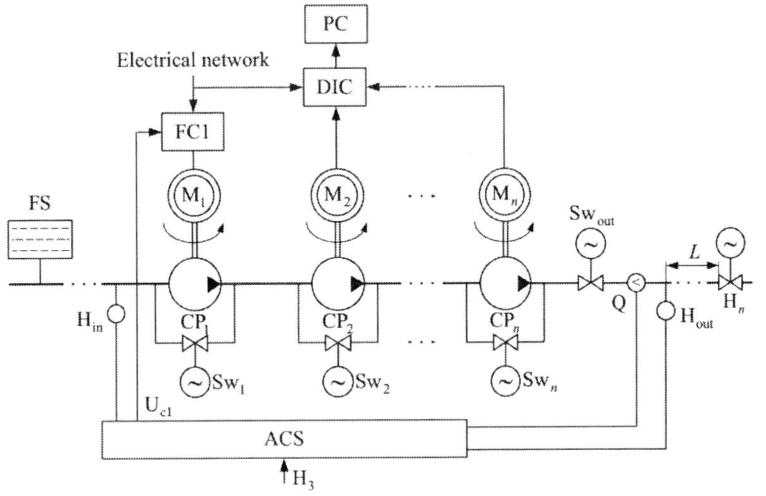

Fig. 1. A Functional Scheme of the Liquid Transmission Systems with One Frequency Converter.

Thus, the development and research of the asynchronous electric drive with the indirect control device of the output variables of the AM is relevant and allows to control the output variables, providing the specified technological parameters.

2 Mathematical model

For the mathematical description of the indirect control of the electromagnetic torque and angular velocity, one can use the expressions given in [5].

The mathematical expressions for the indirect determination of the output variables of the asynchronous motor of electromagnetic torque and angular velocity [7-11] can be represented with the following system of equations (1)

$$\begin{cases} M(t) = \sqrt{3} p_n \left(i_a(t) \int \left[u_b(t) - R_s i_b(t) \right] dt - i_b(t) \int \left[u_a(t) - R_s i_a(t) \right] dt \right); \\ \omega(t) = \omega_m(t) \left[1 + \Delta\omega_{\text{int}}(t) + \Delta\omega_{\text{dif}}(t) \right], \end{cases} \quad (1)$$

Where $\omega_m(t)=\dfrac{\sqrt{3}\left[u_a(t)-(R_s+R'_{r\alpha})i_a(t)\right]}{\int\limits_0^{1/f}\left(R_s\left[i_a(t)+2i_b(t)\right]-\left[u_a(t)+2u_b(t)\right]\right)dt-L_\beta\left[i_a(t)+i_b(t)\right]}$ – measured the instantaneous value of the angular velocity;

$\Delta\omega_{\text{int}}(t)=\dfrac{\int\left[u_a(t)-R_s i_a(t)\right]dt}{T'_r\left[u_a(t)-(R_s+R'_{r\alpha})i_a(t)\right]}$ – the dynamic integral component of the relative value of the angular velocity; $\Delta\omega_{\text{dif}}(t)=\dfrac{L_\beta\dfrac{di_a(t)}{dt}}{u_a(t)-(R_s+R'_{r\alpha})i_a(t)}$ – the dynamic differential component of the relative value of the angular velocity; $R'_{r\alpha}=R'_r\alpha$ – the active reduced resistance of the rotor winding, taking into account α coefficient; R'_r – the active reduced resistance of the rotor; $\alpha=\dfrac{L_s}{L'_r}$ – the coefficient equal to the ratio of the total inductance of the stator winding L_s to the reduced total inductance of the rotor winding L'_r ; $L_\beta=L_\mu\beta-L_s$ – the inductance with account of β coefficient; L_μ – the mutual inductance of the stator and rotor windings; $\beta=\dfrac{L_\mu}{L'_r}$ – the coefficient equal to the ratio of the mutual inductance L_μ to the reduced total inductance of the rotor winding L'_r; $T'_r=\dfrac{L'_r}{R'_r}$ – the rotor time constant.

According to the above equations of the system (1) one can determine the output variables of the asynchronous motor by measuring the voltage, currents of stator phases and data of the electric motor to control their current values.

3 Mathematical modeling

On the basis of the abovementioned mathematical description in Fig. 2 in the software package *MatLab* the simulation model of an asynchronous motor with the indirect control device of the output variables of the asynchronous motor has been made, it allows to carry out studies of the asynchronous electric drive during frequent scalar control excluding PWM of motor supply voltage.

DOI: 10.1051/matecconf/20179101039

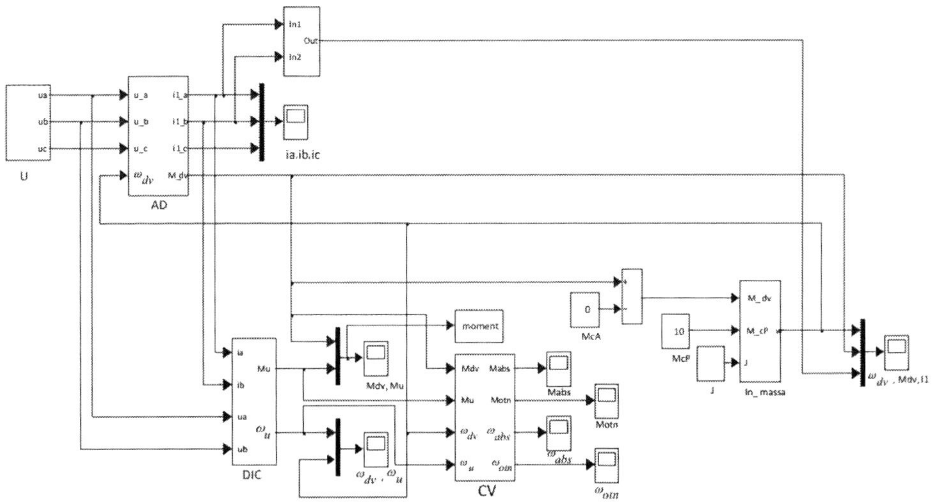

Fig. 2. Simulation model of an asynchronous motor with the indirect control device.

The structure of the simulation model includes superblocks: a model of the asynchronous motor is made based on the mathematical description of the asynchronous motor (AM); phase voltage generator stator windings in a three phase coordinate system (U); inertial mass (In_massa); the indirect control device (DIC) of the electromagnetic torque of the asynchronous motor; a comparator of values obtained of the electromagnetic torque of an asynchronous motor with indirect control devices (CV).

Fig. 3 show the results of simulation studies of the asynchronous motor (AIR90L4) for various values of load, as dependences, respectively, current, speed and torque of time $I(t)$, $M(t)$, $\omega(t)$, which are obtained from an asynchronous motor and indirect control device during frequent scalar control of an asynchronous motor excluding PWM motor supply voltage.

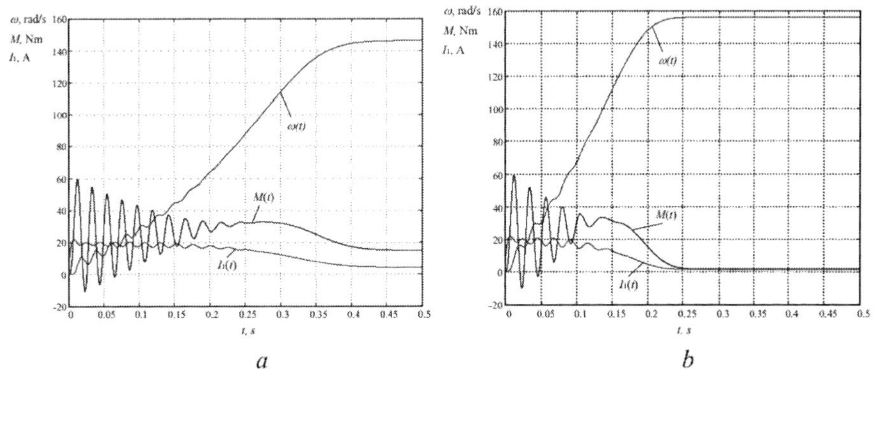

DOI: 10.1051/matecconf/20179101039

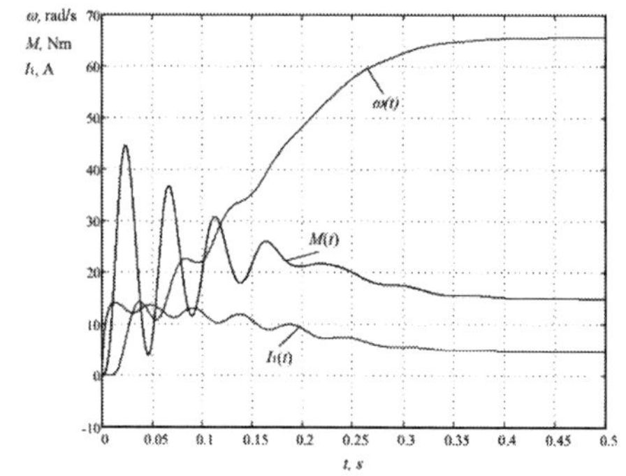

c

Fig. 3. Dependences $I(t)$, $M(t)$, $\omega(t)$, , $a - f = 50$ Hz, $M_l = 10$ Nm; $b - f = 50$ Hz, $M_l = 2$ Nm; $c - f = 25$ Hz, $M_l = 15$ Nm.

From the above-mentioned characteristics, it follows that dependences of the electromagnetic torque and the angular velocity obtained from the output of the asynchronous motor and indirect control device have the convergence permissible in engineering designs.

4 Results

To estimate the results of the control of the output variables of the AM in the comparison unit of the simulation model, the absolute and relative inaccuracies of control of the electromagnetic torque and the angular velocity obtained from the AM and the indirect control device are determined, and their values are summarized in Table 1.

Table 1. Absolute and relative inaccuracies of the control of electromagnetic torque and the angular velocity.

Absolute inaccuracy		**Relative inaccuracy, %**	
ΔM, 10^{-4}, Nm	$\Delta\omega$, rad/s	δ_M, 10^{-2}	δ_ω
– 0,004…0,003	–12…12	–0,1…0.1	–8 … +8

As seen from the table, the relative inaccuracy of control of the electromagnetic torque is less than 1%, and the angular velocity is not more than 10% because of the correction link in the speed calculation unit.

5 Conclusions

The results of modeling obtained in the study of the asynchronous electric drive with the indirect control device, confirm the correctness of the mathematical description of the asynchronous motor with a device of indirect control of output variables.

According to the studies conducted on the models of the asynchronous motor with the device of indirect control, it has been found that the use of the indirect control device in an asynchronous electric drive allows to conduct the continuous monitoring of electromagnetic torque and angular velocity, and confirms the high efficiency of the indirect control device for electric drives of general-purpose machinery carrying out the technological process and requiring the control of its parameters.

References

1. B. Leznov, *Energy conservation and a variable drive in pump units* (Energoatomizdat, Moscow, 2006) [in Russia]
2. V. Galeev, M. Karkachev, V. Kharlamenko, *Main oil pipelines* (Nedra, Moscow, 1988) [in Russia]
3. Yu. Rybalchenko, *Magnetoelastic torque sensors* (Machinostroyenie, Moscow, 1981) [in Russia]
4. M. Leitman, *Automatic measuring the output parameters of electric motors* (Energoatomizdat, Moscow, 1983) [in Russia]
5. Yu. Dementyev, A. Umurzakova and O. Arsentyev, Bulletin of Irkutsk State Technical University, **7**, (2013)
6. Method for measuring the angular velocity of rotation of the three-phase asynchronous motor. (2014, journal, № 23). *Patent for invention, Russia № 2525604*
7. Yu. Dementyev, A.Umurzakova, K. Khatsevski, Dynamics of systems, mechanisms and machines, **1**, (2014)
8. Yu. Dementyev, A. Umurzakova, L. Udut, Fundamental studies, **12**, (2014)
9. Yu. Dementyev, A. Umurzakova, MATEC Web of Conf., **19**, 010027, (2014)
10. The device for measuring the torque of a three-phase asynchronous electric motor. (2013, journal, № 24). *Useful model patent, Russia, № 131874*
11. A. Glazachev, Yu. Dementyev, K. Negodin, A. Umursakova, EPJ Web of Conf., **110**, 01044, 2016

VECTOR CONTROL SYSTEM DESIGN FOR FOUR DEGREE-OF-FREEDOM DYNAMIC FLEXIBLE SIMULATION OF THE VARIABLE-FREQUENCY DRIVE

Sergey N. Kladiev[1*], *Anton* V. Solozhenko[1], and *Alexander* A. Filipas[2]

[1]Tomsk Polytechnic University, 634050, Tomsk, Russia

[2]Seversk Technological Institute of National Research Nuclear University "MEPhI", 636000, Seversk, Russia

Abstract. In the present work we investigate the control system development of the drive simulators to train driver/operator driving skills, taking into account the ever-changing terrain. In order to meet the required response of the four degree-of-freedom motion platform servomotor current studies have been focused on the vector control of the resistance motor angular velocity from the sensor being incremental encoder. In proposed system the standard security of the frequency converter is realized. It leads to overload capacity of two times within minutes determined by servomotor inertia. Further, we represent the algorithms: positional limitation, reliable acceleration and restraint, frequency break. As well as we demonstrate the position switches implement in software. As a result, the control system commands the control of the angular position of the platform in coordinates.

1 Introduction

Modern day motion simulator to train professional driver/operators is a hardware and software system and correspondingly, it has complicated structure. This system is based whether on hydraulic drive using the mobile vehicle simulation taking into account the influence of the ground profile, or on frequency-controlled drive using the motion small platform taking into account the dynamic driving modes to train driver/operator driving skills.

2 Objectives and tasks

- training of the onset of speed and acceleration signals along axis of rotation of the platforms with motion bases, taking into account the ever-changing terrain;
- allowance for the high load inertia of electric drive;

* Corresponding author: kladiev@tpu.ru

- support of the overload capacity of two times determined by servomotor inertia;
- field weakening motor under short-range maintain the torque at the level 0.5 Tn.

3 Subject to control

The motion platform MP-43 is operated by embedded computer through the local area network, Ethernet- and Modbus TCP-compliant based. The inverter FR-A740 Mitsubishi Electric is used as frequency converter.

Using import component and constituent element in this device lead to vendor lock-in. We mean the foreign suppliers. It takes as point of departure for thinking about industry-domestic concept. It has as a consequence the tasks to upgrade an existing motion platform. Mitsubishi Electric FR-A740 power inverters should be replaced by home. The main objective of the import substitution in regard to such simulators is the program development of microprocessor control systems based on Tomsk research production firm "Mechatronica-pro" hardware and software [1].

4 Goal of research

The development of digital microprocessor-based vector control systems for the motion platform drive based on frequency converter FC-01 and CAD MexBIOS Development Studio [1] recommended by embedded software development environment for signal processors Texas Instruments [2].

Hardware component is in the form of the motion platforms DP-43 and four-axis drive. In the present work, we consider one-axis drive constructed as induction motor (IM) integrated into frequency converter FC. We take the parameters of FC output characteristics as technical specification:

Table 1. Technical specification of FC output characteristic.

Characteristics	Motor power	Rated output current for 2.2 kW	Overload capacity (current)	Output frequency range	Frequency settability	Time of acceleration / restraint	Control strategy	Encoder	Communication protocol	Limitation of torque
Value	1.1 kW; 1.5 kW; 2.2 kW	5.5 A	200 % / 60 s	0…100 Hz	0.01Hz	0.1…100 s	magnetic flux vector control in closed loop system in speed [3]	incremental transducer, TTL signal level between 512 and 4096 imp /rev	Modbus TCP/IP	0…200 %

5 The software part

To create vector control system it has additionally to realize algorithms, providing:

- threshold of current acceleration taking into account the dynamics of transient processes (reverse);
- threshold of breaking for the voltage overloading in the DC link (DC Bus);
- displacement constraint of the electric drive output element position (angular displacement) in the range covered;
- short-range maintain the torque at the selected level 0.7Tn in the field weakening range between 50 and 100 Hz [4,5].

Summarized block scheme of vector control system of frequency converter FC-01 [1] includes: *Ramp function* – power-up sensor; *CUR_MOD* – the current-operated model [6,7]; *PI_* – the proportional plus reset controller; *IPARC, ICLARC* – the Clarke and Park inverse transformations; *SAT Irms* – the threshold of effective current (saturation); *SAT Udc* – the safe braking ramp; *Udc* – the voltage in the FC DC link.

According to the structure of the vector control system described above it differ from widely used systems by presence of the blocks SAT Irms, SAT Udc, Brake, ICLARK. These elements handle follow problems: the threshold of current acceleration; the threshold of acceleration for the voltage overloading in the DC link; the frequency break; specifying the value of voltage to power switch using a third harmonic. It was found that it lead to increase output voltage 15% The function of block "Brake" is to define «0» to the proportional plus reset controller of the speed loop when the displacement constraint of the drive output element occurs [5]. The velocity signals input to the block from the control overhead panel (reff SPD), computer command accept (Start), feedback signal for the speed $\omega^{\bullet}$. Further, if the tripping of the limit switches, which is software-based open, the signal (*Start*) put into reset state, forming the logical «1» in the block output (*RS trigger*). Switch *MUX* changes the state «1» and transfers zero assignment to the speed control unit in the block BRAKE output. The stopping and tripping of the FC run the command from the block (*Comparator*). The logical «1» in the block output is formed by comparing the feedback signal for the speed «0» that indicates the drive stopping. Figure 1 present the transient phenomena for the starting of the motor to the minimum ramp control $\omega = 0.013$ p.u. (20 min^{-1}) and the step-response time (application of load $2M_n$=30 H m to motor shaft) at a given time $t = 3$ s. The off-peak speed is 0.008 p.u. (12 min^{-1}). For disturbance the step-response time does not exceed $\Delta t_{pos} = 0.2$ s. The load dump is shown in this figure at the time $t = 4$ s. In addition, the amplitude quantization is equal $\Delta\omega$=0.005 p.u. (7.5 min^{-1}). It is also was investigated the operating of the motor at the rated speed of the motion platform rotation and the step-response time of the motor (application of load to motor shaft $2M_n$ at a given time $t =$ 1,5s). Figure 2 illustrates that the off-peak speed does not exceed 0.06 p.u. (9 min^{-1}), the step-response time is not more than 0.25 seconds.

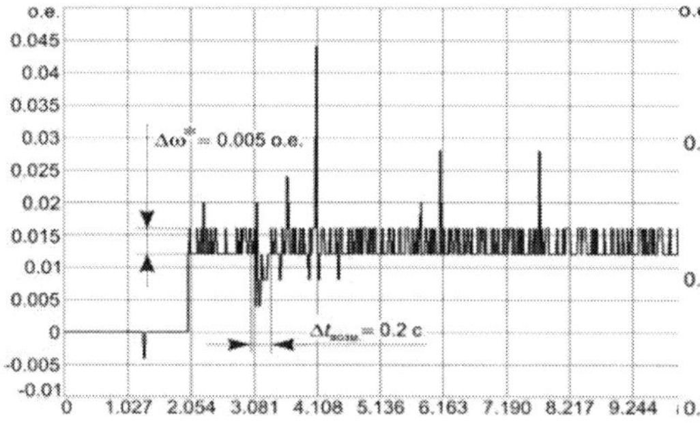

Fig. 1. Starting of the motor idly to the minimum ramp control ω=0.013 p.u. (20 min^{-1}), application of load $2M_n$ to motor shaft at a given time t=3 s.

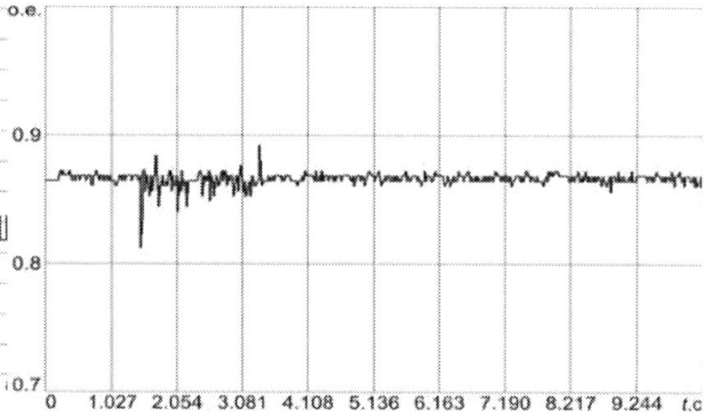

Fig. 2. Transient phenomena for disturbances after application of load $2M_n$ at a given time t=1,5s speed ω=0.86 p.u. (1200 min^{-1}).

Both figures 3, 4 show results of optimization of the reliable acceleration and restraint states when the motor reverses line power sensor. Applied algorithms of asynchronous motor parameters identification[8] and sensorless vector control[9,10] allows the control system to adapt to the dynamic modes of operation depending on the inertia of the mechanical system. There is an obvious advantage that applied control algorithms intended to ensure adapt the control system under dynamic modes according to mechanical system response time. Both threshold parameters which is Isat – the current acceleration, Udc – the voltage of reliable restraint allow threshold of acceleration and jerking of the platform within the prescribed FC limits.

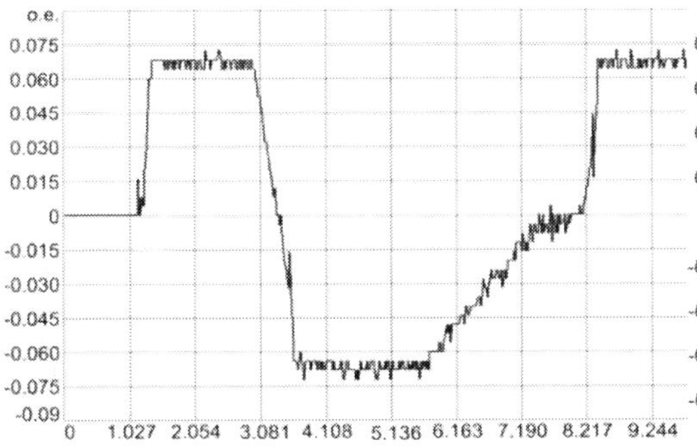

Fig. 3. Acceleration, restraint, reverse on-load $2M_n$; t_z=0.5; ω=10 s^{-1}; I_{sat}=5.5 A; U_{dc}=550 V.

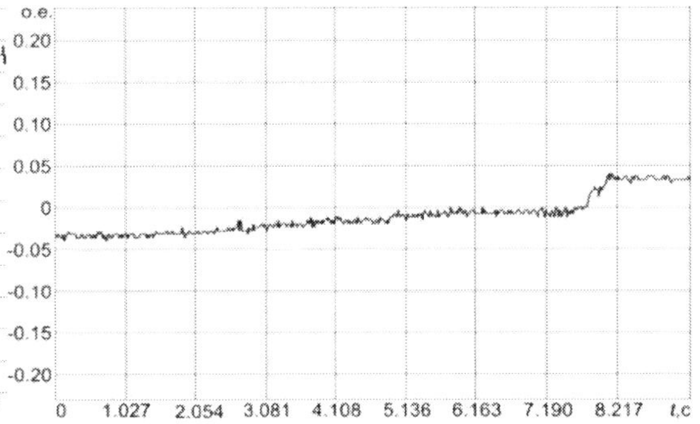

Fig. 4. Reverse on-load $2M_n$; t_z=0.5; ω=10 s^{-1}; I_{sat}=4.5 A; U_{dc}=550 V.

6 Conclusion

This control system takes into account the allowance for the load inertia of electric drive in the process of speed and acceleration setting, training of the onset of speed and acceleration signals along axis of rotation of the platforms with motion bases, taking into account the ever-changing terrain. In addition, it supports the overload capacitance of two times determined by servomotor inertia. Note that besides the technical specification listed above the control system provides the reliable acceleration and restraint modes, the function of frequency break, the control of the angular position of the platform in coordinates.

References

1. http://mechatronica-pro.com/en
2. http://www.ti.com/devnet/docs/catalog/embeddedsoftwarefulldetails.tsp?productId=12220
3. A.S. Anuchin, *Control systems of electric drives: textbook for high schools* (Publishing House MEI, Moscow, 2015) [in Russian]
4. R.T. Schreiner, *Mathematic simulation of the alternating-current drive with semiconductor frequency converters* (URB RAS, Ekaterinburg, 2000) [in Russian]
5. D.W. Novotny, T.A. Lipo, *Vector Control and Dynamics of AC Drives*. (Clarendon Press Oxford, 2003)
6. Peter Vas. *Sensorless vector and direct torque control* (Oxford University Press, 1998)
7. D.V. Robkanov, Y.N. Dementiev, S.N. Kladiev The 8th Russian – Korean International Symposium on Science and technology Proceedings, **1**, 280 (2004)
8. E.V. Bolovin, A.S. Glazyrin, V.N. Brendakov, International Siberian Conference on Russia, 7, 2015 [in Russian]
9. E.A. Kolomiyets; A.Y. Chernyshyov, Int. Sib. Conf. Control Commun., 15287590, 2015
10. V.S., Borisov, N.V. Gusev, P.V. Bykovskih, K.V. Obraztsov, Int. Sib. Conf. Control Commun., 2016

DEVELOPMENT OF SOFT-HARDWARE PLATFORM FOR TRAINING SYSTEM DESIGN OF ELECTROTECHNICAL COMPLEXES AND ELECTRIC DRIVES

Ekaterina A. Koltunova[1], *Kseniya* V. Ikonnikova[1*], and *Sergei* V. Lyapushkin[1]

[1]Tomsk Polytechnic University, 634050, Tomsk, Russia

Abstract. The article presents the results of the development of software and hardware platform as the equipment for the training of children and youth work skills with robotics, allowing in the future to apply this knowledge in practice, implementing automation system for home use. We consider the problems of existing solutions. The main difference is the integration of the proposed fees and extensions into a single set by connecting the connectors and the ability to connect third-party components from different manufacturers, without limiting users. As well as a simplified method using a visual object-oriented programming allows you to immediately engage in the work. Prepared lessons and tasks in the game style simplifies the information and allows you to understand how you can apply one or another technical solution.

1 Introduction

Automation of any manufacturing systems and processes is a key point of technology development [1-2]. Complication of applicable methods demand increasing number of specialists, therefore fast adaptation in studying of any subject helps accelerate the adoption of new information and development in terms of career [3-4]. One, familiar with initial knowledge base in electronics and laws of robotics, introspects deeper in nuances and have no need in extra time for explanation of main theories of physics and electrical engineering. Therefore, a new training tool for the electromechanical systems is needed.

2 Concept of the kit

The most successful way of information flow on circuit design is a kit (usually laboratory works in training stands given in that form). It is necessary to create a variant of kit with open architecture and accessories, which can allow construct robotics systems for carrying out development work in domestic conditions, to solve educational and scientific tasks and to automate scientific experiments. The kit should solve these problems with a view to

* Corresponding author: ikonna@tpu.ru

minimizing the time for its development and acceptable price for mass consumer. And also be able for consistent complication of works in progress.

The concept of that kit built on the basis of the following abstracts:

1. Functional extensibility through the use of additional modules;
2. The ability of organizing a network of several controllers;
3. Establish communication between controller and modern gadgets (smartphone, tablet PC etc.) through support of communication protocols;
4. The ability of uploading an applications for using standard procedures (navigation, audio and video processing, management of complex electromechanical objects);
5. The ability of creating software for controller by user with minimal level of qualification in the field of programming by using graphical programming languages and providing a templates to address specific tasks (for example AC motor control, synchronization of multi-axis motion systems, etc.);
6. The ability of equipping the controllers with additional modules, made by third-party (including user himself) to address the unique challenges.

3 Experimental results

The resulting work is software-hardware platform "MexConStructor" – equipment for schools and youth centers of innovation creation, and for private use.

The kit consists of set of PCBs, several mechanical platforms, sensors, wires and application. Developed set of PCBs with microprocessor STM32F4 (Figure 1) allows user to conveniently use the entire periphery, and has a variety of options. To this was realizable, connectors for different types of interfaces (CAN, I2C, RS-485 etc.) and analog and digital inputs and outputs are applied. A special feature is the expansion-PCB for Arduino-components, which allows connect everything that required for work (Figure 1). Kit designed in one style, allowing user to collect the necessary PCBs in one robotic system.

Fig. 1. Expansion-PCBs.

Proposed a set of sensors and mechanics are widely available and commonly used. Users, who already have their own components with any interface, can adapt it for work (Figure 2)

 DOI: 10.1051/matecconf/20179101041

Fig. 2. Electronic components.

The main competitors of the product is the Arduino platform, robotic systems WEX, LEGO Mindstorm and analogues. Unlike competing technologies, the proposed solution makes it possible to work directly with digital devices without complicated stage communications settings.

4 Conclusion

In addition to that, the kit must provide the control for household appliances (including hi-power), get access via standard Internet tools, set the logic by graphic and text programming languages (with the possibility of easy learning and subsequent complication of programs while operating the equipment).

Cost characteristics are at the level Arduino series controllers and significantly less than robotic kits. Unlike competing products, in the price included the system of a graphical programming MexBIOS, allowing user to create the project firmware simultaneously in 5 languages (with the choice of language by the user for each single project solved subtasks) [5]. The use of such software allows users with minimal knowledge in programming quite easily find a way to create the desired robotic complex programs (followed by the development of other ways of creating programs with increasing complexity and detail tasks.

The platform is open both in terms of hardware implementation and software. The openness of the platform should lead to the emergence of numerous applications precedents with an opportunity to copy the most successful options for the use of third-party users. Mass use ensured by the ease of mastering a specialized programming language MexBIOS, which allows graphical methods and due to built in ready-made templates, to create complex control algorithms easily enough for unskilled users. A training course in the form of a game with increasing complexity is created for children and adolescents. Successful proposed ideas are applicable in practice.

References

1. N. Gusev, S. Lyapushkina, M. Kovalenko, *Integrated automation of technological processes* (Publisher of Tomsk Polytechnic University, Tomsk, 2011)

 DOI: 10.1051/matecconf/20179101041

2. N. Gusev, S. Lyapushkina, M. Kovalenko, *Automation of technological complexes and systems in the industry* (Publisher of Tomsk Polytechnic University, Tomsk, 2011)
3. K. Ikonnikova, L. Ikonnikova, E. Koltunova, Turkish Online Journal of Educational Technology, 557 (2015)
4. K. Ikonnikova, L. Ikonnikova, E. Koltunova, Turkish Online Journal of Educational Technology, 488 (2015)
5. N. Gusev, Bulletin of South Ural State University. Series: Energy, **15**, 59 (2015)

MATHEMATICAL SIMULATION OF THE ASYNCHRONOUS ELECTRIC DRIVE OF PERIODIC MOVEMENT

Vasiliy O. Nagorniy[1], and *Anatoliy* V. Aristov[1,*]

[1]Tomsk Polytechnic University, 634050, Tomsk, Russia

Abstract. The article deals with the approach to construction of mathematical models of electric drives working in the mode of periodic motion. An example of a mathematical model and simulation results are presented.

1 Introduction

The modern requirements for drives of antennas of radar stations [1] or optomechanical scanning devices [2] require profound study of the dynamic processes occurring in the motor drives of periodic motion. Since the full-scale pilot studies increase the financial expenses and periods of development and don't allow to conduct comprehensive researches, the use of mathematical simulation in case of complex researches seems to be the most appropriate [3].

As a system for simulation and research of the electric drive of periodic movement the environment MATLAB/Simulink has been chosen [4, 5].

The article deals with the model of the electric drive of the pulsating movement that is based on a two-phase asynchronous motor (AM), working directly in the mode of periodic motion by the linear phase or balanced amplitude modulation supply voltages to the interruption of one of them in moments of time, when the electromagnetic torque of the AM passes through zero [6]. The complete model of the electric drive consists of electric models of the pulse voltage transformer (PVC) and a two-phase asynchronous motor with a generalized load on the shaft.

The model of the electrical drive of pulsating motion in the mathematical environment, Simulink, implementing phase method when one of the phase windings of the AM from the voltage source is energized through pulse-width voltage converter.

* Corresponding author: aristovav@tpu.ru

DOI: 10.1051/matecconf/20179101042

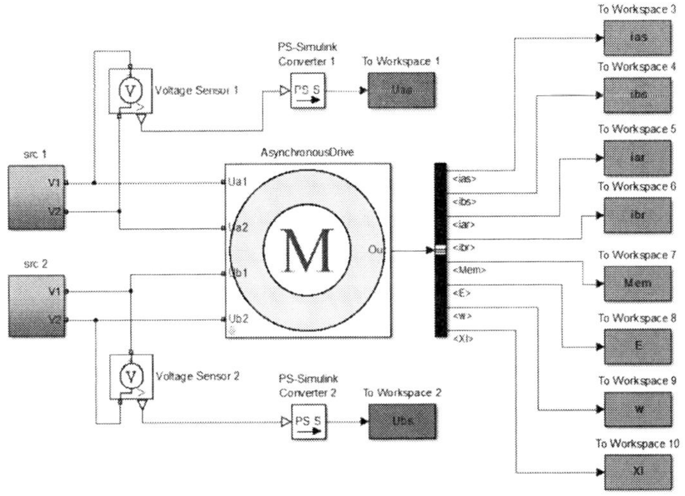

Fig. 1. Electric drive model in the Simulink environment.

In Figure. 1 src1, src2 stand for power supplies of phases of the asynchronous motor; AsynchronousDrive stands for the two-phase asynchronous machine with a load on the shaft. The blocks To Workspace are designed to transfer information about the variables Uas, Ubs (indicate phase stator voltage); ias, ibs (are stator phase currents); iar, ibr (are current rotor phase); Mem (stands for the electromagnetic torque of the electric machine); E (is angular acceleration); w (indicates angular velocity); XI (is the movement the motor shaft) in the MATLAB working environment.

2 Simulation of pulsed voltage converter

Model of a pulse voltage converter (Figure. 2) contains power transistor *VT1*, diode *VD1*, filter inductor *L1*, filter capacitor *C1*, current sensors *UA1*, *UA2*, voltage sensor *UV1*, *Control* – a system of the PVC control.

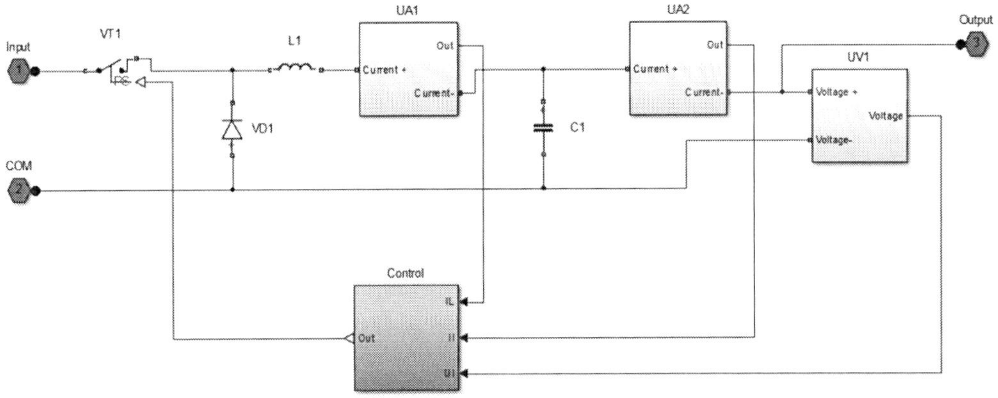

Fig. 2. Model of a pulse voltage converter.

While the PVC is being operated, the values of the inductor current, the load current and load voltage enter the system *Control* (Figure. 3), which controls the power transistor *VT1*.

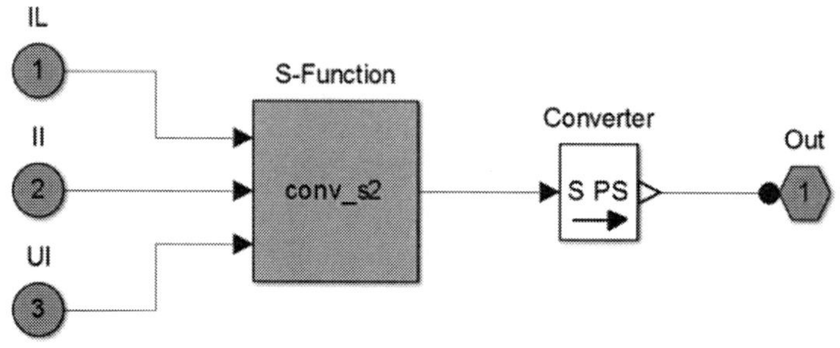

Fig. 3. Pulsed voltage converter control system.

In order to integrate the PVC control systems with the mathematical Simulink model, a special approach was selected in which the control system code is executed on a personal computer [7]. The PVC control system is adapted to Level-2 MATLAB S-function. The language used to create a model management system is 'C' [8, 9]. According to model (Figure. 7) S-function is a dynamic link library (dll) [10], exporting the function mexFunction. The mexFunction is executed every time when the interaction of the Simulink with the S-function is required. If the mathematical model is started the S-function launches a thread (by using the CreateThread function) to run the executable code of the PVC control system.

The model calculation runs in the context of the two threads of the operating system. Since the model threads run in parallel, they are forced to keep a record of working time models. To synchronize the time of the threading model the synchronization cycle is introduced. In this case the leading flow is the thread created in the MATLAB environment, and driven thread is that of the control system.

After a specified period of time the model is sure to complete its work by stopping the thread of the control system. In mathematical models of the PVC as well as in the laboratory prototype we use *the same source code* of the control system, *requiring no modifications in its transfer*.

Using the approach described above, the following the PVC control algorithms were tested:

Proportional-integral-derivative regulation with PWM (pulse-width modulation).

Controlling the balance of required and the stored energy [11].

Controlling the balance of required and the stored energy with PWM [12].

3 Modeling of the electromechanical drive

While investigating the periodic mode it is convenient to represent the electric machine in the form of a certain idealized generalized model with two couples of co-perpendicular windings: on primary and secondary elements [13], which allows, under certain assumptions, to analyze the greatest number of AM of different types working in a mode of periodic movement.

For this model, there are several options for recording a system of differential equations describing its electrical state. The simplest analysis of equations written in the coordinate axes of α, β or d, q, rigidly connected to the stator and the rotor, respectively. For example, the coercion to the axes α, β allows to simplify the mathematical description of the process by eliminating periodic coefficients in the equations, describing the state of the electrical windings in the stator AM.

Under conventional assumptions (the magnetic circuit of the machine is unsaturated, the magnetic flux is sinusoidally distributed in space, eddy currents and edge effects are negligible) [14] the generalized model of the motor, operating in a pulsed mode of motion in the coordinate system α, β is described by the following system of the differential equations:

$$\left.\begin{aligned}
&U_{\alpha s}(t) = i_{\alpha s}R_{\alpha s} + L_{\alpha s}\frac{di_{\alpha s}}{dt} + M_{\alpha}\frac{di_{\alpha r}}{dt};\\
&U_{\beta s}(t) = i_{\beta s}R_{\beta s} + L_{\beta s}\frac{di_{\beta s}}{dt} + M_{\beta}\frac{di_{\beta r}}{dt};\\
&0 = i_{\alpha r}R_r + L_r\frac{di_{\alpha r}}{dt} + M_{\alpha}\frac{di_{\alpha s}}{dt} - p\omega\left(M_{\beta}i_{\beta s} + L_r i_{\beta r}\right);\\
&0 = i_{\beta r}R_r + L_r\frac{di_{\beta r}}{dt} + M_{\beta}\frac{di_{\beta s}}{dt} + p\omega(M_{\alpha}i_{\alpha s} + L_r i_{\alpha r});\\
&M_{em} = k_d\left(M_{\alpha}i_{\alpha s}i_{\beta r} - M_{\beta}i_{\beta s}i_{\alpha r}\right);\\
&L_m\frac{d^2\chi}{dt^2} + R_m\frac{d\chi}{dt} + C_m\chi + M_l = M_{em},
\end{aligned}\right\} \quad (1)$$

where $U_{\alpha s}(t)$, $U_{\beta s}(t)$, $i_{\alpha s}$, $i_{\beta s}$, $i_{\alpha r}$, $i_{\beta r}$ are voltage and current of the phase windings; $R_{\alpha s}$, $R_{\beta s}$, R_r, $L_{\alpha s}$, $L_{\beta s}$, L_r are active resistance and total inductance of the phase windings; M_{α}, M_{β} are mutual inductances between the stator and rotor windings on coordinate axes α, β; ω - the rate of change of the generalized coordinates of the movable element; M_{em}, M_l are generalized electromagnetic force and the load; p is the number of pairs of poles of the electric machine; k_d is a generalized power factor; L_m is an inertia load factor; R_m is a damping load factor; C_m is a positional load factor; χ is the relocation of the output element of the drive.

For angular movements $k_d = 1$, and for the rectilinear $k_d = \pi^2 / \tau^2$; where τ is a pole pitch. Now we arrange the system of equations (**1**) as a Level-2 MATLAB S-Function. Let us examine the internal structure of the mask *AsynchronousDrive* (Figure. 4).

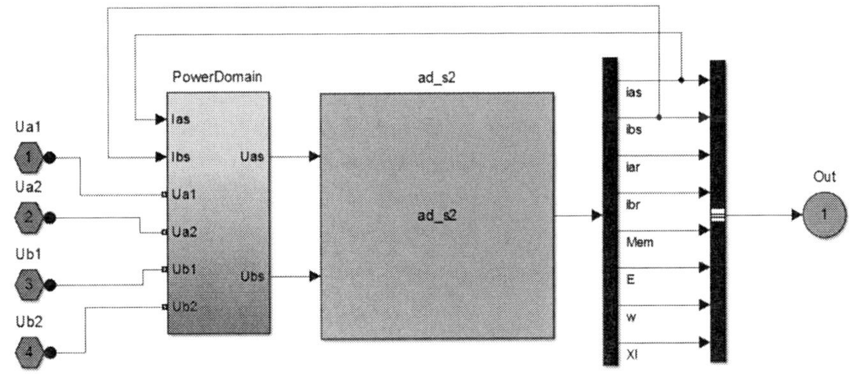

Fig. 4. The internal structure *AsynchronousDrive* mask.

The *PowerDomain* subsystem converts the input electrical quantities in dimensionless quantities MATLAB/Simulink. Block *ad_s2* – block MATLAB S-Function, described in the MATLAB language.

While in operation Simulink environment solver periodically requests the current values of the derivatives of internal states ($i_{\alpha s}$, $i_{\beta s}$, $i_{\alpha r}$, $i_{\beta r}$, ω, χ) of S-function, which are calculated

 DOI: 10.1051/matecconf/20179101042

in accordance with **Ошибка! Источник ссылки не найден.**, and updates the state values selected according to the solver.

Such a construction of a mathematical model allows using of different solvers for the same the mathematical description of the model, and changing them if necessary. Further for mathematical simulation the solver *ode*45 (*Dormand-Prince*) [15, 16] was used.

Conclusion

The implementation of the proposed approach allows using the algorithm worked out on the mathematical model of a PVC in the control system without additional changes.

Running the algorithm on a personal computer simplifies the process of debugging and enables to usc thc powerful visualization tools of the various CAD systems.

The application of Simulink environment allows using a visual, graphical representation of a mathematical model; it is easy to alter both the model itself and its parameters.

The construction of mathematical models in MATLAB/Simulink enables to use a variety of solvers for one and the same mathematical description of the model, and change them if necessary.

The power of one of the windings of AM with the DC voltage eliminates the high-frequency pulsation electromagnetic torque that inevitably arises when balanced amplitude or phase modulation of supply voltages are being used and thereby provides a higher energy performance of the drive.

References

1. V.O. Nagorniy, A.V. Aristov, Herald SibSAU, **16**, 97 (2015)
2. A.V. Aristov, V.O. Nagorniy, Reports of the Tomsk state University of control systems and electronics, 2, 60 (2013)
3. B.G. Lyubarskiy, E.S. Ryabov, *MATLAB : Materials V Intern. scientific. conf.*, 404 (2011)
4. Simulink – Simulation and Model-Based Design. URL: http://www.mathworks.com/products/simulink/
5. V.P. Dyaconov, *Simulink 4. Special handbook* (Piter, St. Petersburg, 2001) [in Russian]
6. A.V. Aristov, Energy, **313**, 107 (2008)
7. V.O. Nagorniy, A.V. Aristov, A.M. Gavrilov, Electromechanical energy converters: materials VII Intern. scientific and engineering. Conf., 201 (2015).
8. B. Kernighan, D. Ritchie, *The C Programming Language* (Williams, 2007)
9. ISO / IEC 9899: 2011 – Information technology – Programming languages
10. What is a DLL? URL: https://support.microsoft.com/ru-ru/kb/815065
11. A.F. Lekarev, V.O. Nagorniy, A.V. Aristov, VI Intern. scientific and engineering. conf., 192 (2013)
12. Yu.M. Kazantsev, A.F. Lekarev, Devices and systems. Management, control, diagnostics, **5**, 17 (2009)
13. I.P. Kopylov, *Electromechanical energy converters* (Energy, St. Petersburg, 1973) [in Russian]
14. A.K. Gromov, A.G. Ershov, A.V. Likhacheva, Electromechanical energy conversion: tutorial. Ivanovo, 2008.
15. J.R. Dormand, Numerical Methods for Differential Equations: A Computational Approach. Boca Raton: CRC Press, 1996.
16. A.E. Novikov, Development of algorithms of variable structure for solving stiff problems: Dis. Cand. Physics and Mathematics. Sciences. Krasnoyarsk, 2014.

MATEC Web of Conferences **91**, 01043 (2017) DOI: 10.1051/matecconf/20179101043
Smart Grids 2016

PRESSURE PUMP POWER CONTROL IN THE PRIMARY CIRCUIT OF THE HEAT EXCHANGE SYSTEM

Aleksandr Shilin[1,*], *Viktor* Bukreev[1], and *Sergey* Prohorov[1]

[1]Tomsk Polytechnic University, 634050, Tomsk, Russia

Abstract. In this paper we consider the problem of speed in hot water systems where highly efficient plate heat exchanger is used. Especially marked the problem which is connected with long transition drive of constant speed exceeding the time of the heat exchanger accumulative tank emptying more than twice. As a regulating element in the heat exchange system there was proposed to use asynchronous electric drive of pressure pump in the primary circuit of the heat exchanger. For correct use of such electric drive we solved the problem of control object mathematical model synthesis, which has non-linear properties, in particular, the transfer coefficient of the circuit can vary in more than 6 times. At the same time there was revealed the dependence of the transfer coefficient on the motor speed, which must be considered in the controller synthesis. In conclusion we suggested the solutions of regulators synthesis tasks with customizable settings for speed and switchable structure between relay λ and PI regulators.

1 Introduction

In hot water systems (HWS) as a control device the valve with a controlled rod position is traditionally used. The very position of the rod is set by driving the constant speed with three-point control [1]. The block diagram of such a system is shown in Fig. 1, where in addition to the valve we have a high-efficiency heat exchanger and circulation pumps in both primary circuits of flow. As a coolant in most cases ordinary water is used. The control of such systems is performed on programmable logic controllers (PLCs). The principle of hot water system control is as follows: the water temperature T_{pd3}, which is supplied to the consumer, depends on the water temperature in the first circuit, which in turn depends on the ratio of the volume q_i of the entering coolant from input to the volume q_1 circulating in the first circuit. In its turn the volume of the incoming coolant depends on the position of the controlled stem valve h, which is set by the drive. To use traditional PID controls it is enough to organize two circuits [2] the first one sets the valve position, the second one, including PID, calculates the specified position of the rod on the mismatch between the target T_{set} and the measured temperature T_{pd2} values.

* Corresponding author: shilin@tpu.ru

In such a configuration it is possible to use ready-made libraries FBD of blocks [3] PLC producers [4], including the basic laws of dynamic objects control and self-tuning procedures of closed contours. It would seem that in this field there is no subject management problem, but with the advent of highly efficient plate heat exchanger there appeared the speed problem which has not been previously considered in heat consumption systems.

The use of modern automatic control theory in the field of optimal control for speed may give little effect in performance, which cannot solve the problem of hot water temperature dips at the time of a sharp change in consumption. Therefore, usually the system includes a storage tank, allowing to increase the inertia of the control object commensurate with the time of transition. However, it significantly increases the cost and size of the design of automatic hot water supply control.

The article explores a radical solution to increase the speed of HWS, where as the actuator pressure pump with frequency controlled asynchronous motor is used instead of the valve with a slow drive.

2 Problem statement

As the actuating element it is supposed to use frequency motor control of pressure pump included in the counter pressure drop of inlet and reverse coolant coming from the supplier. The block diagram and the mathematical model will have several basic differences that should be considered in the synthesis of a closed system. Therefore, for the implementation of the proposed control system it is necessary to perform the synthesis of a mathematical model, to determine the impact on the model and the structure of differences between the traditional models. To analyze the use of the traditional PID controller. To propose alternative methods to PID controller which can improve performance of the system as a whole. To perform numerical simulation and analysis of the results.

3 The structure of the object and its mathematical model

The mathematical model of the DHW control system with AC electric drive can be built using the methodology presented in [5], where the coolant mixing process is calculated using the equivalent circuit. In this case, the element of the structure mixing and the equivalent circuit can be represented as shown in Fig.1.

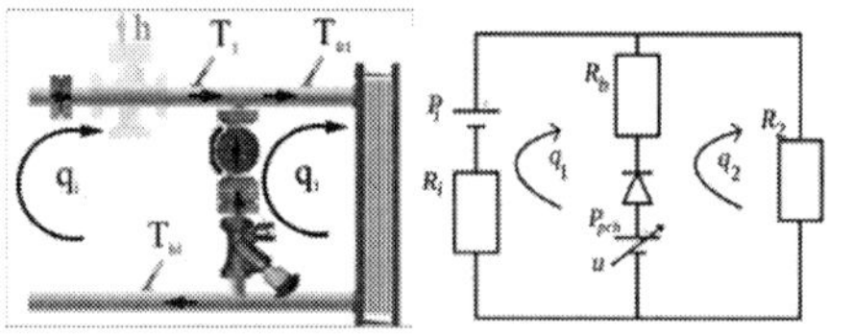

Fig. 1. Block diagram and equivalent circuit of coolant mixing in the frequency controlled drive.

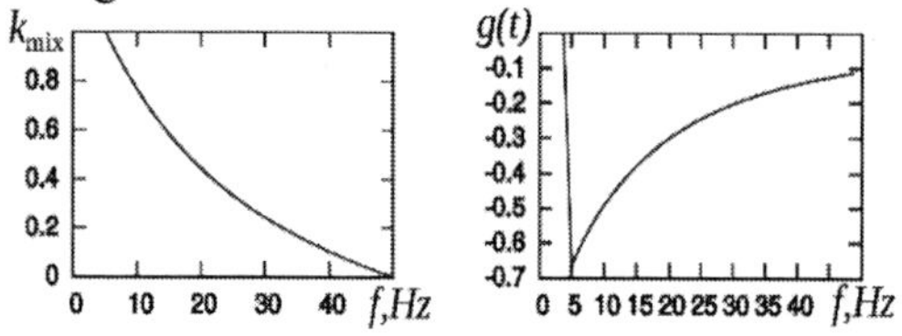

Fig. 2. Static dependence of mixing coefficient and transfer coefficient.

Kirchhoff equations will be as follows:

$$\begin{cases} P_i - P_{ch}(f) = q_i(R_i - R_b) - q_2 R_b \\ P_{ch}(f) = q_i(R_i + R_b) - q_i R_b \end{cases} \quad (1)$$

where P_i – the differential pressure generated by the central thermal point (CTP); R_i – equivalent output hydraulic resistance of CTP; P_{pch} – pressure drop on the pressure pump; R_2 – hydraulic resistance of the heat exchanger input circuit; R_b – hydraulic resistance of balancing valve.

It is required to find the mixing coefficient $k_{sm}(f)$ as a function depending on the converter frequency $f \in (5;\ 50)$. The result of calculating the coefficient of the mixing speed is a numerical solution of the nonlinear system of Kirchhoff's equations (1) on the characteristic points of the equilibrium state

From mixing characteristics, it can be seen (Fig. 2) that the control object has non-linear properties influencing on the transfer coefficient of the control object for different equilibrium points, which can be changed in six times. The presence of the equivalent output hydraulic resistance from the supplier point has a significant e ect on the nonlinear properties of the mixing coefficient. The pressure pump should be chosen so that it would be able to form a pressure difference P_{pch} greater than P_i at least in three times.

To form a complete mathematical model of management we can use the results published in [5], the block diagram of which after inclusion frequency control differs from the traditional only in a fragment, shown in Fig. 1, a.

For the management method synthesis, it is advisable to use an equivalent model of the second order, which can be obtained using the methods presented in [6]. The procedure of obtaining a mathematical model of the second order is virtually not different, except the type of the nonlinear dependence of the mixing coefficient, which has now an inverse monotonic function of depending on the engine speed, so the model has a negative coefficient of the management transfer.

$$\dot{x}_1 = T_0^{-1}(g_{sm}(f_{pch})k_u(x,t)u_{set}(t-\tau) - x_1) \tag{2}$$

where τ – equivalent delay of the elements with a short inertia, T_o – inertia moment. The coefficient k_u has a complex dependence on the parameters of the entire system and is presented in [6]. The transmission coefficient g_{sm} has negative $g_{sm} < 0$. This fact should be taken into account in the implementation of the system management.

When using the PID controller for a given system, there appear some problems associated with the specific application of the frequency converter and the limits on rate of change of differential pressure in the circuits of the plate heat exchanger. In particular, there is a limit to the rate of frequency change f_{pL} and flux vector-clutch changing, which depend on the load on the motor shaft and other conditions [7–9].

In this case the component of the PD regulator suggests the possibility of an instantaneous change of the control voltage u_{set} coming to the inverter input. The actual discrepancy between the actual shaft speed and a predetermined controller output make the restriction of the closed control system speed of the linear system on that should be considered when controller synthesis.

In this case more reliable equivalent mathematical model corresponds to the expression

$$\begin{cases} \dot{x}_1 = (k_{mix}(x_2)k_u(x,t) - x_1)/T_0 \\ \dot{x}_1 = u(t-\tau)/T_{dr} \end{cases} \tag{3}$$

where $u(t)$ – the calculated control and the variable x_2 corresponds to f_{pch} on the condition that the time T_{pr} exceeds the allowable time of rise speed and available for use in the control calculation. In this case the management $u(t)$ is imposed with limitation (-1; 1).

 DOI: 10.1051/matecconf/20179101043

4 Using the high-speed relay controller

From the standpoint of high-speed optimality, it is advisable to use a -control [10], as its dynamic and optimal in performance properties are best suited to such systems [11]. It is possible to take advantage of softening condition of the control signal limiting. In the synthesis of -control the management is considered as three-position relay belonging to the set (1; 0; +1), but in this case it is possible to use a continuous management value at the integrator input bounded with domain [1::: + 1]. To assess the control quality there were chosen two methods of calculating control values implementation which were realized in X-cos environment [4].

In the first case, λ-controller was used in the form given in the primary source [5] for the traditional HWS system in Fig. 4.

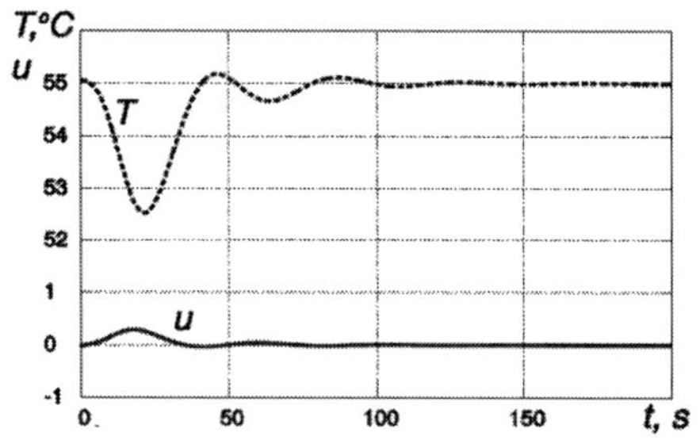

Fig 3. The transition process with the PI controller and frequency control.

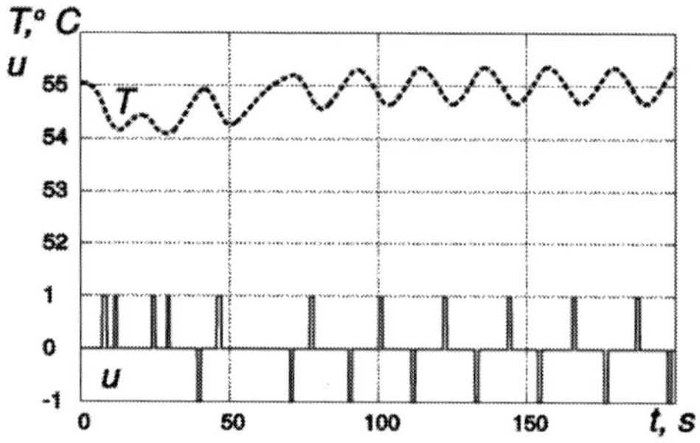

Fig. 4. The transition process with the λ-control.

The simulation results, where as the control element the drive frequency is taken, are given in Fig. 3 which shows that the temperature failure time is almost two times less than the simulation results, where a valve on drive with a constant speed is used.

Using λ-control in a frequency control system is shown in the Fig. 4, where the temperature failure does not exceed one degree. The main disadvantage of this structure is the presence of minor fluctuations in the vicinity of the equilibrium point of the closed system. However, this disadvantage can be eliminated using a controller with a switchable structure, which forms the control in the vicinity of the equilibrium point by means of the linear regulator, built on the basis of the correction link with the transfer function. As a result, the obtained closed systems have the best speed for the given control object.

As an example and basis of further development we can consider PLC [4], built on the microprocessor STM32 of the series F1 [12], which allows the FBD-programming in a graphical environment and includes a control in its library. The microcontroller used in the PLC has a hard-tools [13] key control of the frequency converter for induction motor.

So, in presence of the means of FBD-programs compiling and the development of additional library item, which will perform the configuration of hard-means of frequency control, it is possible to create inexpensive hot water management system based on a microcontroller with significant functionality borrowed from the programming environment [13]. Constructive solutions are available in [7], and the necessary calculations of the power unit of the pump motor control can be taken both from [8] and [9].

5 Conclusion

In this paper we describe a fundamentally new approach to domestic hot water system management using electric drive; the direction of this research has become urgent in connection with a significant decline in prices for frequency converters. In the framework of the tasks there has been solved the problem of synthesis of a mathematical model suitable for research and development of asynchronous motor control methods.

It should be clarified that there are some unsolved scientific and practical problems in the way of further development. The function of transfer ratio of the engine speed is obtained from fairly simplified calculations to assess the range of variation of its value, taking into account mainly the principles
of operation of the system. At the same time the search for approximating functions could allow to consider the model of the object as deterministic, and there is a possibility to develop methods for the control synthesis with adjustable parameters in agreement with the current value of the calculated speed.

Using the principle of management with the switchable control structure between high-speed relay and linear asymptotically stable ones requires a systematic analysis of the stability of such systems. The obtained results of numerical experiments give evidence of such systems operability, therefore the search for solutions in the field of their analysis will provide valuable theoretical results.

References

1. A. Klyuev, A. Kolesnikov, *Optimization of automatic speed control systems* (Energoizdat, Moscow, 1982)
2. J. Shen, H. Chiang, 5th Asian Control Conference, **1**, 472 (2004)
3. Scilab enterprises, open source software for numerical computation, (2015), http://www.scilab. Org
4. 3s-smart software solutions, the complete software suite for automation technology (2015), http://www.codesys.com
5. A. Shilin, V. Bukreev, Communications in Computer and Information Science, **487**, 387 (2014)
6. A. Shilin, V. Bukreev, Thermal Engineering, **61**, 741 (2014)
7. K. Buczek, J. Bartman, A. Koziorowska, Przeglad Elektrotechniczny, **83**, 9 (2007)
8. A. Koziorowska, J. Bartman, International Journal of Numerical Modelling: Electronic Net-works, Devices and Fields, **25**, 103 (2012)
9. M. Ioannides, IEEE Transactions on Energy Conversion, **19**, 469 (2004)
10. A. Shilin, V. Bukreev, Devices and systems. Management, monitoring, diagnostics, Moscow, **10**, 45 (2014)
11. L. Rozonoer, Autom. Remote Control, **64**, 1237 (2003)
12. Stmicroelectronics, reference manual, stm32f103xx arm-based 32-bit mcus (2016), www.st.com/ resource/en/reference_manual/cd00171190.pdf
13. Programmable logic controller west-02 (2015), http://npowest.ru/index.php/produktsiya/ programmiruemye-logicheskie-kontrollery

SHUT-OFF VALVES OF ELECTRIC DRIVE WITH THYRISTOR VOLTAGE REGULATORS AND SOFTWARE IMPLEMENTATION OF THE INDIRECT DETERMINATION OF MOMENT

Anton I. Vlasov[1], *Vladislav.* Yu. Dementyev[2], *Alexandr.* S. Karakylov[1,*], and *Leonid.* S. Udut[1]

[1] Tomsk Polytechnic University, 634050, Tomsk, Russia
[2] Specialist of 1st class of production unit of "Gazprom Transgaz Tomsk", Tomsk, Russia

Abstract. In this paper we consider the shut-off valve of electric drive with thyristor voltage regulator and microprocessor control which ensures a constant indirect control of torque value on the electric drive as well as its limitation in the case of the creation of an emergency on the main oil pipeline. It was found that the use of the integrated microprocessor control systems with software implementation calculating of given torque in the system of induction motor - thyristor voltage regulator, allows you to get the best performance of electric drive shut-off valve with the restrictions on the maximum current, torque whith deviations of voltage parameters and improve the safety of the process of oil pumping.

1 Introduction

The development of the fuel and energy complex of our country is largely dependents on the operation of pipeline transport, in particular the main oil pipelines. Main oil pipeline is the cheapest means of delivery of oil and oil products from their place of production to refineries. For effective work of the oil pipeline and constant control of the of the oil transportation used shut-off valves, the main purpose of which – to block fluid flow in a pipe by locking valve body, and again let the medium depending on the requirements of the process, according to an accepted conduit. The drive of shut-off valves was designed to move the locking bolt body in end positions, as well as the establishment, where necessary, efforts to achieve the desired tightness of the seal.

At the moment, as the motorized shut-off valves apply electric drive with a frequency converter and a thyristor voltage regulator. The main advantages of the inverter are the high control accuracy, large range. Application of the frequency converter is not limited, so at the modernization and introduction of new technologies they are used everywhere [1]. However, the mechanisms that do not require a large adjustment range, but need a soft start and accurate stop, possible to use a thyristor voltage regulators of trimmer. The main advantage of thyristor voltage compared with a frequency converter is its low cost and high reliability [2, 4].

* Corresponding author: karakulov@tpu.ru

Ensuring the safety of technological process of oil transporting is only possible with the high reliability of the drive The failures occur primarily in the power control unit coz of external factors (low supply voltage, imbalance voltage of one phase, and etc.). Therefore, the creation of the electric shut-off valves to operate at a reduced supply voltage, as well as asymmetrical modes of electric grid operation, is today an urgent problem [3, 4].

The purpose of this article is to develop electric drive with thyristor voltage regulator and microprocessor control provides constant control the value of output torque of the drive as well as its limitation in the case of the creation of an emergency on the main oil pipeline, for example, the electric shut-off valve at lower supply voltage, as well as the unbalance in one phase of the mains supply.

2 Study technique

To carry out such a task often used computational methods based primarily on modeling modes of operation of the induction motor (IM). However, they are computing algorithm repeated with high frequency in real time and require significant resources of the microprocessor system. In addition, the accuracy of these methods depends on the accuracy of determining IM parameters. The use of an alternative method for the indirect determination of output torque IM [4, 5], implemented on the basis of microprocessor control unit with thyristor voltage regulator, allows the use of less powerful microprocessors, making the system electric shut-off valve cheaper, and therefore more competitive on the drive technology market shut-off valves. As is well known [6, 7], in static mods, torque on the shaft of induction motor can be represented as a function of the current values of the stator current (with electric drive movement) or as a function of the steering angle of the thyristors voltage. And for each of the stator voltages, the data will look different curves, so we are dealing with the group characteristics. Therefore, if in three dimensions on the axes postpone current, voltage and torque we get a surface that describes the time induction motor as a function of two variables: current and voltage (Fig. 1, A). Consequently, the same three-dimensional surface we can get if we represent time as a function of voltage and controls the angle of the thyristors by thyristor voltage regulator (Fig. 1 b)

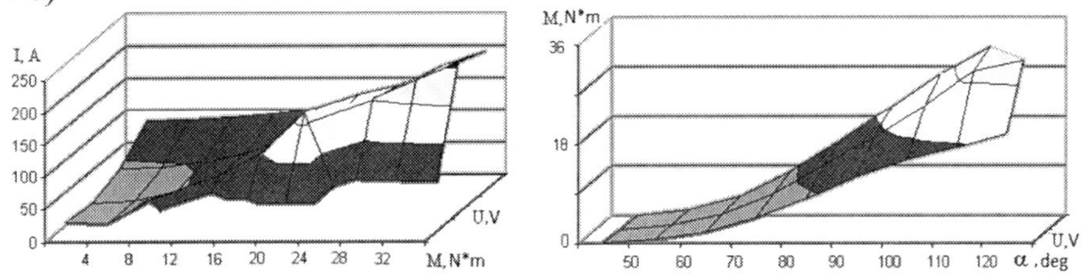

Fig. 1. Dependence of the torque developed by induction motor:a) the stator current and suply voltage b) opening angle of thyristors and the supply voltage.

The asymmetry (imbalance) electric phase grid causes fluctuations of currents and voltages, which affects on the output signal extracted from the current and voltage sensors, which will be complex. In this case it is necessary to interpolate information from current and voltage sensors and with the help of three-dimensional surfaces to determine torque setpoint, providing the maximum possible speed of the shut-off valve body.

Fig. 2 shows a graphical interpretation of the algorithm for torque calculating, which used in the microprocessor control unit - thyristor voltage regulator. Here: the characteristic described by the equation M_2=kI_2+b matches to the values of the torque at nominal supply voltage, by the equation M_1=kI_1+b - the values of the torque when an imbalance in

 DOI: 10.1051/matecconf/20179101044

one of the phases of the grid and the equation M=kI+b - the values at the torque of any supply condition.

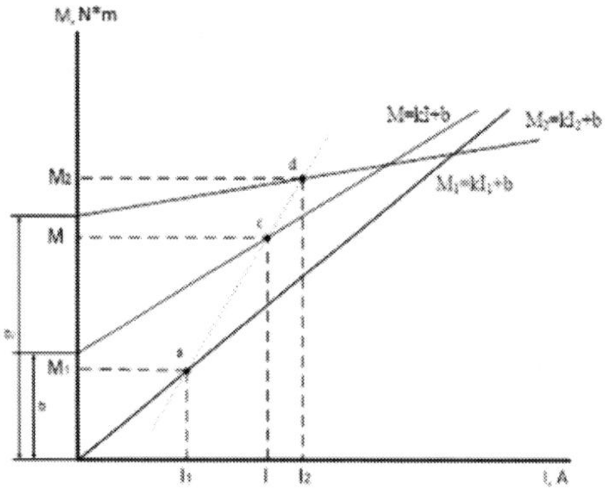

Fig. 2. Graphical interpretation of calculation of the set torque by the microprocessor control unit.

If, at a certain time, it is necessary to have information about the current value (point "c" in Fig. 2) matching to a given torque on the linear characteristic M=kI+b through the predetermined point of a straight line intersecting characteristics $M_2=kI_2+b$ and $M_1=kI_1+b$ at points "a" and "d", which are known. Using the method of interpolating rectangular, one can determine the coordinate point "a" (in this case, the required current value I) according to following equation $I=M_1+\frac{M_2-M_1}{I_2-I_1}\cdot(M-M_1)$.

The parameter b characteristics of the torque is the ratio of the imbalance voltage of the nominal voltage and matches necessary coordinate of torque value on the y-axis of the torque characteristic.

3 Conclusion

The use of the integrated microprocessor control systems with software implementation calculating with a given torque in the system of induction motor – thyristor voltage regulator allows you to get the best performance shut-off valve electric drive with with the restrictions on the Overcurrent, the torque when deviations of parameters voltage and improve the safety of the process of oil pumping.

References

1. Y.N. Dementyev, K.N. Negodin, N.V. Koyain, L.S. Udut, Proceedings of the Higher Educational Institutions. Electromechanics, **4**, 49 (2016)
2. V.A. Shybenko, N.Y. Braslavsky, *Thyristor asynchronous electric drive with phase control* (Energy, St. Petersburg, 1972) [in Russian]
3. A.S. Karakulov, the fourth Scientific Conference Modern Equipment and the Automation System, 63 (2004)
4. A.S. Karakulov, V.V. Krasilyev, S.V. Langraf, A.Y. Marjanov, the V Interdisciplinary Scientific-Technical Conference Automation and Advanced Technologies, (2005)
5. Y.N. Dementyev, A.D. Umurzakova, MATEC Web of Conf.,**19**, 01027 (2014)
6. Y.N. Dementiev, A.Y. Chernyshev, I.A. Chernyshev, *Electric drive* (Yurayt, Moscow, 2016) [in Russian]
7. R.F. Bekishev, Y.N. Dementiev, *Power* (Yurayt, Moscow, 2016) [in Russian]

ENERGY SAVING CONTROL OF SERIES EXCITATION MOTOR

Yuri N. Dementyev[1,*], *Kirill* N. Negodin[1], *Min Tu* Aung[2], and *Viktor* I. Susdorf[2]

[1]National Research Tomsk Polytechnic University, 634050, Tomsk, Russia
[2] State Technical University, Komsomolsk-na-Amure, Russia

Abstract. Losses optimization in transients with series excitation motor (SEM) control just by control, at that controlled value is input voltage of SEM, is considered in this paper. Algorithm of optimal control of SEM, that allows to define SEM control optimality in relation of production by method of dynamic programming with eliminations of current and voltage, is suggested. It is defined, that optimal control in SEM transients in relation of production is same as optimal control with energy losses minimum

1 Introduction

The problem of controlling series-wound motor in terms of optimization of its energy still remains relevance for researchers to these days. Although the widespread use of series-wound motor in the traction electric drive, electrical hand tools and household appliances, theoretical investigations are extremely rare. We can notice the works of Russian scientists devoted to the theoretical and practical improvement of energy performance of electric drives with single phase collector series excitation motor: V.P. Klimov, N.F. Elias, V.G. Kagan, G.V. Lebedeva, L.I. Malinina, Y.P. Petrov, M.H. Bellman and also T. Fujimaki, G.K. Dubey and others [1,2,3,4,5], who correctly point out that the classical variation methods for finding the optimal control of series-wound motor does not provide direct solutions. One approach for finding the optimal control of collector series excitation motor is described in [2, 4]. Specifically, the following divergent objectives for controlling drives were considered:

a) maximum productivity by limiting the electric heat losses in the motor and specified motion time;
b) minimum loss at specified productivity and time;
c) maximum performance with constraints on the phase coordinates movement and heat loss;

It is known that the principle of reciprocity isoperimetric problem, they have a common solution.

2 Study technique

* Corresponding authors: dementev@tpu.ru

Modern technical equipment allows providing regulation of the input electrical parameters of the electric motor in such a way as to provide a current waveform and its spectral composition to minimize the energy loss associated with the resolution of control. Based on this, consider the control of series-wound motor with "smooth" power supply. Losses from hysteretic currents, mechanical losses and other kinds of losses, the struggle of which is the improvement of technical equipments, in our opinion, quite well – we leave out the present article and ask the optimization of losses in transient states only by control. The control parameter is obviously the input voltage of the motor.

Classically the optimal control problem is solved for speed control drive in the increment of speed, in the formulation of minimum losses: it is required to find a function *i(t)*, providing a minimum loss of Q.

When using the Pontryagin maximum principle, in general case the solution of the problem can be obtained in the form [3, 6, 7]:

$$i(t) = \begin{cases} +i_{max} & where \quad o \le t \le t_1 \\ \sqrt{M_c} & where \quad t_1 \le T \end{cases} \tag{1}$$

The parameters of transient process depend on the limits of dynamic properties of drive and mainly depend on the restrictions imposed on it (especially the maximum permissible value of the current, the voltage and current slew rate), which determine the permissible rate of acceleration – deceleration.Since the electromagnetic torque and the heat losses are proportional to the square of the motor current, then in equation 1 logically substitute the maximum current value. This corresponds to the optimality criterion, and allows proceeding to consideration of the physics of energy processes, where the energy loss due to heating depend on the time: $Q = \Delta P.t = i^2.t$.

Due to the fact that the rectangular dynamic torque cannot be ensured in practice, we find the optimal control of series-wound motor from the point of view of performance using dynamic programming with constraints on voltage and current. Algorithm for solving the problem is shown in Figure 1 (symbols are common).

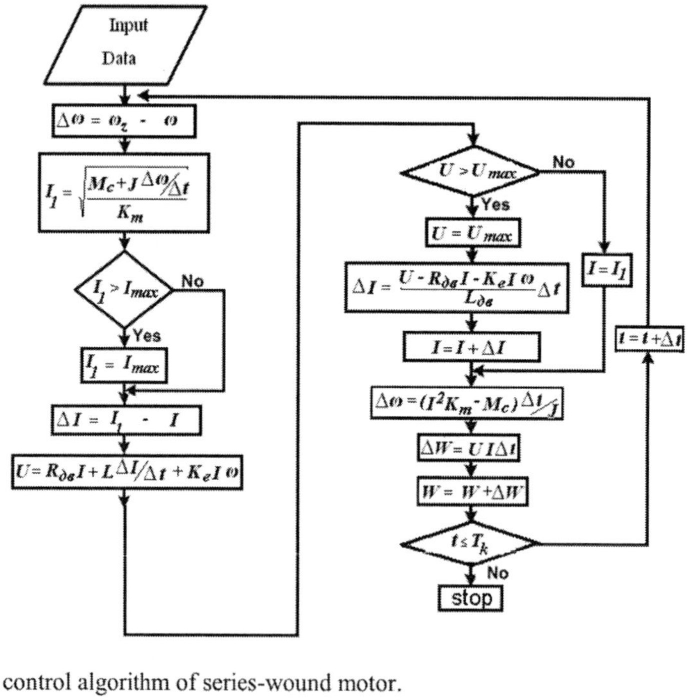

Fig. 1. Optimal control algorithm of series-wound motor.

DOI: 10.1051/matecconf/20179101045

3 Experimental results

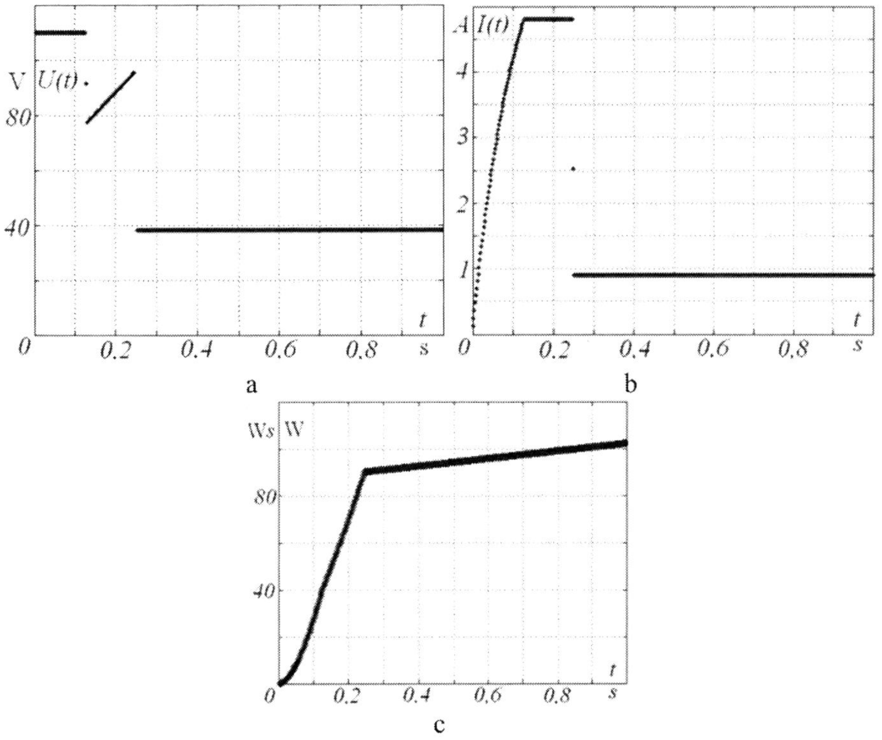

Fig. 2. The result of the calculation is shown in Figure 2 (a, b, c).

For comparison, figures 3. a and b shows the same characteristics of the same motor for direct start in the absence of restrictions.

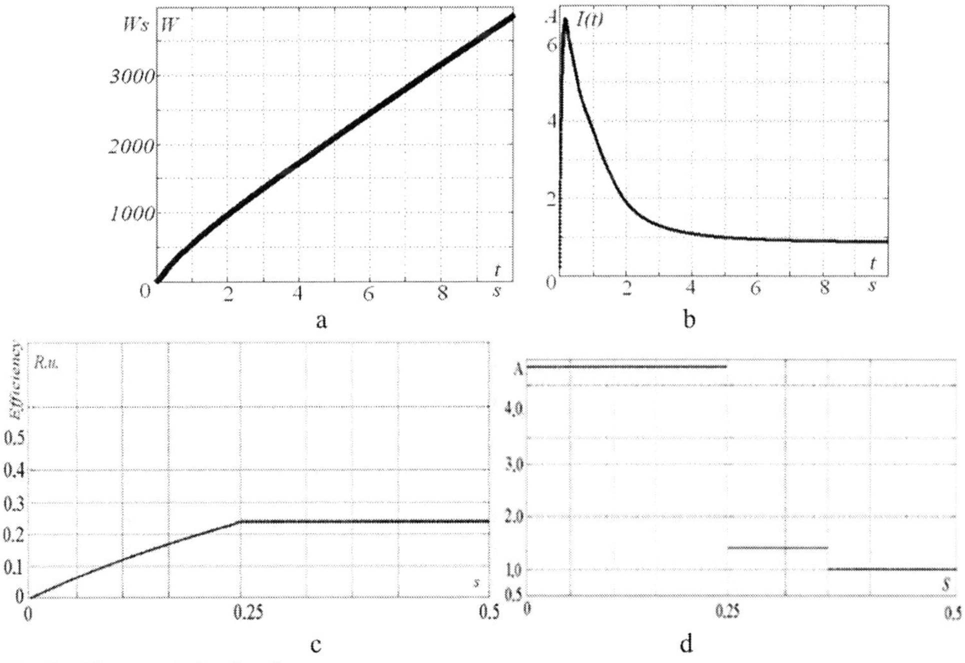

Fig. 3. Characteristics for direct start.

Comparative analysis of the characteristics shown in figure 3 and 4 shows that the time of transition (9 s vs 0.25 s) as consumed during the transition process of electric energy (3250 W vs 90 W) differs in 36 times.

Figure 3 c and d shows the current and efficiency curve of electric drive, which reduces the load on the shaft by 50 % for 0.3 seconds in the transient process. From the characteristics it shows that the maximum efficiency is achieved when the output speed is stable, so the duration of the transient process negatively impact on the energy sector. In this plot with a maximum current value corresponds to the maximum rate of speed change and hence minimizes the operating time with a relatively low efficiency.

The above illustration is consistent with the known results of research [2, 3, 6, 8, 9], and allows concluding that for the series-wound motor in the transient optimum control in terms of performance is the optimal control of minimum energy losses.

4 Conclusion

According to the experimental results, the following conclusions can be made:

- Created a new algorithm of synthesis of optimal control with constraints on the coordinates of movement.
- Set numeric energy efficiency optimal control of engine serial excitation

References

1. T. S. Bukanova, L. A. Steshina, A. B. Savinih, Mehatronika, avtomatizaciya, upravlenie, **10**, 48 (2009)
2. A. S. Meshkov, Nauchno-issledovatelskih rabot finalistov konkursa aspirantov I molodih uchenih v oblasti energosberezheniya v promishlennosti, g. Novocherkassk, 49 (2010)
3. A. E. Alehin, Yu. N. Dementyev, B. I. Susdorf, DC Electric Drive with series excitation motor. License RUS 1676418 (1990)
4. A. S. Meshkov, V. I. Susdorf, Y. N. Dementyev , D. A. Kaftasyev, MATEC Web of Conferences, **19**, (2014)
5. B. V. Suhinin, V. V. Surkov, S. A. Ciruh, E. I. Feofilov, *Optimal in accuracy (speedwork, energyefficiency) control of electromechanical objects*, (Learning edition, Tula, 2014) [in Russian]
6. Yu. A. Shurigin, Elektronnie sredstva I sistemi upravleniya, 8 (2003) [in Russian]
7. M. V. Pronin, *Electric drives and systems with electric machines and semiconductor converters*, (Spb. 2004) [in Russian]
8. A. G. Ivanov, G. A. Belov, A. G. Sergeev, *Control system of semiconductor converters*, (2010) [in Russian]
9. A.E. Alehin, Yu. N. Dementyev, B.I. Susdorf, 6th Conferenc II, Budapest, 564 (1990)

A NEW TORQUE CONTROL SYSTEM OF PERMANENT MAGNET SYNCHRONOUS MOTOR

Andrey Evstratov [1,*], *Valery* Zavyalov[1], *Alexander* Grigoryev[2],and *Irina* Semykina[1,2]

[1] Kuzbass State Technical University, 650000 Kemerovo, Russia
[2] Tomsk Polytechnic University, 634050 Tomsk, Russia

Abstract. The article describes a new approach to control of permanent magnet synchronous motor drive based on the analysis of the electromechanical transformation. The proposed control system provides quick response and low ripple of the motor torque and flux. To synthesis this control system, the authors put the electromagnetic torque and the modulus of stator flux vector as controlled values and use the Lyapunov's second method. In addition, the stator voltage constriction and ability of low-pass filtration are taken into account. The investigation of the proposed control system has carried out with the simulation and the experimental research which have confirmed that the proposed control system correspond to all above-mentioned control tasks and the permanent magnet synchronous motor controlled under this system may be recommended to use in robotics.

1 Introduction

A permanent magnet synchronous motor (PMSM) is a type of electric motor having a low mass and inertia and also small time constants that nowadays become widely used as a input force source, for example in robotics.

There are many researches dealing with PMSM control system [1-17] that may be classified by optimization criteria: the minimum time [4, 5, 7, 8, 10-12, 17], the minimum loss [2, 16], the minimum torque ripple [1, 3, 6, 9, 17] and the minimum flux ripple [11, 17]. The new torque control system of PMSM with quick response and low ripple, which match well with robotics tasks, is given below.

2 The control approach

To obtain the controlling of PMSM dynamic state and the energy optimization of motor, the control system regulates the electromagnetic torque T and the modulus of stator flux vector ψ_s:

* Corresponding author: eaegemf@mail.ru

$$\mathbf{Y}^* - \mathbf{Y} \to 0,\ \mathbf{Y} = \begin{bmatrix} y_1 \\ y_2 \end{bmatrix} = \begin{bmatrix} T \\ \psi_s \end{bmatrix}.$$

13.5 liters), was manufactured from optical glass of thickness 5 mm. A solution of castor oil in ethyl alcohol in different percent ratios and glycerin were used as liquids; soft soap was used as surfactant.

$$V = [\mathbf{Y}^* - \mathbf{Y}]^T \mathbf{1} [\mathbf{Y}^* - \mathbf{Y}] = \sum_{i=1}^{2} (y_i^* - y_i)^2 . \tag{1}$$

By virtue of the fact that V is the positive definite function, the negative value of its derivative dV/dt certifies $V \to 0$ when $t \to \infty$ meaning $\mathbf{Y} \to \mathbf{Y}^*$. The model of PMSM with nonsalient pole in the d,q reference frame is defined the follow equation [13]:

$$\begin{gathered} u_{sd} = \frac{d\psi_{sd}}{dt} + i_{sd}R + p\omega\psi_{sq};\ u_{sq} = \frac{d\psi_{sq}}{dt} + i_{sq}R - p\omega\psi_{sd}; \\ \psi_{sd} = Li_{sd} + \lambda;\ \psi_{sq} = Li_{sq};\ M = (3/2)p\lambda i_{sq};\ J\frac{d\omega}{dt} = T - T_l, \end{gathered} \tag{2}$$

where λ is the rotor flux; u_{sd}, u_{sq}, i_{sd}, i_{sq}, ψ_{sd}, ψ_{sq} are the vector components of the stator voltage the stator current and the stator flux correspondingly; L is the stator inductance; R is the stator resistance; p is the number of pole pair; T_l is the load torque; J is the rotor inertia; ω is the rotor speed. So, the conditions of dV/dt negativity is obtained by means expression of dV/dt from (1) using (2):

$$\frac{(T^* - T)}{T_n}\left(\frac{3}{2}pe_{sq}\frac{\lambda}{L}\right) + \frac{(\psi_s^* - \psi_s)}{\psi_{sn}}\left(e_{sd}\frac{Li_{sd} + \lambda}{\psi_s} + e_{sq}\frac{Li_{sq}}{\psi_s}\right) > 0, \tag{3}$$

where e_{sd}, e_{sq} are the vector components of the stator EMF calculated as $e_{sd} = d\psi_{sd}/dt$, $e_{sq} = d\psi_{sq}/dt$; T_n, ψ_{sn} are the nominal values of the electromagnetic torque and the modulus of stator flux vector.

On the assumption with the stator voltage is constraint, the applying of the positive maximum voltage $u_{sd\max}$ and $u_{sq\max}$ force the stator EMF e_{sd} and e_{sq} to be positive, and the opposite is true. Given that point the analysis of (3) leads to the conclusion to satisfy the inequality the PMSM control system should form the stator voltage as:

$$\begin{gathered} u_{sq} = \begin{cases} u_{sq\max},\ if\ \dfrac{(T^* - T)}{T_n}\dfrac{3}{2}p\dfrac{\lambda}{L} + \dfrac{(\psi_s^* - \psi_s)}{\psi_{sn}}\dfrac{Li_{sq}}{\psi_s} > 0; \\ -u_{sq\max},\ if\ \dfrac{(T^* - T)}{T_n}\dfrac{3}{2}p\dfrac{\lambda}{L} + \dfrac{(\psi_s^* - \psi_s)}{\psi_{sn}}\dfrac{Li_{sq}}{\psi_s} < 0; \end{cases} \\ u_{sd} = \begin{cases} u_{sd\max},\ if\ \dfrac{(\psi_s^* - \psi_s)}{\psi_{sn}}\dfrac{Li_{sd} + \lambda}{\psi_s} > 0; \\ -u_{sd\max},\ if\ \dfrac{(\psi_s^* - \psi_s)}{\psi_{sn}}\dfrac{Li_{sd} + \lambda}{\psi_s} < 0. \end{cases} \end{gathered} \tag{4}$$

3 The control system investigation

To investigate the dynamic performance of proposed control system the simulation and experimental research was carried out. The object of research was PMSM with capacity of

DOI: 10.1051/matecconf/20179101046

2.2 kW and the next parameters: L=0.022 H, J=0.055 kg·m^2, λ=0.393 Wb, R=2.7 Ω. This PMSM was supplied by pulse width modulated voltage source transistor converter with the switching frequency of 5 kHz.

The simulation result of the control system with (4) showed the mean values of T and ψ_s was controlled with the required accuracy, however there was high-level ripples in the amount 32.68 % from T_n and 5.5 % from ψ_{sn}. Such ripples is caused of the voltage converter use only $u_{sd\max}$ и $u_{sq\max}$ and not to apply intermediate values of the stator voltage.

To decrease the torque ripples the low-pass filter was put in the control system. Fig. 1 shows its final structure, and fig. 2 displays the results of its simulation that testify the ripple reduction to 16.39 % for electromagnetic torque and 1.39 % for the modulus of stator flux vector. In addition, it have to be noted the low-pass filter was not affect on the transient time, which is 0.005 s for the PMSM with non-exited stator in the initial state.

The experimental research was carried out to compare results with simulation.

The experimental installation consist of the PMSM with MBS-FC01 frequency converter, TRB-5K torque sensor and the loading direct current motor. MBS-FC01 frequency converter is the coverless pulse width modulated voltage source transistor converter with open software supported platform based on TMS320F28335 microcontroller by Texas Instruments and the intelligent power unit. The converter's software powered by MexBIOS real-time operating system and MexBIOS Development Studio graphical programming environment.

To examine the behaviors obtained via simulation PMSM was tested the next way: loaded start and start under no-load conditions, step injection of load torque after starting and reversion of motion (fig. 3).

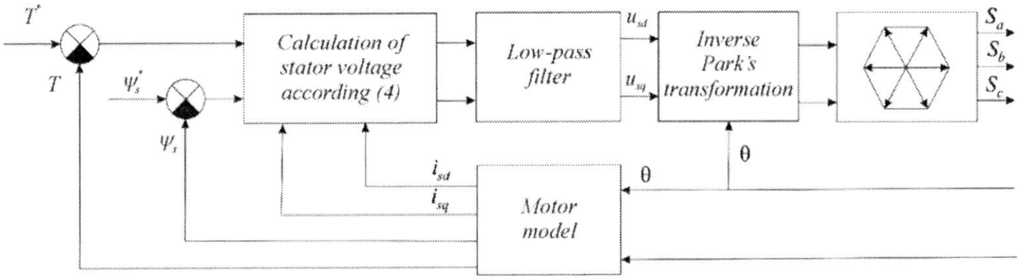

Fig. 1. The structure of proposed control system.

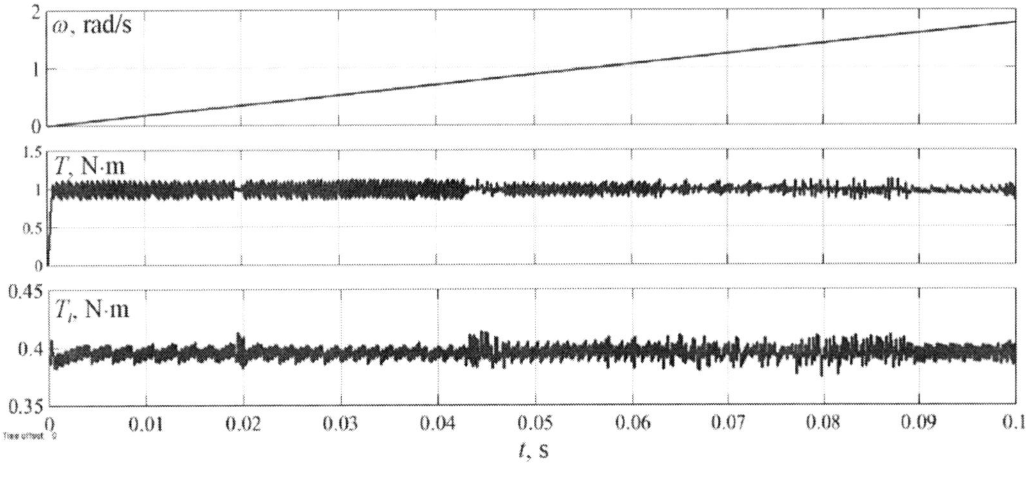

Fig. 2. The simulation results of proposed control system.

 DOI: 10.1051/matecconf/20179101046

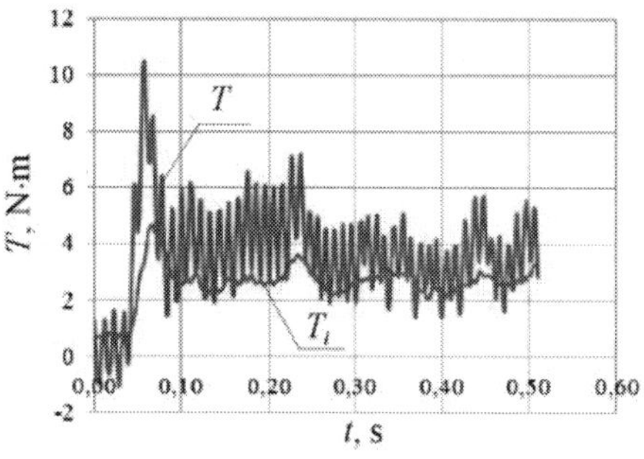

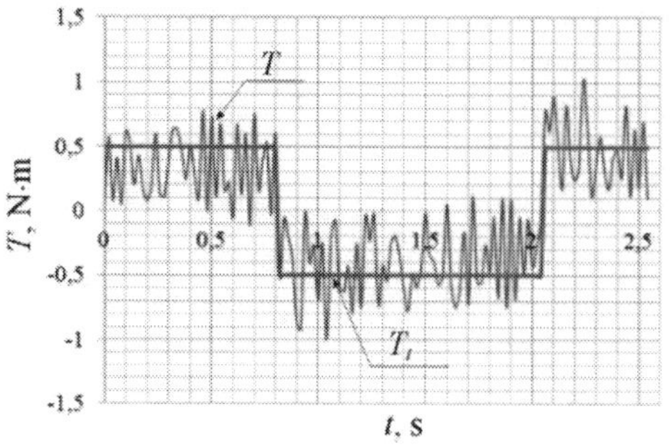

Fig. 3. The results of experimental research: a – loaded start; b – reverse.

4 Conclusion

The results depicted on fig. 3 have high-level ripples, however this ripples caused not the control system but the noises from analog-digital conversion of sensors' data explained by crosslinks. The carried research confirmed the proposed control system correspond to all above-mentioned tasks, and PMSM controlled under it may be recommended to use in robotics.

References

1. T. D. Batzel, K. Y. Lee, IEEE Trans. Energy Convers. **13**, 257 (1998)
2. C. Bizot , J. Brottes , M. Lungeanu , B. Poulsen , D. Séra , M. Sørensen, Sensorless Control for PMSM. Power Electronics and Drives, Institute of Energy Technology, Aalborg University, Denmark, (2003)
3. S. Bouchiker, G. A. Capolino, M. Poloujadoff, IEEE Transactions Power Electronics **13**, 1089 (1998)
4. H. Chaoui, P. Sicard, IEEE Trans. Ind. Electron. **59**,1123 (2012)
5. K. Chikh, A. Saad, M. Khafallah, D. Yousfi, Canadian Journal on Electrical and Electronics Engineering **2** , 400 (2011)
6. M. L. Corradini, G. Ippoliti, S. Longhi , G. Orlando,Sliding Mode Approach for Robust **59** , 1096 (2012)
7. Gabriel Noriega, Miguel Strefezza, Wseas Transactions. Electronics **11**, 2452 (2007)
8. Jian Wang, Honghua Wang, Tianhang Lu, Dehong Teng, ICEMS, 110 (2011)
9. Kyeong-Hwa Kim, In-Cheol Baik, Gun-Woo Moon, Myung-Joong Youn, IEEE Trans, on Control System technology **7**, 630 (1997)
10. M. Kadjoudj, S. Taibi, N. Golea, International Journal of Sciences and Techniques of Automatic control &computer engineering **1**, 167 (2007)
11. Y. A. Mohamed, IEEE Transactions Power Electronics **22**, 2042 (2007)
12. Y .A. Mohamed, IEEE Transactions Industrial Electronics **54**, (1981)
13. P. Pillay, R. Krishnan, IEEE Transactions Industry Applications **25**, 265 (1989)
14. T. Takeshita, N. Matsui, EMF Constant Identifier. IECON **1**, 8 (1994)
15. C Edwards, S Spurgeon. Sliding Mode Control: *Theory And Application. CRC Press Reference*, (Taylor & Francis, London, 1998)

ESTIMATING ACCURACY OF EXPERIMENTALLY OBTAINED POWER SYSTEM ZIP LOAD MODELS

Aleksey V. Pankratov[1*], *Igor* E. Kopylov[1], *Alexander* S. Tavlintsev[2], and *Mikhail* A. Kondrashov[1], and *Li* Wang[3]

[1] Tomsk Polytechnic University, 634050 Tomsk, Russia
[2] Ural Federal University, 620002 Ekaterinburg, Russia
[3] National Cheng Kung University, Tainan City 701, Taiwan

Abstract. Experimentally obtained ZIP load models are used for the power system analysis. Estimating accuracy of such models is presented in this paper. The dependence of ZIP load model error on the range in which voltage changes during the experiment, the number of measured values of voltage and power, standard deviation of the load change random process have been obtained using the computing experiment. The results could be used to estimate the accuracy of ZIP load models or to design the experiment to determine ZIP load models with prescribed accuracy.

1 Introduction

Load models are necessary for power system study. The impact of the load model on voltage stability is demonstrated in [1, 2]. Different load model types and international industry practice on load modeling are described in [3]. Generally all existing load models can be classified in two groups: static load models [1, 4] and dynamic load models [5, 6]. Static load model in the form of parabolic curve contains three components: constant resistance, constant current and constant power. This model was called as ZIP model. ZIP load model is widely used [3].

Load rate and structure vary over time [4,6]. It leads to the necessity of periodic model parameters reassessment. For large load centers the static load models are determined by means of special experimental researches. Despite the large number of studies, devoted to the measurement based load model identification [1, 4, 6, 7], the problem of obtained models accuracy remains actual till the present day.

The research objective is to calculate the accuracy of ZIP load models obtained experimentally depending on the parameters of the experiment. The parameters of the experiment influencing the accuracy of ZIP load models are the following: the range in which voltage changes during the experiment (Δu %); the number of measured values of voltage and power (N); a standard deviation of the load change random process (σ_p %).

* Corresponding author: pank@tpu.ru

2 Research methods

The difficulty of estimating the accuracy of the ZIP load models lies in the fact that the standard which the result obtained in the experiment could be compared with is lacking. So the authors have decided to use a computing experiment instead of a full-scale one.

The initial reference static load model was defined in relative units by the 2nd degree polynomial (1) with known coefficients a'_0, a'_1, a'_2:

$$P_{ORIG}(U) = a'_0 + a'_1 \cdot U + a'_2 \cdot U^2. \tag{1}$$

Arrays of voltage values $U_{(i)}$ and real power $P_{(i)}$, where i=1,2,...,N is the index of reading, were determined in the computer experiment. The values $U_{(i)}$ have been obtained in relative units on the assumption that voltage increases regularly during the experiment within the limits $\pm\Delta u_{\%}/2$ from the rated voltage according to (2):

$$U_{(i)} = 1 - \frac{\Delta u_{\%}}{2 \cdot 100\%} + i \cdot \frac{\Delta u_{\%}}{N \cdot 100\%}. \tag{2}$$

The influence of the random error of measurements and random fluctuations of the load power on the accuracy of the obtained static load models may be generalized by taking into account one random normally distributed value with a standard deviation $\sigma_{p\%}$:

$$P_{(i)} = (a'_0 + a'_1 \cdot U_{(i)} + a'_2 \cdot U_{(i)}^2) \cdot error(\sigma_{p\%}), \tag{3}$$

where $error(\sigma_{p\%})$ is the random normally distributed value with the mean value of 1.

Approximation initial data $U_{(i)}$ and $P_{(i)}$ by 2nd degree polynomial using ordinary least squares method gives the experimental static load model with new coefficients a_0, a_1, a_2:

$$P_{TEST}(U) = a_0 + a_1 \cdot U + a_2 \cdot U^2. \tag{4}$$

The results of one experiment are shown in fig. 1.

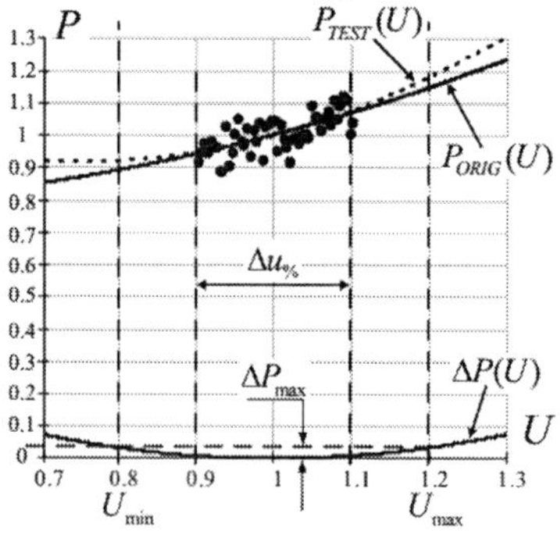

Fig. 1. Comparison of reference and experimental static load models.

The voltage was varied within the $\Delta u_{\%}$ range during the experiment. A comparison of the reference static load model $P_{ORIG}(U)$ and the experimental static load model $P_{TEST}(U)$ is performed within the predetermined range $(U_{min} \div U_{max})$ that was chosen on the basis of the voltage variation deviation under normal and post-emergency conditions (0.8/1.2 from the rated voltage). The maximum deviation of the models $\Delta P(U)=|P_{ORIG}(U)-P_{TEST}(U)|$ within the predetermined range $(U_{min} \div U_{max})$ of ΔP_{max} was used to obtain the numerical value of the error of the static load models.

The random values $P_{(i)}$ were used as the initial data so the obtained result ΔP_{max} is also random with the same values $\Delta u_{\%}$, N, and $\sigma_{p\ \%}$. The results of the set of experiments are presented in fig. 2.

The values ΔP_{max} obtained in the set of S=1000 computing experiment are shown by dots in fig. 2a. The probability-distribution function of the values ΔP_{max} is shown in fig. 2b. The value of the quantile Δ_{max} was taken as the final value of the static load model's error. This value will not exceed the probability of 0.95.

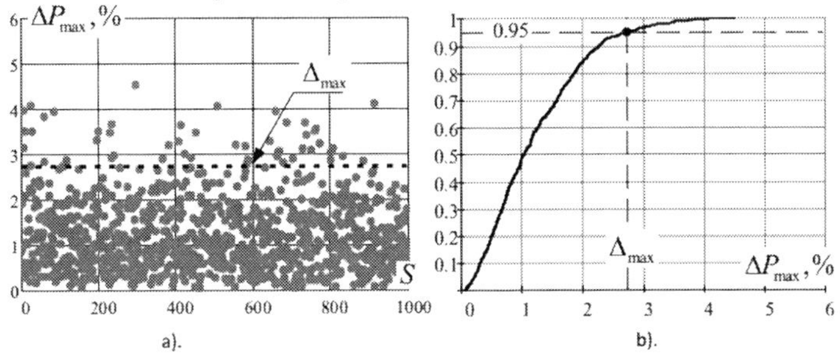

Fig. 2. a) set of computing experiments; b) probability-distribution function.

3 Obtained results

The described sets of experiments were carried out for the different values $\Delta u_{\%}$, N and $\sigma_{p\ \%}$. The dependences of the static load model's error Δ_{max} on the number of measurements N for various values $\Delta u_{\%}$, with $\sigma_{p\%}$=1 % are shown in fig. 3.

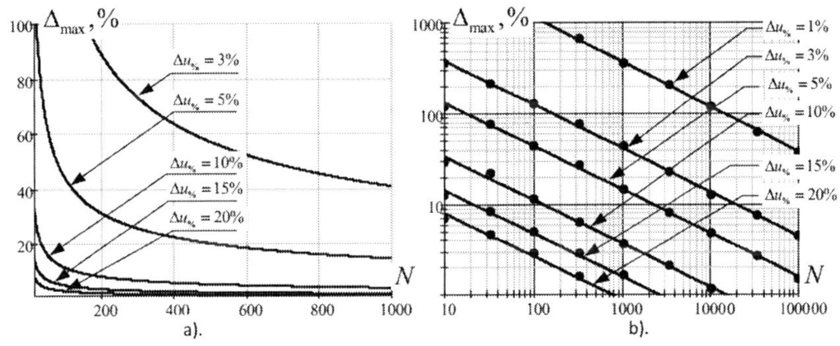

Fig. 3. Values of the static load model's error with σ_p%=1%: a)linear scales; b) logarithmic scales.

These dependences are linear on the logarithmic scales (fig. 3b), so they can be approximated by the linear functions. The same linear functions can be obtained for $\Delta_{max}(\Delta u_{\%})$ and $\Delta_{max}(\sigma_{p\ \%})$ dependences. Finally all results are generalized by equation

$$\Delta_{\max} = 118 \cdot \frac{\sigma_p\%}{(\Delta u\%)^2 \cdot \sqrt{N}} \cdot 100\%. \tag{5}$$

Thus, given the known values of the range of the voltage variation $\Delta u_{\%}$, the number of measured values N and the standard deviation of load power $\sigma_{p\ \%}$, the resulting error of the static load model will not exceed the value Δ_{max} calculated from (5), with the probability of 0.95. For example, with $\sigma_{p\ \%}$=1 % and $\Delta u_{\%}$=20 %, and N=10^3, the error of the static load models will not exceed the value Δ_{max}=0.93%, calculated from (5), with the probability of 0.95. If $\sigma_{p\,\%}$=10%, $\Delta u_{\,\%}$=5 %, N=10^3, then Δ_{max}=149 %.

The inverse problem (design of the experiment for determining the static load models with the prescribed accuracy) could be solved using the obtained results. For example, if the accuracy of the static load model does not exceed 10% and the tolerance range of the voltage variation is $\Delta u_{\,\%}$=15 %, then with $\sigma_{p\,\%}$=10 %, solving (5) for N, we will obtain

$$N = \left(\frac{118 \cdot \sigma_{p\%} \cdot 100\%}{(\Delta u_{\%})^2 \cdot \Delta_{\max}} \right)^2 = \left(\frac{118 \cdot 10 \cdot 100}{15^2 \cdot 10} \right)^2 = 2750. \tag{6}$$

Therefore, under these conditions, for the measurements to obtain the required accuracy it is necessary that N must be equal to no less than 2750. If $\sigma_{p\,\%}$=15 %, it is necessary $N \geq 6188$ measurements to obtain the same accuracy.

4 Conclusion

The described researches have shown that the parameters of the experiment influence significantly the accuracy of the obtained ZIP load models. The error can vary from part to thousand of percent, depending on the range of the voltage variation $\Delta u_{\,\%}$, the number of measured values N and the standard deviation of load power $\sigma_{p\,\%}$.

The detected dependence of the static load model's error on the parameters of the experiment can be applied to solve two main problems:

- estimating the experimentally obtained static load models accuracy;
- planning the experiment for determining static load models with the prescribed accuracy.

The work was supported by Act 211 Government of the Russian Federation, contract № 02.A03.21.0006 and the Ministry of Education and Science of the Russian Federation (in the framework of state assignment, №13.1928.2014/K (project №1928)).

References

1. C.A. Baone, S. Veda, Y. Pan, W. Premerlani, J. Dai, A. Johnson, IEEE Power and Energy Society General Meeting, **2015**, 7285681 (2015)
2. J.N.S. Mrudveeka, S.V.N.L. Lalitha, M. Ramamoorty, Indian J. Sci. Technol., **9** (23), 92514 (2016)
3. J. Milanović, K. Yamashita, S. Martínez Villanueva, S. Djokić, and L. Korunović, IEEE Transactions on Power Systems, **28**, 3038 (2013)
4. A. Bokhari, A. Alkan, R. Dogan, M. Diaz-Aguilo, F. De Leon, D. Czarkowski, Z. Zabar, L. Birenbaum, A. Noel, R.E. Uosef, IEEE Transactions on Power Delivery, **29**, 1372 (2014)
5. E. Polykarpou, E. Kyriakides, 18th Mediterranean Electrotechnical Conference, **2016**, 7495363 (2016)
6. E.O. Kontis, A.I. Chrysochos, G.K. Papagiannis, T.A. Papadopoulos, IEEE Eindhoven PowerTech, **2015**, 7232800 (2015)
7. X. Zhang, C. Lu, Y. Han, S. Yu, J. Wang, H. Huang, Y. Su, IEEE Power and Energy Society General Meeting, **2015**, 7285620 (2015)

MATHEMATICAL SIMULATION OF THE TRANSFORMER DIFFERENTIAL PROTECTION, BASED ON THE TIME-IMPULSE PRINCIPLE

Mikhail V. Andreev[1,*], *Yuri* S. Borovikov[1], *Nikolay* Yu. Ruban[1], and *Aleksey* V. Khlebov[1]

[1]Tomsk Polytechnic University, 634050, Tomsk, Russia

Abstract. The modern electric power systems (EPS) are large, multi-parameter, stiff, non-linear and dynamic systems. Control and protection of such systems are extremely complex and urgent tasks. It is confirmed by the statistic of accidents in EPS: about 25 % of severe accidents are caused by wrong actions of relay protection (RP) and automation of EPS. The 30 % of these incorrect actions are caused by inadequate setting. The goal of this work is to solve the problem of RPs setting through the usage of detailed models that take into account specific features of the RPs and processes in transducers. The paper focused on the solution of the problem for the transformer differential protection (TDP), based on the impulse-time principle – DZT-21/DZT-23. The presented research results confirm the effectiveness of the usage of the appropriate mathematical models for the problem solution. The presented fragments are a part of a single project aimed at the problem' solution for all the RPs devices.

1 Introduction

Due to unity and continuity of generation, distribution and consumption of electrical energy all equipment involved in the process: primary movers, generators, transformers, power transmission lines, electrical consumers and etc., including relay protection (RP) constantly interconnected in normal, emergency and post-emergency operational regimes.

Since the vast majority of this equipment is a dynamic elements, mostly non-linear and with very significant range of time constants (> 1000), any modern electric power system (EPS) [1] is a large, multi-parameter, stiff, non-linear, dynamic system. Research, design and utilization of such systems are very complex problems due to the difficulty of determining the influence of transients on the operation of RPs and the associated consequences for the operational regimes of the equipment and the EPS as a whole. According to the generalized statistics of accidents in the global electric power industry about 25 % [2] of severe accidents are the result of incorrect actions of RP and emergency automation. In the 50 – 70 % of cases wrong action of RP leads to the development of emergencies to severe system crash.

Because of development of the EPS simulation tools and, especially, creation in the Tomsk Polytechnic University of Hybrid Real-Time Power System Simulator (HRTSim)

* Corresponding author: andreevmv@tpu.ru

[3, 4], which in this case is a source of primary information about the processes in the power system, became possible full investigation of the indicated problem, identification of its existence reasons and development on the basis of the results of these studies concept for solution of this problem.

2 Mathematical simulation of relay protection

The mathematical modelling is currently the only way to obtain information for the full spectrum of the processes and regimes of EPS. Depending on the specific tasks used models of the main and auxiliary equipment of power systems, including RP, varying details. In most projects, devoted to the evaluation of EPS stability, protection and automation reproduced as simplified models using the basic functions, such as a comparison with a threshold, timing and sending a command to circuit breakers. However, there are other approaches to the modelling of the RPs:

- Detailed modelling of physical processes in the RPs with the system of differential-algebraic equations. This approach is mainly applied to study of electromechanical and electronic protection. The main drawback, according to the authors opinion, is the neglecting the measurement transducers (MT), which are mostly determine the nature of the identified processes.
- Detailed modelling of RPs operational algorithms. This approach is most popular at the moment, due to the fact that the researchers focused on the simulation of digital RPs (DRP) actively displacing other elemental bases. Here again, there are different approaches to the creation of such models. In some cases, models reproduce only a digital part of DRPs that implements the algorithm of its operation, in others – take into account a measurement and converting section, which includes primary and intermediate MT, as well as frequency filters in the form of dynamic models varying details. The complexity of development of the full models of specific DRPs caused by unavailability, in most cases, information about particular devices. Due to it a model designed on a base of data from manufacturer's guidelines.

Based on the analysis briefly presented above, it should be noted that the vast majority of projects for EPS modelling RPs models are not implemented properly, i.e. without taking into account of key software and hardware features of specific implementations and processes in the primary and intermediate MT. That is ultimately reduces the adequacy of the EPS model and simulation results.

Proposed by the authors and shown in a several publications, for example in [5, 6], approach aimed at creating universal all-regimes mathematical models of RPs, taking into account the mentioned features and processes. Such models can be used to solve the following scientific and practical problems: 1) investigation of the influence of the spectrum of transients in electric power systems on the operation of relay protection and identification of specific laws and theories; 2) reliable analysis and identification of the causes of incorrect actions of protection associated with the operation of the elements in a specific RP, development of recommendations for their improvement or modification; 3) testing and adequate setting of RP for the specific operational conditions in EPS.

This article presents fragments of the research results of the above problems on the example of transformer differential protection (TDP), based on the time-impulse principle – DZT-21/DZT-23.

3 Experimental results

On the basis of the approach, which was indicated in the previous section, a universal all-regimes mathematical model of DZT-21/DZT-23 is created. The phrase "all-regimes" means what a one model used for simulation the full spectrum of normal and abnormal regimes of EPS operation. Because of the limited volume of the article below shown a fragment of the model – second harmonic braking circuit (SHBC) of DZT-21/DZT-23: equivalent circuit (Fig. 1a), directed graph (Fig. 1b) on the basis of the equivalent circuit, transfer function (1) and the differential equation (2).

The transfer function:

$$W_{secHarm}(p) = U_{secHarm}(p) \,/\, U_{Filter}(p) = (K_1\,\mathrm{p} + K_2) \,/\, (K_3\,\mathrm{p} + K_4) \quad (1)$$

where K_1, K_2, K_3, K_4 – coefficients determined by RLC-parameters of SHBC.

The differential equation:

$$d(K_3\, u_{secHarm}(t) - K_1\, u_{Filter}(t)) \,/\, dt + K_4\, u_{secHarm}(t) - K_2\, u_{Filter}(t) = 0 \quad (2)$$

Implementation of the synthesized mathematical model of DZT done through transforming the model presented in the form of a system of differential equations into a program code, integrated into the software of HRTSim. Another option also was done – autonomous software with capability of downloading information about the processes in the EPS from simulators. The main advantage of the first option – it is an opportunity of real-time tracing the response of EPS on the action of RPs and analyzing consequences.

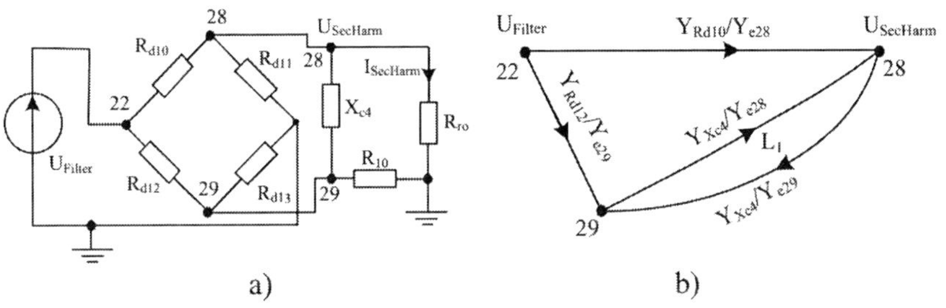

Fig. 1. Equivalent circuit (a) and graph (b) for second harmonic braking circuit: UTAV2 – transreactor output voltage; Z_{load} – load impedance; U_{Filter}, I_{Filter} – second harmonic filter output voltage and current; X_L – inductive resistance.

Currently two basic approaches to validation of RPs mathematical models are applied: 1) comparison of results of operation of mathematical models and real devices; 2) comparison of characteristics of particular elements models with theoretical information available in open sources. The first approach is difficult to implement due to difficulty of information extraction from particular nodes of the real device, as well as the lack of access to the equipment. For validation of the DZT-21/DZT-23 model the second approach was applied.

The spectrum of regimes for model testing includes all possible internal and external short-circuit (SC), as well as inrush current regime. Below presented are fragments (Fig. 2 and 3) of the research of the DZT-21/DZT-23 mathematical model in regime of external three-phase SC on the 220 kV bus-bar – the highest voltage (HV) of protected autotransformer.

According to the oscillograms (Fig. 3b) a incorrect tripping of BC complect of protection is occurred. This fact show a ouput signal of tripping-time-delay element of operating element.

DOI: 10.1051/matecconf/20179101048

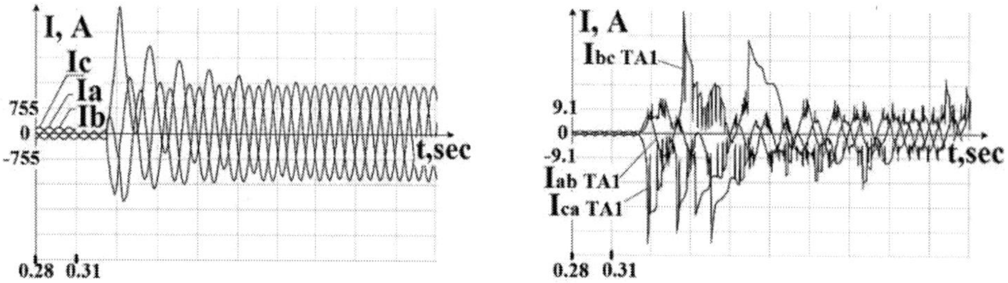

a) b)

Fig. 2. a) Oscillograms of phase primary currents of the HV side of the autotransformer; b) oscillograms of linear current at the output of current transformers on the HV side.

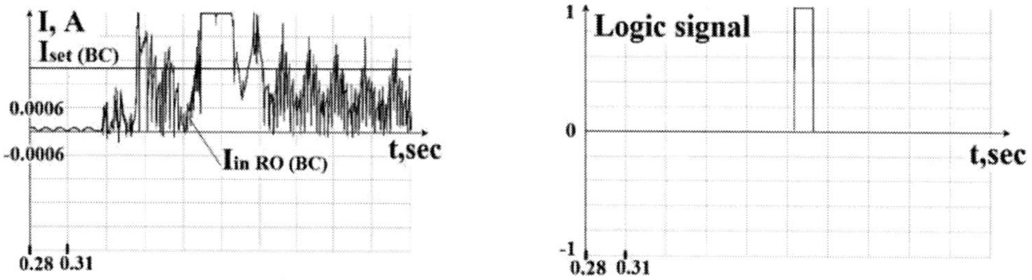

a) b)

Fig. 3. a) Oscillograms of input currents of operating element of BC complect; b) oscillograms of output signal of tripping-time-delay element of BC complect.

4 Conclusions

The done studies allow analysing the errors of specific elements of the TDP, testing the setting and correct them based on the analysis of RP operation in different regimes. The research results allow confirming the fact of a significant impact of transients in EPS on the TDP operation and make some conclusions about the effectiveness of time-impulse principle and effectiveness of the protection work as a whole. The presented results of studies in combination with others, obtaining of which is planned in the near future, will form the basis of the concept for solution of the problem indicated at the beginning.

The reported study was funded by RFBR, according to the research project No. 16-38-60043 mol_a_dk.

References

1. M. Mazurov, A. Nikolaev, N. Lozinova, IEEE Russia Power Tech, 1 (2005)
2. A. Kuzmichev, E. Konovalova, S. Sakhorov, A. Zakharenkov, Relay protection and automatization, **1**, 60 (2012)
3. A. Gusev, A. Suvorov, A. Sulaymanov, IOP Conf. Ser.: Mater. Sci. Eng., **93**, 1 (2015)
4. R. Ufa, V. Sulaymanova, Yu. Borovikov, MATEC Web of Conf., **37**, 01061 (2015)
5. Yu. Borovikov, A. Sulaymanov, M. Andreev, IOP Conf. Ser.: Mater. Sci. Eng., **93**, 1 (2015)
6. A. Suvorov, Y. Borovikov, A. Gusev, A. Sulaymanov, M. Andreev, N. Ruban, R. Ufa, Electrical Engineering (to be published)

MATEC Web of Conferences **91**, 01049 (2017) DOI: 10.1051/matecconf/20179101049
Smart Grids 2016

RESEARCH OF THE POWER PLANT OPERATIONAL MODES

Nina M. Koismynina[1,*], and *Timur* T. Valiev[1]

[1]Tomsk Polytechnic University, 634050, Tomsk, Russia

Abstract. In this article the algorithm of the power plant operational modes research is offered. According to this algorithm the program for the modes analysis and connection power transformers choice is developed. The program can be used as educational means for studying of the power plant electric part, at the same time basic data are provided. Also the program can be used for the analysis of the working power plants modes. Checks of the entered data completeness and a choice correctness of the operational modes are provided in the program; in all cases of a deviation from the correct decisions to the user the relevant information is given.

1 Introduction

The most important problems of power supply system functioning and management are the equipment structure choice and the power plants modes analysis [1]. The electric power isdeveloped at power plants of a different type depending on technological process: thermal, nuclear, hydraulic. According to the Russian Federation Energy Ministry in 2016 year 65.4 1% of the developed power were the corresponds to the thermal power plants, 15.4 % – to the hydraulic, 19.3 % – to the nuclear power plants. Thermal and nuclear power plants turbogenerators with the drive from steam turbines are established. Powerful thermal and nuclear power plants as the installed turbines are condensation.

Feature of condensation power plants block diagrams is power development from generators to power transformers and to voltage switchgear [2]. As a rule, condensation power plants have two voltage switchgear: high voltage switchgear (HV switchgear) and the middle voltage switchgear (MV switchgear). The electric power on HV switchgear power lines is transferred to substations HV switchgear. Substations electric power will be transformed to the required level of distributing devices voltage – low voltage switchgear (LV switchgear). The powerful substations can have the middle voltage switchgear (MV switchgear) [2]. The example of the condensation power plants block diagram with three winding connection transformers is presented in Fig.1. The power plant has three turbogenerators (G_1, G_2, G_3). The G_1 generator is connected to a lowest voltage winding of the T_1 transformer which works at HV switchgear. The G_2 generator is connected to a lowest voltage winding of the T_2 transformer which works at MV switchgear. The G_3 generator is connected to a lowest voltage winding of the T_3 transformer which works at MV switchgear. Communication between HV switchgear and MV switchgear is carried out

* Corresponding author: kosm_nm@tpu.ru

by three winding transformers T_4 and T_5. At certain combinations of voltage instead of the three winding transformers autotransformers are used; in this article this case isn't considered. The HV switchgear has connection with a power supply system (in drawing it is designated – a source S) and external consumers loading by the total power of S_{HV}. The MV switchgear has middle voltage external consumers loading of the total power of S_{MV}. Power supply of power plant internal consumers (own needs) is carried out from blocks the generator – transformer and is designated as c.n; reserve food of internal consumers from the connection transformers is also shown.

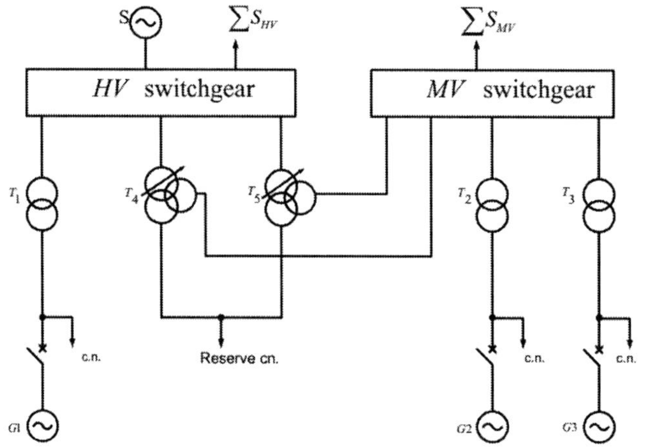

Fig. 1. The power plant with the connection transformer block diagram.

2 Experimental setup and study technique

At power plant the following operational modes are possible.

Load modes: the maximum load of MV switchgear, the minimum load mode of MV switchgear, the maximum load mode of HV switchgear, the minimum load HV switchgear. Repair mode: removal from work of one connection transformers. Postemergency mode: shutdown of one block generator – the transformer from switchgear (HV or MV) leads to emergence of postemergency mode as this shutdown is possible only at block equipment accident and it removal from work. Each of the described modes (or their combinations) is characterized by the balances capacities of power plant characteristic point: turbogenerators stator winding, switchgear (Fig. 2), neutral points of three winding transformers equivalent circuits.

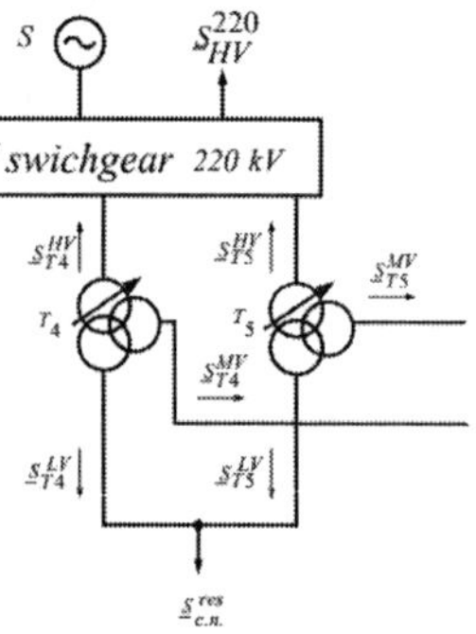

Fig. 2. HV switchgear point.

For research only those operational modes at which there is a change of connection transformers overflows are chosen. The required rated power the three winding power transformers is made after definition the capacities balances equations and calculation of power overflows: $S_{RATE} = \max(S_{LW}, S_{MV}, S_{HV})$, where S_{LW} – power overflow on a low voltage winding; S_{MV} – power overflow on a middle voltage winding; S_{HV} – power overflow on a high voltage winding.

In practice of the power plant modes analysis it is accepted not to consider losses of power transformers. In this case for condensation power plant with connection transformers above the given formula becomes simpler as $S_{MV} = S_{HV}$ – at not accounting of power transformers losses and here is also a working food of internal consumers. Therefore, $S_{RATE} = S_{LW}$.

Similar calculations are carried out for each operational mode, and the largest required power for the connection transformer is defined.

3 Experimental results

The program for research of power plant operational modes is carried on Electric Power Systems Department of National Research Tomsk Polytechnic University Power Engineering Institute. This development is further work on re-equipment of the department laboratory modern software [3].

The program is written in the Delphi programming language which has a number of advantage in comparison with similar products: speed of development; low requirements to resources to computers; building due to embedding new a component and tools; possibility of development of new components and tools own means; successful study of hierarchy of objects [4].

The program consists of the next windows: the description, the basic data input, the block diagram formation, the operational mode choice, the calculation results (the program work protocol with his subsequent preservation in the file form). The description window contains the program name and assignment, the executed functions, data on the developer and the research supervisor. The input generator parameters window contains the fields: rated active power, stator winding voltage, power factor, electric power expense on power plant own needs, number of generators (Fig. 3). The data input window on loadings contains the fields: maximum active power of one loading line, power factor, coefficient of simultaneity of loading maxima, number of lines, voltage; for HV loading only voltage is requested.

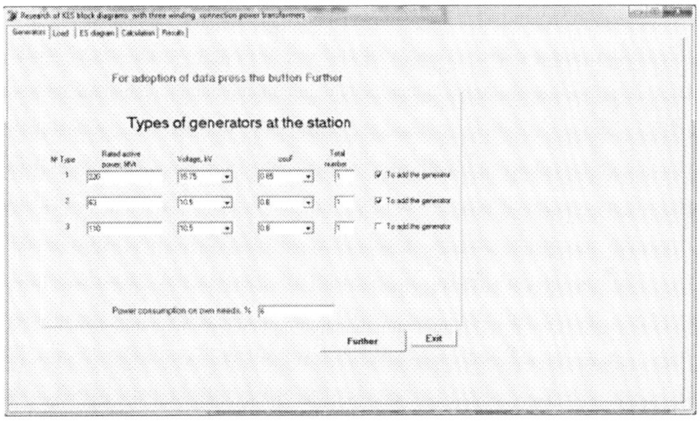

Fig. 3. Input window of power plant generators parameters.

 DOI: 10.1051/matecconf/20179101049

For formation of the power plant block diagram the user needs to execute input of connection transformers number (from 1 to 3) and the number of generator-transformer blocks connected to MV switchgear and HV switchgear (Fig.4).

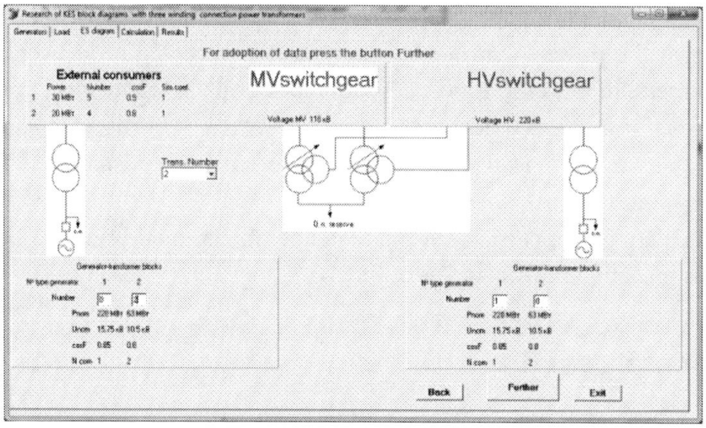

Fig. 4. Window of the power plant block diagram formation.

The window of the station operational mode choice is described below. In this window the user chooses station mode, and for postemergency operation mode obligatory specifies type of the switched-off generator and a power expense on own needs for maintenance of a disconnected equipment normal state.

For the aid to the user checks in equipment nominal parameters and also power plant structure and the studied operational modes are provided in the program. So, generators rated power, own needs (internal consumers) power expense coefficients; power factors for external consumers; simultaneity coefficients of loadings maxima for external consumers; value of switchgear voltage; the block generators number, used in the block diagram; feasibility of the operational mode set users for the concrete block diagram are checked.

4 Conclusion

The presented material contains the concentrated statement of operational power plant mode and calculation algorithm of power connection power transformers overflows

On this basis the program allowing according to the entered data is developed: to create the power plant block diagram; to set the operational mode; to calculate connection transformers power overflows; to create the calculation protocol; to keep calculation data in the file; to study reference methodical material on this subject.

Now work on introduction of the developed program for educational and research tasks performance in TPU is conducted.

References

1. Y. Gusev, Thermal Engineering, **3**, 62 (2015)
2. B. Starshinov, M. Piratorov, M. Kozinova, *Electric part of power plants and substations* (MEI, Moscow, 2015) [in Russian]
3. I. Tsoy, N. Kosmynina, IOP Conference Series: Earth and Environmental Science, **6**, 27 (2015)
4. A. Arkhangelsk, *Programming in Delphi: the textbook according to classical versions* (Binom, Moscow, 2008) [in Russian]

SIMULATION OF A HIGH-FREQUENCY LINK OF PHASE COMPARISON PROTECTION OF TRANSMISSION LINES FOR OPTIMIZATION ITS SETTINGS

Nikolay Yu. Ruban[1,*], *Almaz* O. Sulaymanov[1], *Ruslan* A. Ufa[1], and *Igor* A. Razzhivin[1]

[1]Tomsk Polytechnic University, 634050, Tomsk, Russia

Abstract. The objective of this work is to determine the influence of high-frequency link (HFL) on functioning of phase comparison protection (PCP) through the usage of detailed models that take into account specific features of the PCP and processes in HFL. The article describes the case of the most common variant of the high frequency signal transmission by using a scheme 'phase-to-ground'. To verify suggested solution the developed mathematical model of HFL has been studied. Presented simulation results confirm the effectiveness of the proposed approach for solving the problem.

1 Introduction

One of the main reasons for wrong actions of the phase comparison protection (PCP) is inappropriate setting for the particular implementation and operating conditions. It is related to over-generalized view of the errors generated by measuring transformers and actually by apparatus of the used protection, as well as the use of insufficiently reliable primary regime information [1]. This means that, in accordance with the existing methods for calculating relay protection (RP) settings, accounting the actual conditions of the functioning of the particular implementation of protection is carried out very simplistically - through a variety of generalized adjusting coefficients. The latter is due to the fact that the software systems used to determine the RP settings in which the calculation is made in strict accordance with existing procedures, as well as systems for modelling regimes and processes in electric power systems (EPS) is used several limitations and assumptions. This is connected with a numerical method for solving differential equations [2, 3], and therefore reproducible primary regime information and its transformation by particular implementations of PCP contains errors.

At the same time, the main sources of errors and consequently, the main cause of incorrect PCP settings are its measuring and transmitting parts.

Transmitting part (TP) is a high-frequency component of the protection, and regardless of the type of execution comprises: a part of transmission line (PL), linking half-sets at opposite ends of the protected line; high-frequency chokes including a reactor (R),

* Corresponding author: rubanny@tpu.ru

 DOI: 10.1051/matecconf/20179101050

equivalent capacitance (E) and inductivity; coupling capacitors (CC); connecting filters (CF) and high-frequency cables (HFC) for each end of the protected line, respectively. The above elements of TP constitute an electric circuit comprising RLC elements that significantly affect the transmitted signal. In addition to the mentioned elements, in its structure also high-frequency signal receivers and transmitters (RT) are included, together they combine a high frequency link (HFL) of PCP.

2 Mathematical simulation of relay protection

Thus for PCP modelling is also advisable to take into account the impact on its functioning HFL, the elements of which inevitably have an impact on the transmitted signal.

Figure 1 shows the equivalent circuit of HFL.

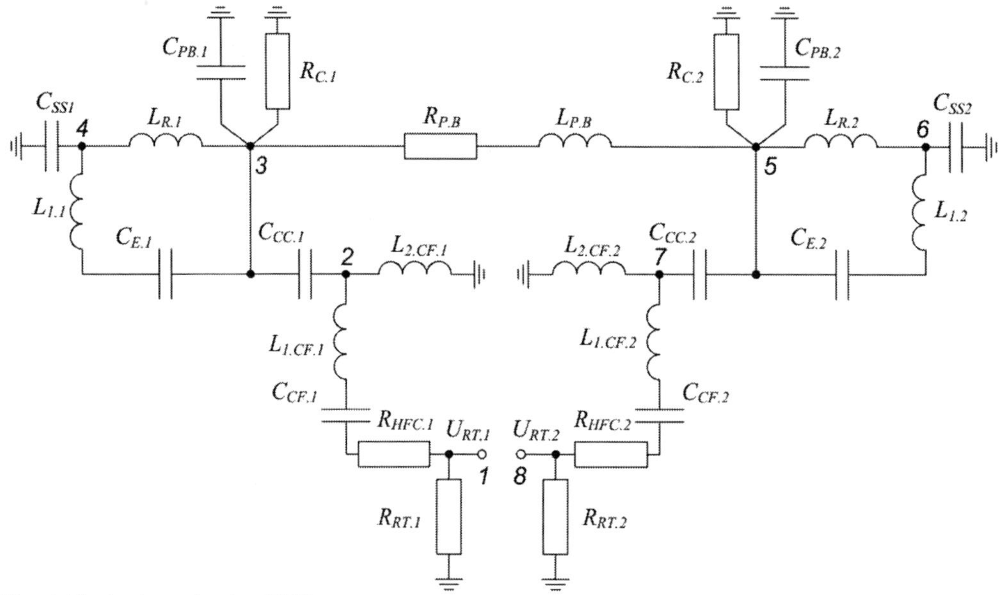

Fig. 1. Equivalent circuit of HFL.

On the basis of the equivalent circuit considering the relationship HFL functional blocks, which is carried out by the determination of appropriate equivalent load resistance of these blocks, is formed the differential equation describing the processes in HFL.

$$\frac{d^{15}\left(u_{OUT.HFL1}(t)\cdot A_{2HFL1}\right)}{dt^{15}}+\frac{d^{14}\left(u_{OUT.HFL1}(t)\cdot B_{2HFL1}\right)}{dt^{14}}+\frac{d^{13}\left(u_{OUT.HFL1}(t)\cdot C_{2HFL1}\right)}{dt^{13}}+$$
$$+\frac{d^{12}\left(u_{OUT.HFL1}(t)\cdot D_{2HFL1}\right)}{dt^{12}}+\frac{d^{11}\left(u_{OUT.HFL1}(t)\cdot E_{2HFL1}\right)}{dt^{11}}+\frac{d^{10}\left(u_{OUT.HFL1}(t)\cdot F_{2HFL1}-u_{IN.HFL1}(t)\cdot F_{1HFL1}\right)}{dt^{10}}+$$
$$+\frac{d^{9}\left(u_{OUT.HFL1}(t)\cdot G_{2HFL1}\right)}{dt^{9}}+\frac{d^{8}\left(u_{OUT.HFL1}(t)\cdot H_{2HFL1}-u_{IN.HFL1}(t)\cdot H_{1HFL1}\right)}{dt^{8}}+\frac{d^{7}\left(u_{OUT.HFL1}(t)\cdot I_{2HFL1}\right)}{dt^{7}}+$$
$$+\frac{d^{6}\left(u_{OUT.HFL1}(t)\cdot J_{2HFL1}-u_{IN.HFL1}(t)\cdot J_{1HFL1}\right)}{dt^{6}}+\frac{d^{5}\left(u_{OUT.HFL1}(t)\cdot K_{2HFL1}\right)}{dt^{5}}+\frac{d^{4}\left(u_{OUT.HFL1}(t)\cdot L_{2HFL1}\right)}{dt^{4}}+$$
$$+\frac{d^{3}\left(u_{OUT.HFL1}(t)\cdot M_{2HFL1}\right)}{dt^{3}}+\frac{d^{2}\left(u_{OUT.HFL1}(t)\cdot N_{2HFL1}\right)}{dt^{2}}+\frac{d\left(u_{OUT.HFL1}(t)\cdot O_{2HFL1}\right)}{dt}+u_{OUT.HFL1}(t)\cdot P_{2HFL1}=0$$

Factors F_{1HFL1}, H_{1HFL1}, J_{1HFL1} and $A_{2HFL1}-P_{2HFL1}$ are presented in general form and are not given in the article because of the large size.

3 Experimental results

As a preliminary tests of studies with a programs MathCAD and MATLAB Simulink are determined amplitude-frequency and phase-frequency characteristics of the HFL mathematical model necessary to assess the amplitude and phase errors for the particular frequency of the converted signal. They are shown in Figure 2.

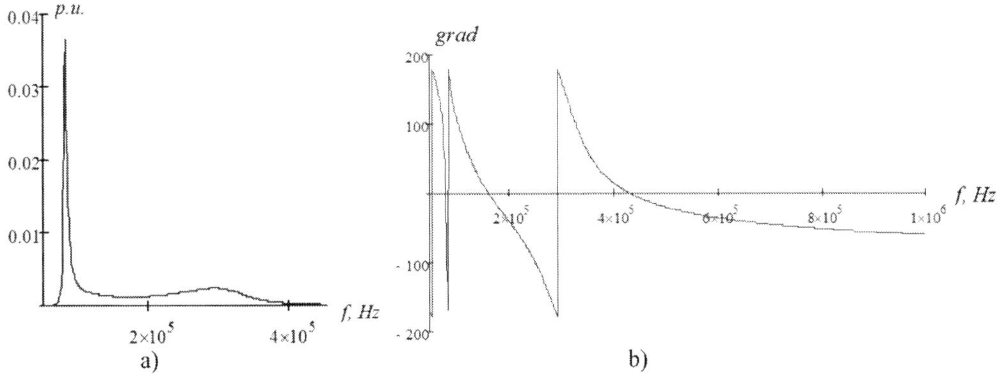

Fig. 2. Amplitude-frequency (a) and phase-frequency (b) characteristics of the HFL mathematical model.

After the pre-tests studies obtained mathematical model is converted into a program code and integrated into the specialized program for PCP simulation – DPPmodels [4, 5], previously developed by the author. The program allows to set the parameters of HFL and with the integrated oscilloscope to analyze the impact of changes in these parameters on the transmitted signal.

For example for double ended lines setpoint of PCP blocking angle should be selected as short as possible considering signal propagation delay and the necessary reserve. The value of reserve on the blocking angle is recommended to choose a minimum of 30° using the same half-sets of PCP. The propagation delay of the signal is assumed to be 6° for every 100 km of line, that is typical for HFL signal transmission. Often, however, this value is chosen significantly enhanced. Using the developed tool for PCP modeling, including HFT, allows to carry out a substantiated calculation of reserve angle. Example oscillograms confirming features of the program DPPmodels are shown in Figure 3.

The oscillograms presents HF signals at the transmitter output and the receiver input, logical signals at the input and output of the phase comparison element of the first half-set of PCP.

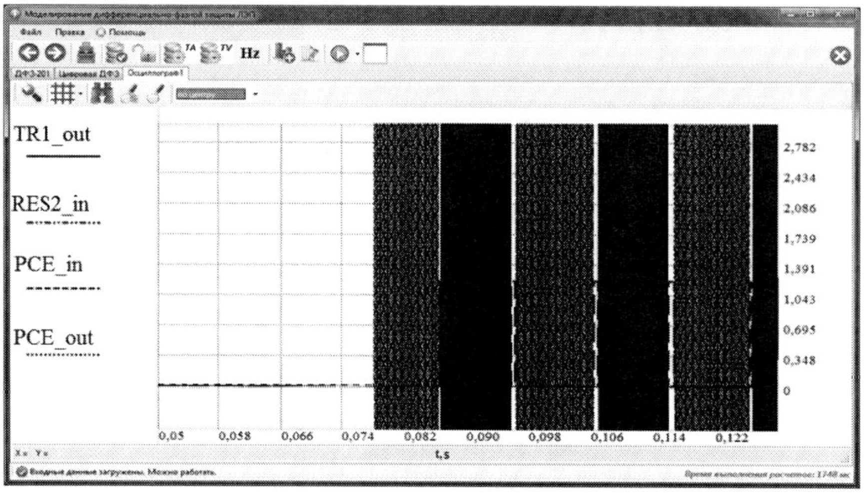

Fig. 3. Oscillograms of HF and logical signals.

Oscillograms confirm the ability to accurately determine the offset between the HF signals, limited by value of blocking angle.

4 Conclusions

The synthesized model provides adequate reproduction of processes in a high-frequency part of the PCP. Thus it is achieved the possibility of determining and accounting of the actual errors of the HFL elements formed by a particular spectrum of input regime values. At the same time errors definition of the transmission signal through the HFL becomes particularly urgent in connection with the tendency to increase the use of FACTS in Russian networks with different capacitive filter-compensating devices, connection of which has an impact on transient processes [6]. The results confirmed the possibility of using the proposed approach for the determination of errors in a HFL.

The work was supported by the Ministry of Education and Science of the Russian Federation under the governmental grant "Science" №3901 (Research and development of the hybrid model of the back-to-back high-voltage direct current transmission system).

References

1. M. Andreev, A. Sulaymanov, A. Gusev, 13th IET International Conference on Developments in Power System Protection, 539 (2016)
2. G. Dahlquist, Math. Scand., **4**, 33 (1956)
3. J. D. Lambert, *Computational methods in ordinary differential equations* (John Wiley and Sons, Ltd., New Jersey, 1973)
4. N. Ruban, A. Gusev, A. Pishchulin, MATEC Web Conf., **37**, 01046 (2015)
5. A. Suvorov, Y. Borovikov, A. Gusev, A. Sulaymanov, M. Andreev, N. Ruban, R. Ufa, Electrical Engineering (to be published)
6. L. Balyberdin, E. Zmaznov, Yu. Kraichik, B. Krasnova, N. Lozinova, M. Mazurov, G. Prochan, Power Technology and Engineering, **38**, 119 (2004)

SOFTWARE AND HARDWARE COMPLEX FOR SETTING OF AUTOMATIC EXCITATION REGULATORS OF TURBOGENERATORS

Venera A. Sulaymanova[1,*], *Yuri* S. Borovikov[1], *Mikhail* V. Andreev[1], and *Aleksey* A. Suvorov[1]

[1]Tomsk Polytechnic University, 634050, Tomsk, Russia

Abstract. The motivation of the presented research is based on the needs for development of new methods and tools for research setting problem of automatic excitation regulators of turbogenerators. Simulation tools must meet the requirements of reproduction processes reliability in all elements of electric power system. The developed block diagram of the adequate mathematical model of automatic excitation regulator is presented. The simulation results of electric power system scheme confirm the adequacy of the reproduction processes of functioning of automatic excitation regulator and a generator.

1 Introduction

In view of the interconnection of all elements of electric power system (EPS) in a single continuous process of generation, transmission, distribution and consumption of electric power, any changes of mode parameters (load change, connection and disconnection of equipment, short circuits and others) influence on functioning of the power equipment and EPS as a whole. The basic elements of maintaining required voltage level, increasing the stability of EPS generators operating in parallel and damping of oscillation are automatic excitation regulators (AER) of generators. At the present time used AER with power system stabilizer (PSS) have a large number of tuning factors, and their choice has impact on reliable and stable EPS functioning. In this way, setting of AER is an important practical problem, which for many years many research groups engaged and engage.

Setting must be implemented for each generator considering of scheme-regime operational conditions in a scheme with its characteristics and parameters, and also EPS reaction on setting. Great importance in the study of this problem has reliability of used mathematical models of all EPS elements. Technical requirements for the excitation system (ES) and AER PSS of synchronous generators, and also a method for certification and setting of AER PSS are defined by the standard of Joint-stock Company "System Operator of the United Power System" [1].

* Corresponding author: venera20@tpu.ru

2 Mathematical model of excitation system with an automatic excitation regulator with power system stabilizer

Using developed in Tomsk Polytechnic University Hybrid Real-Time Simulator (HRTSim) [2, 3] of EPS is one of the most preferred tools for solving the problem of adequate setting of AER. HRTSim of EPS is the soft- and hard-ware complex, comprising set of universal specialized processors for all kinds of power equipment of simulated EPS and control system.

In HRTSim of EPS the mathematical model of a synchronous machine represents full Park's-Gorev's system equations increased accuracy, in cooperation with the equations of forming three-phase system A, B, C and mutual conversion systems variables d, q and A, B, C. Required reliability of simulation is provided by increasing number of simulated damping circuits (three d-axis and four q-axis) and accounting the frequency dependence of their parameters [4].

For adequate reproduction processes in EPS also it's necessary to have detailed mathematical models of the auxiliary EPS equipment: ES, primary engines, driven mechanisms and others. Developed structure of hybrid processor HRTSim of EPS [5] allows to realize any kinds of AER. The block diagram of the mathematical model of ES with AER PSS is presented in figure 1.

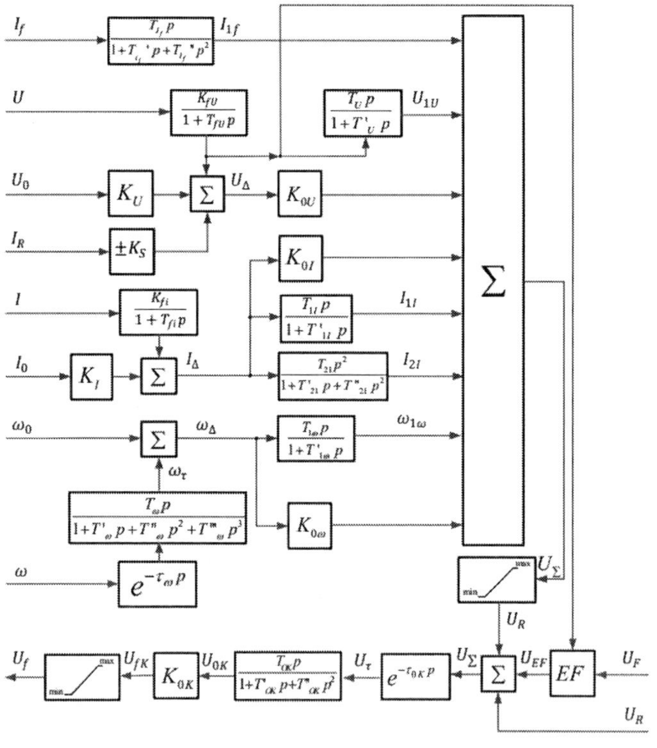

Fig. 1. Block diagram of the mathematical model of ES with different kinds of AER PSS.

Presented block diagram takes into account significant factors essential for adequate reproduction, namely the specifics of such ES and AER and inherent control channels, tuning factors, parameters and defining differential equations transfer functions: K_S – a static factor of voltage corrector of the stator U, K_U – threshold factor of stator voltage U, K_I – a compounding factor for the complete stator current I, EF – different types of

 DOI: 10.1051/matecconf/20179101051

excitation forcing, K_{0U} – a gain factor of control channel on the stator voltage deviation U_{Δ}, $K_{0\omega}$ – a factor of the control channel on the frequency deviation ω_{Δ}, blocked when the limit value, $W(p)=\dfrac{K_{fU}}{1+T_{fU}p}$ and $W(p)=\dfrac{K_{fi}}{1+T_{fi}p}$ – transfer functions of the rectifiers and voltage filters U and stator current I, $W(p)=\dfrac{T_U}{1+T'_U p}$ – the transfer function of the control channel on the derivative of the stator voltage U_{1U}, $W(p)=\dfrac{T_{I_f}}{1+T'_{I_f}p+T''_{I_f}p^2}$ – a transfer function of the control channel on the derivative of the excitation current I_{1f}, $W(p)=\dfrac{T_{1\omega}}{1+T'_{1\omega}p}$ –a transfer function of the control channel on the derivative of the frequency, $W(p)=e^{-\tau_{\omega}p}\dfrac{T_{\omega}p}{1+T'_{\omega}p+T''_{\omega}p^2+T'''_{\omega}p^3}$ – a transfer function of the measurement unit and frequency conversion to form its deviation ω_{Δ}, $W(p)=e^{-\tau_{OK}p}\dfrac{T_{OK}p}{1+T'_{OK}p+T''_{OK}p^2}$ – a transfer function of total excitation control channel, including the rectifier, exciter and others, K_{0I} – a gain factor of control channel on the stator current deviation I_{Δ} from set value I_0, $W(p)=\dfrac{T_{1I}p}{1+T'_{1I}p}$ – a transfer function of the control channel on the derivative of the stator current I_1, $W(p)=\dfrac{T_{2I}p^2}{1+T'_{2I}p+T''_{2I}p^2}$ – a transfer function of the control channel on the second derivative of the stator current I_{2I}.

In figure 2 the fragment of simulated scheme of Tomsk EPS in HRTSim is given.

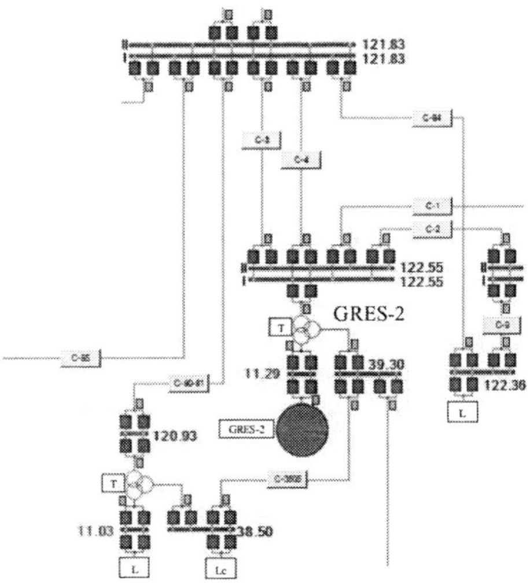

Fig. 2. Fragment of scheme of Tomsk EPS: GRES-2 – thermal power plant, L – load, Lc – controlled load, T – transformer, C – power transmission line.

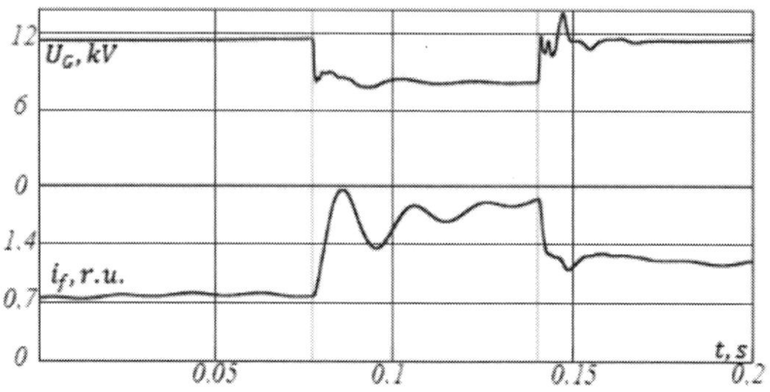

Fig. 3. Oscillograms of automatic voltage regulation process at the generator terminal GRES-2 for three-phase short circuit at the end of the line C-84.

The results of the experiment perform oscillograms of automatic voltage regulation by AER GRES-2 for three-phase short circuit at the end of the line C-84 (figure 3). Implemented in HRTSim of EPS oscillograms confirm the adequacy of the reproduction processes of functioning of AER PSS and generator, that allow to use this complex for solve setting of AER problem, as well as for other tasks, such as research regimes of EPS with the presence of compensating devices [6].

3 Conclusion

Efficiency of solution of AER setting problem depends on used mathematical models of AER, generators and all the other elements of the simulated EPS. Developed HRTSim allows to use detailed models of AER and other equipment for the problems of research and design of EPS. Developed and tested adaptable mathematical model of ES, allowing to simulate different types of AER PSS and ensuring reliable reproduction of the processes occurring in this equipment.

The work was supported by the Ministry of Education and Science of the Russian Federation under the governmental grant "Science" № 3901 (Research and development of the hybrid model of the back-to-back high-voltage direct current transmission system).

References

1. Standard of organization "Requirements for excitation systems and automatic excitation regulators with power system stabilizer of synchronous generators"
2. A. Suvorov, Y. Borovikov, A. Gusev, A. Sulaymanov, M. Andreev, N. Ruban, R. Ufa, Electrical Engineering (to be published)
3. N. Y. Ruban, A. S. Gusev, A. Y. Pishchulin, MATEC Web of Conf., **37**, 01046 (2015)
4. V. M. Yakutov, Electricity, **2**, 23 (1992) [in Russian]
5. Y. S. Borovikov, A. S. Gusev, A. O. Sulaymanov, R. A. Ufa, ISCAI, 148 (2014)
6. L. L. Balyberdin, E. Yu. Zmaznov, Yu. S. Kraichik, B. P. Krasnova, N. G. Lozinova, M. I. Mazurov, G. G. Prochan, Power Technology and Engineering, **38**, 119 (2004)

APPROACH TO CREATING AN INFORMATION-CONTROL SYSTEM OF HYBRID POWER SYSTEM SIMULATOR

Aleksey A. Suvorov[1,*], *Alexander* S. Gusev[1], *Almaz* O. Sulaymanov[1], and *Olga* V. Suslova[2]

[1] Tomsk Polytechnic University, 634050, Tomsk, Russia
[2] Scientific and Technical Center of Unified Power System JSC, 194223, Saint-Petersburg, Russia

Abstract. The results of development and experimental investigations of information and control hybrid simulator system of conventional and smart grids are considered in the paper. The fragments of the given system experimental investigation and its software tools are provided. Performed results illustrate the possibilities and features offered to a user and necessary for solution of complex power system design, research and exploitation tasks.

1 Introduction

Obvious condition of the reliable and effective solution of complex challenges of design, research and operation of modern electrical power systems (EPS), according to their operation, is existence of a possibility obtaining, including in real time, full and reliable information about the processes in the equipment and EPS in all possible standard, emergency and post-emergency operational regimes. The mathematical modelling of EPS is the main way of obtaining this information in view of the known specifics and complexity of EPS. At the same time full and reliable mathematical model of any real EPS, taking into account a possible simplification, always contains very stiff non-linear system of the differential equations of high order, which satisfactory solution is improbable by means of numerical methods, according to the discretization technique of ordinary differential equations [1]. Therefore in all digital tools for calculation regimes of EPS, inevitably used these methods, are applied significant simplifications and limitations of mathematical models and EPS in whole. Moreover, in all cases the methodical error of the solution is unknown and may be accumulated. Meanwhile, mentioned known simplifications and limitations become even less acceptable for modelling smart grids [2, 3].

2 The information-control system structure of hybrid power system simulator

* Corresponding author: suvorovaa@tpu.ru

As the designated reason has fundamental character and within unilateral especially numerical simulation of EPS isn't removable, the alternative way to simulation of EPS is hybrid approach. Developed according to this approach the concept and tools of adequate simulation in real time of EPS, including smart grids, are considered in [4-6]. Hybrid Real-Time Power System Simulator (HRTSim) is developed according to this concept. The information-control system of HRTSim consists of adapted specialized processors (SP). The adaptable set of SHP is combined by the three-phase switch according to the topology of the modelled power system, this set provides continuous methodically accurate solution in real time and with the guaranteed instrumental error of full three-phase mathematical models of all types significant power equipment and EPS in whole, including smart grids.

All information-control functions, including modelling of relay protection, emergency automatic, etc., are carried out in each SHP by means of the microprocessor unit (MPU), which includes central processing unit (CPU). CPU is connected by the local area network (LAN) between themselves and with the HRTSim Server. In this way, the set of MPU and specialized software tools (SST) constitute the information-control system (ICS) of HRTSim. For users – clients, as a part of ICS, the modern automated user's workstations (WS), established without restrictions on external computers, are developed.

Depending on specifics functions and necessary resources for its, the processing of its may be possible in MPU, Server or WS. Therefore, ICS represents multilevel program and hardware system:

- at the level of the Server are carried out the information interactions with MPU SP, WS and all system functions, including modelling of various system and centralized means of automatic equipment, etc., and also various scenarios of modelling are realized;
- at the level of MPU SP interaction on LAN of their CP and Server, and also CP and peripheral processors (PP) MPU is provided by the processors of analog-digital converting and processors of series and shunt digitally controlled three phase switches;
- at the level of WS are realized external the interactive, automatic and combined information interactions with his Server, which can be also various external devices: the existing tools of simulation of power systems, relay protection devices, supervisory control and data acquisition (SCADA) and etc.

The Fig. 1 is demonstrated the structure of multilevel ICS.

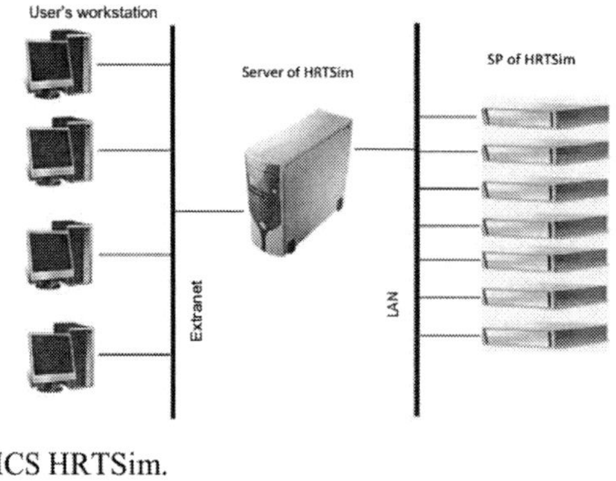

Fig. 1. Block diagram of ICS HRTSim.

According to the block diagram, the HRTSim Server contains two network interface cards. By means of one of them via network switches and LAN control information interactions with all CP are carried out. Within broadcasting interactions are provided: modes of the loader or working CPU, initialization of the MPU databases, synchronization of MPU and Server, various regimes of simulation. Tasks and changes of parameters and coefficients of mathematical models in SP, inquiries and import of simulation results for their converting and display by various software devices: digital, chart, vector, tabular, oscillographic, etc. on specialized dynamic monitoring and control panel (DMCP) are carried out by local interactions, also for implementation of various local and system scenarios of simulation, management means of power systems. Hardware interactions of the CPU with the Server are carried out by means of the Ethernet driver.

3 Software tools of hybrid power system simulator information-control system

All interactions in ICS are carried out in the form of quasiparallel processes. At the same time the interaction with WS is carried out via TCP/IP, this protocol uses the squeezed for acceleration XML messages, which are coded and converted into BASE64. Each user connects to the Server by means of login and password. Any external devices with various data's formats interact with the Server by means of the interfaces created for this purpose. Special SST for user's WS are developed, which is the main tool for interactive using of HRTSim. All program of this SST for WS are professionally focused and structured:

1. DMCP represent the graphic forms used for display of schemes of the equipment, energy area and their regime state, and also for interactive management in real time by parameters, settings, longitudinally cross three-phase and incremental commutation. Realization of this properties and opportunities is provided with the corresponding program management information devices placed on DMCP: digital, chart, vector, tabular, indicator, etc. Moreover, the edition by users of the existing DMCP and creation new are provided in SST for WS. Basic purpose of DMCP is display of the set regimes and processes in the equipment and interactive control of them. Examples of DMCP are given in Fig. 2.

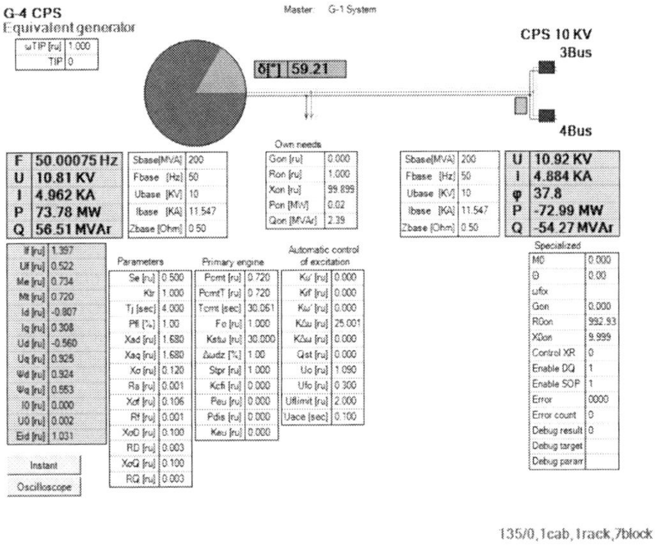

Fig. 2. DMCP of the G-4 generator of thermal power plant.

2. Multipurpose and multibeam oscilloscope represent the program devices used for display of the processes proceeding in the equipment, EPS and also for their in-depth analysis. Processes, their list and combinations in one or several windows of oscilloscopes are chosen by the user. Two types of oscilloscopes are provided in SST for user's WS:

- continuously functioning for display and the analysis of the quasi-steady state and slow transition processes;
- functioning on the set time interval for fast transition processes: electromagnetic, including switching overvoltage, etc.

Examples of using the mentioned program tools are given in Fig. 3.

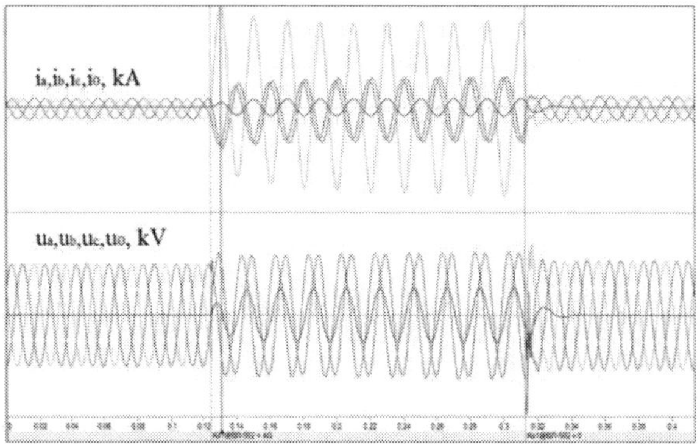

Fig. 3. Single-phase (AG) short-circuit 220 kV transmission line: i_a, i_b, i_c – phase currents; i_0 – zero sequence current; u_a, u_b, u_c – phase voltage; u_0 – zero sequence voltage.

3. Scenarios of simulation represent the program procedures made by the user at discretion with use of the standard symbols and equipment names, parameters, etc. modelled EPS, used for automatic implementation of various scenarios of simulation. In view of essential difference between quasi-steady state and transition processes, two types of the scenarios are provided in SST of WS: quasi-static and dynamic. Scenarios of a quasi-static are intended for changing the necessary set modes. Scenarios of dynamics are intended for reproduction of various abnormal regimes and processes.

4 Conclusions

1. The developed ICS of HRTSim provides necessary set of management information properties and opportunities for the solution of complex challenges of design, research and operation of EPS, including smart grids.
2. The results of experimental research the developed ICS created on the basis of modern IT technologies have confirmed reliability and efficiency of functioning of these tools in the user's WS and in HRTSim in general.

The work was supported by Ministry of Education and Science of the Russian Federation under government grant "Science", Project 3901: Research and development of hybrid model of back-to-back high voltage direct current transmission system.

References

1. G. Hall, J. Watt, *Modern Numerical Methods for Ordinary Differential Equations* (Oxford University Press, USA, 1976)
2. K. Suslov, S. Solodusha, D. Gerasimov, Smartgreens 2015, 195 (2015)
3. M. Mazurov, A. Nikolaev, N. Lozinova, IEEE Russia Power Tech, 4524569 (2005)
4. A. Gusev, A. Suvorov, A. Sulaymanov, MATEC Web of Conf. **37**, 01023 (2015)
5. A. Suvorov, Y. Borovikov, A. Gusev, A. Sulaymanov, M. Andreev, N. Ruban, R. Ufa, Electrical Engineering (to be published)
6. Y. Borovikov, A. Gusev, A. Sulaymanov, R. Ufa, A. Vasilev, M. Andreev, N. Ruban, A. Suvorov, IEEE Trans. on Smart Grid, **7**, 2242 (2016)

MATEC Web of Conferences **91**, 01053 (2017) DOI: 10.1051/matecconf/20179101053
Smart Grids 2016

HYBRID MODELS OF ALTERNATIVE CURRENT FILTER FOR HVDC

Ruslan A. Ufa[1, *], *Alexander* S. Gusev[1], *Alexey* S. Vasilev[1], and *Natalya* G. Lozinova[2]

[1] Tomsk Polytechnic University, 634050 Tomsk, Russia
[2] JSC High Voltage Direct Current Power Transmission Research Institute, 194223, St.-Petersburg, Russia

Abstract. Based on a hybrid simulation concept of HVDC, the developed hybrid AC filter models, providing the sufficiently full and adequate modeling of all single continuous spectrum of quasi-steady-state and transient processes in the filter, are presented. The obtained results suggest that usage of the hybrid simulation approach is carried out a methodically accurate with guaranteed instrumental error solution of differential equation systems of mathematical models of HVDC.

1 Introduction

The HVDC (high voltage direct current) systems are widely used in electric power system (EPS) with its technical and economic advantages [1, 2]. However, operation specific of convertor stations of HVDC is the cause of harmonics which reduce the power quality and disturb the communication system [3, 4]. Therefore, AC (alternative current) filters are necessary in convertor station to compensate harmonics, supply the reactive power consumed by the converter and mitigate current and voltage distortion in the EPS. AC filters, employing combination of resistors, capacitors and inductors, are used in three types of filter configuration – single tuned, double tuned and high-pass filters.

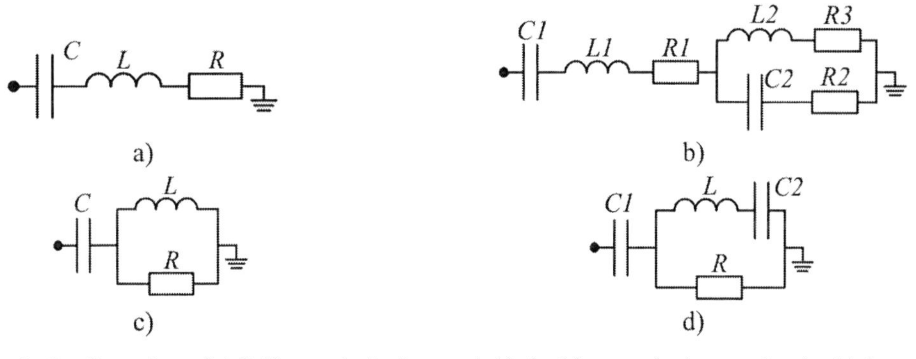

Fig. 1. Configuration of AC filters: a) single tuned; b) double turned; c) second-order high pass; d) high-pass “C” type.

* Corresponding author: hecn@tpu.ru

Single tuned AC filters is a series connected RCL circuit (fig.1a) that are tuned at one harmonic frequency (in general, low order characteristic harmonics). Double tuned AC filters (fig.1b) are used to out two discrete frequencies (for example 5th and 7th or 11th and 13th harmonics). The main advantages of employing double tuned AC filters over the single tuned AC filters are: one reactor subjected to full line voltage, occupying less space, and needing to only one switchgear etc [3, 4]. The damped filter or high-pass filter (fig.1c, d) offers low impedance over a broad band of frequencies. The high-pass filters are designed to damp more than one harmonic, for example, a filter tuned at 24th harmonic will give low impedance for both 23rd and 25th harmonic and for higher order harmonics. But the losses are higher than tuned filters.

In this paper the fragments of design of double turned and second-order high pass AC filter models for hybrid model of HVDC, presented in [2, 5], are introduced.

2 Development of AC filter model

According the basic principles of hybrid simulation concept for realization of HVDC model [6, 7] all power equipment of HVDC, which do not contain any switching elements, are described by mathematical models implemented via hybrid coprocessors. This approach is carried out the methodically accurate with guaranteed instrumental error solution of differential equation systems of mathematical models of HVDC by means of the continuous implicit integration method.

There are following steps for development of AC filter model: 1) formation and synthesis of hybrid mathematical model of the filter, providing a sufficiently full and adequate modeling of a single continuous spectrum of quasi-steady-state and transient processes in all normal, emergency and post-emergency conditions of filter operation; 2) usage of the continuous implicit integration method that carried out the methodically accurate with guaranteed instrumental error solution of differential equation systems of the model of filter; 3) the development of the specialized parallel digital-to-analog structures of the continuous implicit integration method, providing the methodically accurate solution of differential equation systems of the model of filter in real time and on an unlimited range.

2.1 The double turned AC filter model

The equivalent circuit and differential equation system of the double turned AC filter is presented in figure 2.

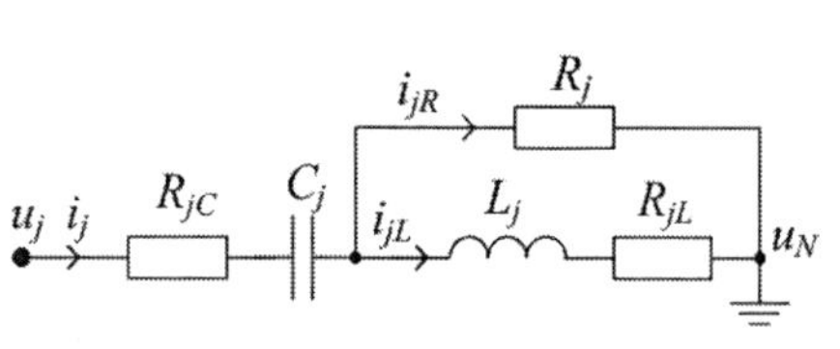

a)

$$\begin{cases} i_{jR} = \dfrac{1}{R_j} \cdot \left(u_j - i_j \cdot R_{jC} - u_{jC} - u_N \right) \\ \dfrac{di_{jL}}{dt} = \dfrac{1}{L_j} \cdot \left(i_{jR} \cdot R_j - i_{jL} \cdot R_{jL} \right) \\ i_j = i_{jL} + i_{jR} \\ \dfrac{du_{jC}}{dt} = \dfrac{1}{C_j} \cdot i_j \end{cases}$$

b)

Fig. 2. The equivalent circuit (a) and differential equation system (b) of the double turned AC filter, where j=A,B,C – phase of filter; R_{jC} and C_j – active resistance and capacity of capacitor bank; R_{jL} and L_j – active resistance and leakage inductance; R_j – resistor; u_N – neutral voltage, u_{jC} – voltage on reactive resistance of capacitor bank.

The developed specialized parallel digital-to-analog structure of the continuous implicit integration method, providing the methodically accurate solution of the differential

 DOI: 10.1051/matecconf/20179101053

equation system of the filter in real time and on an unlimited range and conversion of mathematical variable data input-output currents via *u/i* converters in corresponding model physical variables are introduced:

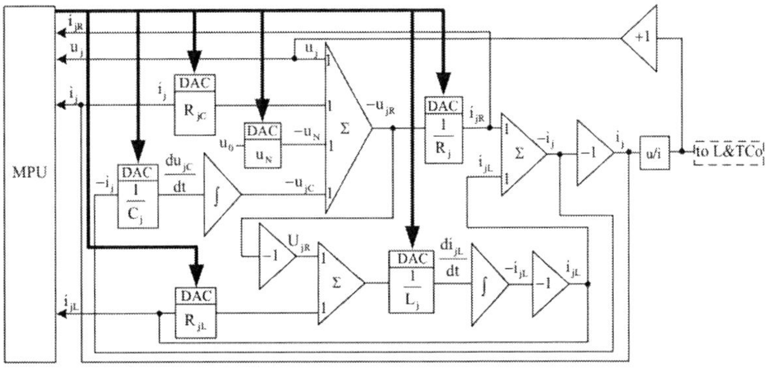

Fig. 3. The specialized parallel digital-to-analog structure of realization of the AC filter model, where all elements carrying out mathematical operations are based on integrated operational amplifiers in inverting gate, DAC – digital-analog converter, MPU – microprocessor unit, L&TCo – longitudinal-transverse commutators, u_0 – reference voltage of DAC [2].

To confirm the adequacy of the developed model of the AC filter a frequency characteristics are obtained in Multisim 11.0 software and compared with the mathematical (base on eq.1) and theoretical frequency characteristics (based on the equivalent circuit of the double turned AC filter – figure 2a):

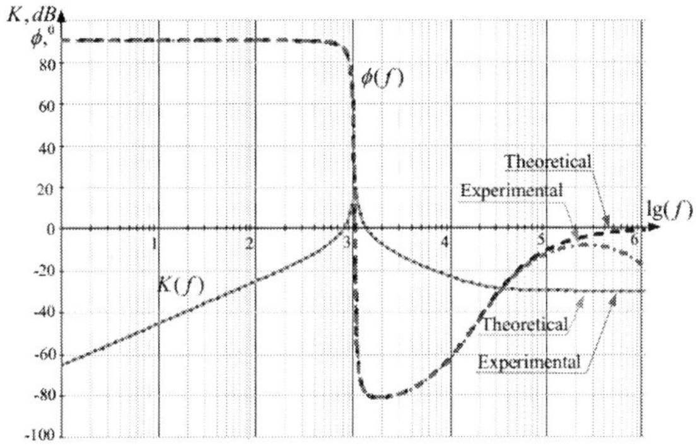

Fig. 4. Bode magnitude and phase plots of the double turned AC filter.

The obtained frequency characteristics confirm the correctness of filter tuning on the frequency 1050 kHz (PWM switching frequency). The maximum value of the suppression ratio and phase reversal are observed at this frequency.

2.2 The second-order high pass AC filter model

The equivalent circuit of the AC filter tuned to suppress the 5th and 7th harmonics and analyze of model' frequency characteristic are presented in figure 5.

The obtained frequency characteristics confirm the correctness of filter tuning on the frequency 1050 kHz (PWM switching frequency). The maximum value of the suppression ratio and phase reversal are observed at this frequency. The obtained frequency characteristics confirm the correctness of filter tuning on the suppressing the 5th and 7th harmonics.

 DOI: 10.1051/matecconf/20179101053

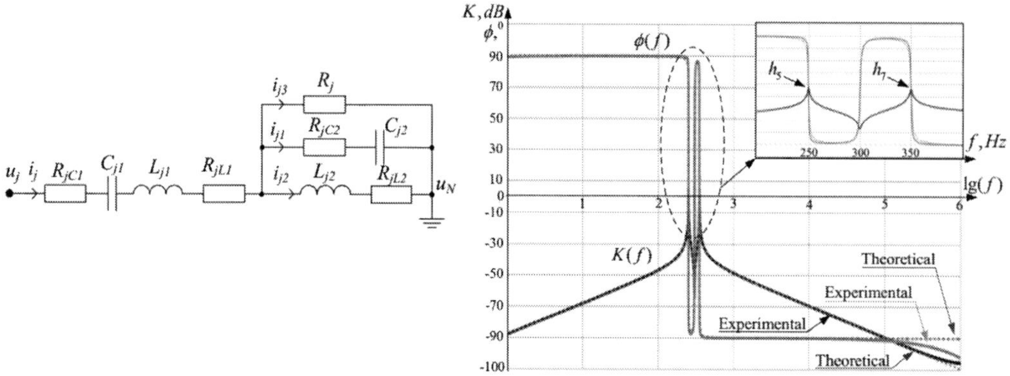

a) b)

Fig. 5. The equivalent circuit (a) and bode magnitude and phase plots (b) of the AC filter tuned to suppress the 5th and 7th harmonics.

Presented results suggest the following conclusions:

- the developed hybrid models of AC filter provide the sufficiently full and adequate modeling of all single continuous spectrum of quasi-steady-state and transient processes in all normal, emergency and post-emergency conditions of operation of the filter;
- usage of the hybrid simulation approach is carried out a methodically accurate with guaranteed instrumental error solution of differential equation systems of mathematical models of HVDC by means of the continuous implicit integration method.

The work was supported by Ministry of Education and Science of the Russian Federation under government grant «Science», project №3901: Research and development of hybrid model of back-to-back high voltage direct current transmission system.

References

1. M. I. Mazurov, A. V. Nikolaev, N. G. Lozinova, *Improvement of voltage quality in AC network by use of STATCOM* (IEEE Russia Power Tech, IEEE, 2005)
2. Yu. S. Borovikov, A. S. Gusev, A. O. Sulaymanov, R. A. Ufa, A. S. Vasilev, M. V. Andreev, N. Yu. Ruban, A. A. Suvorov, *A Hybrid Simulation Model for VSC HVDC* (IEEE Transactions on Smart Grid, IEEE, 2016)
3. M. A. Zamani, M. Mohseni, 9th International Conference on Electrical Power Quality and Utilisation, IEEE, 6, (2007)
4. L. Puming, H. Quanrui, Transmission and Distribution Conference and Exposition, IEEE, 6, (2008)
5. N. Yu. Ruban, A. S. Gusev, V. A. Sulaymanova, IOP Conf. Ser.: Mater. Sci. Eng., **93**, 6, (2015)
6. M. V. Andreev, A. O. Sulaymanov, A. A. Suvorov, MATEC Web of Conf., **37**, 01001, (2015)
7. A. Suvorov, Y. Borovikov, A. Gusev, A. Sulaymanov, M. Andreev, N. Ruban, R. Ufa, Electrical Engineering (to be published)

MATEC Web of Conferences **91**, 01054 (2017) DOI: 10.1051/matecconf/20179101054
Smart Grids 2016

SMART GRID TECHNOLOGIES: TRENDS AND PERSPECTIVES

Anna Buran[1,*], *Gavriil* Nizkodubov[1], and *Ilya* Pryahin[1]

[1]National Research Tomsk Polytechnic University, 634050 Tomsk, Russia

Abstract. The study explores the role of Smart Grid technologies, giving sustainable, reliable and safe energy. These technologies will provide an economic benefit to the government, the investors and the consumers. The paper overviews global trends and perspectives of using Smart Grid technologies in Russia and in other countries. The Smart Grid concept is especially important for Russia, because there are many power supply problems in the energy sector. This study is expected to be important for researchers and engineers studying Smart Grid technologies.

1 Introduction

Any country has a slightly different set of reasons for pursuing smart grid (SG) technologies and programs and a slightly different definition of what constitutes a smart grid. The reason is the energy sector has one of the strategic roles in the development of national industrial complexes, and the world economy as a whole [1]. Generally, smart grids coordinate the needs and capabilities of all generators, grid operators, end-users and electricity market stakeholders to operate all parts of the system as efficiently as possible, minimizing costs and environmental impacts while maximizing system reliability, resilience and stability [2]. The term SG has a lot of definitions, because each country has its own SG strategy and concept. For example, in the European opinion, Smart Grid is a concept of fully-integrated, self-healing and self-regulated electric power system, which has a network topology and includes all generating sources, transmission and distribution networks, all types of electrical energy consumers - driven by a single network of automated devices in real time. At the same time, in the opinion of the USA, SG is a set of organizational changes, new models of the processes, decisions in the field of information technology, as well as innovation in the field of automated control systems and supervisory control in power engineering. Smart Grids can be easily described as "Transactive". This means that the SG has the following characteristics. It will enable new products, services, and markets, as well as active participation by consumers. Also, it will provide power quality for the digital economy and optimize asset utilization. What is more important SG can anticipate and respond to system disturbances. Last feature makes it absolutely secure and means that the SG will operate resiliently against attack and natural disaster. All above mentioned characteristics proved that SG is more effective in comparison with a standard network.

* Corresponding author: aburan@tpu.ru

2 Analysis of using Smart Grid technologies in the East

Smart Grid technologies offer greater choice between energy providers, products and services to consumers, as well as, greater transparency and control over energy consumption. As for vendors, they will have opportunity for a machine-to-machine platform that can assist multiple industries to create new products and services to take to market; also they can reduce the cost of delivery through mass deployments. The government has real opportunities for GDP uplift and green-collar job creation, moreover, spending efficiency, increased by providing options to rationalize national infrastructure investments. In case of SG implementation utilities can provide grid efficiency, reliability and understanding of power flows which increased enabling operational/maintenance savings. They will have new ways to improve customer service and create new revenue channels, like new opportunities to transition from a commodity provider to a service provider and to evolve the operating model and lower operating costs.

In table 1 the concept of Smart Grid implementation in the South Korea is presented. South Korea prefers National Export Strategy and state-owned monopoly in the capacity of the market model. This means, that government of South Korea has the major influence on the strategic choice and implementation of the SG concept in this country.

Table 1. Smart Grid concept in South Korea.

Strategy	National Export Strategy
Country	South Korea
Local industry drivers	Development of an industrial complex to export smart grid technologies and solutions globally; A green economic growth agenda; Integrate with other initiatives – intelligent city, electric vehicles, renewable, transnational supergrid
Market Model	State-owned monopoly
Regulatory Incentives for Smart Grid Pilots	International knowledge sharing programs, e.g. the Korea-Illinois Smart Grid Collaboration Program; National smart grid work flowchart
IP development focus	Transmission and distribution solutions; Storage technology; Electric vehicles;
SG maturity	Low – medium
Local industry challenges	Regional regulatory markets may not be strongly conducive limiting ability to develop innovation in a local context

South Korea prefers National Export Strategy and state-owned monopoly in the capacity of the market model. This means, that government of South Korea has the major influence on the strategic choice and implementation of the SG concept in this country. The South Korean government is planning to reduce CO2 emissions by 2020. The goal seems to be challenging as Korean plants and factories doubled their emissions over the last decade. With this goal Seoul is trying to be a model for other countries, including, such as India and China, being developing countries under the Kyoto Protocol.

Korea pursues sustainable development, dealing with climate change. It shifts toward a low carbon economy and a society. Following this plan, Korea launches a SG project to achieve green growth in an effective way. The main idea of the project was to enhance the Korean industry.

The market model of the SG concept is integrated and the country pursues the policy that is aimed at improving the consumer experience. There are a lot of major companies, which exert an influence on the SG implementation: like GE, FreeDM, ABB group and others. The ABB group should be singled out, because, nowadays, this company produces a lot of SG devices and equipment and it realizes at least a third part of all SG projects in the USA. It has some particular projects, like Stockholm Royal Seaport and SG Got land. As well as major projects like "Smart Grid Center of Excellence", which will be created in the near future in Raleigh, state North Carolina. An initial investment of the company in this project will be about $10 million. This center will include a testing and development laboratory, a verification center and a demonstration center. One of the important functions of this center will be collaboration with the same center of the company FreeDM and invention and testing of new devices in the USA and in the whole world.

3. Analysis of of using Smart Grid technologies in Russia

The Smart Grid concept is urgent for Russia, because of problems in the energy sector, such as the ever more unreliable electric power grids, the necessity of energy supply in many time zones, and the obsolete single-direction energy transmission system. The energy resources in Russia are frequently wasted. Different sources show that losses during the energy distribution in Russia are significantly higher than in the European countries [2]. However, due to the specific economic situation in Russia the Smart Grid concept has become more popular there. The conversion of the Unified National (All-Russia) Electric Grid (UNEG) into the Smart Grid format allows increasing the system reliability of power grid complex, reduces the capital investment into the building of new objects, receives the technical and economical efficiency from the disposition of the UNEG objects in 8 time zones, and decreases the impact of the energy sector on the environment [3]. It's necessary to underline that in the Russian Federation the major power companies are federal, this means, they are regulated by the government and power market trends are chosen by them. One of these companies is the Federal Grid Company of Unified Energy System (FGC of UES). This company, like most power companies in Russia has chosen the SG concept of "Intelligent System" implementation. The advanced investment of capital in the SG technologies will make RUR 952,4 [3,4]. Moreover, this company tries to support and finance national equipment and material manufacturers; they have a special program for them. Also, some pilot projects in Siberia, in the Far East, in North West are realized with the support of this company. In the near future in these regions, there will be mass use of innovation SG techniques: storage batteries with high power, static reactive compensators, modern systems of equipment monitoring and diagnostics. For the full review, another example of SG implementation in our country will be given. It's the Research and Development Center of Smart Grid, which includes National Research Tomsk Polytechnic University; the Institute of Power Engineering; The Energy Research Institute of the Russian Academy of Sciences. National Research Tomsk Polytechnic University is presented by Scientific-research laboratory "Simulation of power systems", which made the invention called "Hybrid fully-variable simulated package of distribution network in real time" in the past. This innovation intends for continuous and high-precision modeling of power systems in real time. What is more important, it has the following specifics: unbounded interval of modeling, modeling in real time, high level of adequacy, profound mathematical model of elements, modeling of the process, which happens to the Power System at the same time. Considering depreciation of electrical equipment and constantly growing demands in the building of new energy facilities, the first step on the way of smart grid implementation is not an intellectualization of existing objects, but building new objects with advanced manufacturing sciences of the power industry. As an example, one step of building smart grids is the project of the Federal Grid Company (FGC) which involves the creation of a single automated technological management system. Such a system is oriented to avoid emergency operation, create an online monitoring system and provide smart diagnostics of equipment state. The Smart Grid technology requires the development and integration of a whole complex of innovative equipment and technologies: FACTS, PMU, WAMS, WAPS, WACTS technologies, powerful storage systems, electric vehicles, renewable and hybrid power supply systems based on them.

4 Conclusion

Smart grid technologies can be defined as self-sufficient systems that can find solutions to problems quickly in an available system that reduces the workforce and targets quality electricity to all consumers [5, 6]. Moreover, the Smart Grid (SG), are likely to bring out subsequently enormous forms of social change [7] in a modern world. The Smart Grid concept is especially important for Russia, as the integration of new technological advances in the energy systems and information and communication technologies [8] can help solve many of power supply problems in the energy sector of our country.

References

1. V.M. Rostovtseva, G.A. Nizkodubov, MATEC Web Conf. **72**, 01096 (2016)
2. B.B. Kobec, I.O. Volkova, *Innovative development of electrical engineering based on the concept "Smart Grid"* (IAC Energy, Moscow, 2010) [in Russian]
3. Federal Grid Company of Unified Energy System., http://www.fsk-ees.ru/eng/
4. Smart Grid: China Leads Top Ten Countries in Smart Grid Federal Stimulus Investments, Zpryme Reports, Zpryme Research & Consulting press release (2010) http://zpryme.com/news-room/smart-grid-china-leads-top-ten-countries-in-smart-grid-federal-stimulus-investments-zpryme- reports.html
5. R. Bayindir, I. Colak, G, Fulli, K. Demirta, Renewable Sustainable Energy Rev. **66**, 499 (2016)
6. S. Maier, Reininghaus District. Energy, Sustainability Society **6**, 1 (2016)
7. O. Ellabban, H. Abu-Rub, Renewable Sustainable Energy Rev. **65**, 1285 (2016)
8. Camarinha-Matos, M. Luís, Renewable Sustainable Energy Rev. **65**, 283 (2016)

DEVELOPING TRANSLATION COMPETENCE IN THERMAL POWER ENGINEERING STUDENTS

Ksenia Girfanova[1,*], *Nikolay* Kachalov[1], and *Inna* Cheremisina Harrer[1]

[1]National Research Tomsk Polytechnic University, 634050 Tomsk, Russia

Abstract. An engineering graduate today should be able to apply a variety of professional and the so-called deep learning competences including communication and translation competence. The paper gives a review of the notion of translation competence and presents key pre-requisites and creative tasks that enable thermal power engineering students of the National Research Tomsk Polytechnic University, Russia, develop and master translation competence.

1 Introduction

An engineering graduate of the 21st century should demonstrate a wide variety of skills and competences to be eligible to fit in the modern context of the international labour market. New pedagogies for deep learning that are available in the 21st century are aimed at training specialists prepared to work creatively in the global economic community and respond to the unpredictable challenges of the time for the sake of the stable and sustainable development [1, 2].

Deep learning skills (also termed as 21st century skills and competencies) are composed of four groups of skills, which include ways of working: communication and collaboration; ways of thinking: critical thinking, problem solving and metacognitive skills; tools for working: information and communication literacy; and ways of living in the world: global citizenship and civil responsibility, as well as cultural awareness and competence [3]. Communication is understood in a very broad sense, including communication in the native language as well as in the second (or foreign) language, which supposes using both languages as a means of communication and for the purposes of direct and inverse translation. In this case the translator plays an important social function of the mediator between the two languages and two cultures.

In 1997 the PACTE research group (Process in the Acquisition of Translation Competence and Evaluation) was formed in Spain with the purpose to investigate the notion of translation competence and evaluate how translation competence can be acquired based on experimental studies involving the process of written translation [4]. In 2000 the group formulated that translation competence is understood as a system of knowledge and skills needed to be able to translate [5]. Later the notion ′translation competence′ was deepened by the following important characteristics constituting it. Firstly, it is seen as expert knowledge and not possessed by all bilinguals; secondly, it is basically procedural

* Corresponding author: kcheremisina@tpu.ru

knowledge (and not declarative); thirdly, it is made up of various interrelated sub-competences; and fourthly, the strategic component is very important [6]. In translating practice most processes should be automatic, and the strategy guides the translator in choosing the right and appropriate way of solving translation problems.

Applying the PACTE model of translation competence to the reality of written translation fulfilled by thermal power engineering students, we should focus on the following key aspects (also termed as translation sub-competences) that provide efficient acquisition of translation competence. To be bilingually fluent the translator needs to be more or less expert in Russian and English and demonstrate pragmatic, socio-linguistic, textual and lexical-grammatical knowledge of both languages. Taken together the above components constitute the bilingual sub-competence. Subject knowledge that is deep understanding of the professional area as well as theory and practice of translation, the knowledge of two cultures, and eventually general encyclopedic knowledge provide the basis for the extra-linguistic sub-competence. Closely linked with the second sub-competence stands the instrumental sub-competence that involves the use of referential and documentation sources (dictionaries, encyclopedias, electronic corpora, terminological databases, etc) and application of information and communication technologies. The strategic sub-competence embraces the all above sub-competences and ensures the effectiveness of the translation process. Finally, the emphasis in PACTE′s model is made on the transfer sub-competence, which includes understanding of the source text and re-expressing its idea in the target language.

The model of translation competence developed by the PACTE research group goes in parallel with the standards elaborated and being implemented in the professional translation activity in Europe. According to the European standard requirements translator's professional competences include translating competence; linguistic and textual competence in the source language and the target language; research competence, information acquisition and processing; cultural competence; and technical competence [7]. The article evaluates the key pre-requisites that are needed to develop translation competence in thermal power engineering students and gives an overview of the creative tasks that promote to master the acquired translation competence at a more advanced level of study.

2 Key pre-requisites for developing translation competence in thermal power engineering students

To become an efficient translator the student should possess good knowledge and skills in both languages – Russian and English. Adequate command of the native language is the key pre-requisite in acquiring translation competence. Mastering the knowledge of lexical and grammar system, enriching general and special terminological lexicon and, as a result, developing the ability to extract the necessary information from the text and functionally and stylistically relevantly transfer it into the translating language are initial milestones in the process of shaping translation competence. The task is to be achieved in the context of special courses in stylistics of the Russian language and the culture of speech being part of the bachelor′s cycle of educational program in thermal power engineering, or as a standard course in the curriculum of the additional educational program ′Translator in the area of professional communication′ available for undergraduate and graduate students at the Department of Foreign Languages, Institute of Power Engineering at the National Research Tomsk Polytechnic University. As an efficient example at this stage can be practical exercises aimed at paraphrasing sentences and coherent passages in the native language without changing or modifying the content, the main idea and the style of the passage under study. As an alternative the sentence or the set of sentences can be restructured, but not changed contextually. The message and the function of the passage should be preserved and transferred by adequate means of the native language.

At a later stage these exercises are offered to students in the English language. The task becomes more sophisticated, and students are to analyse and evaluate that the content of the source passage is preserved in the target language, and the language means used to translate the text are adequate to the stylistic norm and functionally accepted in the context. Didactically this stage appears to be more complicated than exercises in the native language. The attention should be focused not only on the subject matter of the passage under translation, but also (and more important) on the process of decoding the information and understanding its sense. The suggested set of exercises is aimed at providing students practice in extracting the content centre (nucleus) of the passage in the interrelation of words and phrases constituting it and apply adequate means of translating language to transfer its sense completely and precisely.

Another important constituent for developing translation competence is a solid terminological background of the translator's profile, active operation of the widely used terms and terminological combinations in the area of thermal power engineering. The preference should be given to those terminological items which are frequently and widely used, are exact and precise in expression; also the emphasis should be made on their equivalents in the Russian language. The practice of enriching active terminological base is directly linked with abundant use of referential sources, terminological dictionaries, glossaries of special terms, electronic corpora and other online specialized services [8, 9, 10, 11].

To prove the importance of good command of terms we demonstrate a relatively rare phenomenon in the terminological sphere when 2 terms or terminological expressions can be translated absolutely identically, we mean the following terminological units – *thermal power engineering* and *heat power engineering*. By comparing their translation into Russian in several dictionaries we may still feel the slight difference in the scope of interpretation. In Collins Free Online Translator and in Lingvo Online Dictionary *thermal power engineering* is translated as 'тепловая энергетика' and *heat power engineering* – as 'теплоэнергетика'. The Electronic Dictionary Multitran contains the following translation of the above terms: *thermal power* – 'теплоэнергетика'; 'тепловая энергетика'; tech. 'тепловая мощность (энергосистемы)'; 'теплотворная способность'; 'тепловая энергия', and the terminological expression *heat power engineering* is translated as 'теплоэнергетика'.

The value of the correct use of terminological items can be proved by comparing very close in meaning terms in the area of power engineering, namely the synonymic row *energy-power-capacity*. The Electronic Dictionary Multitran provides the following translation of these terms: *energy* – gen. энергия, сила, мощность, энергоносители; econ. электроэнергия; el.tech. потребленная электроэнергия; электропотребление; *power* – gen. производительность; способность; tech. источник энергии; el. питание; *capacity* – gen. ёмкость; вместимость; объём; пропускная способность; электрическая ёмкость; tech. нагрузка; выработка; el.tech. (электрическая) ёмкость; установленная мощность (электростанции); energ. (электрическая) мощность; (электрическая) производительность; (электрическая) пропускная способность; (электрическая) установленная мощность.

In Lingvo Online Dictionary *энергия* means 'energy' in physics and technically 'power'. Examples: *cumulative energy* – 'совокупная энергия', *biomass energy* – 'энергия биомассы'. Also in Lingvo Online Dictionary *power* – 1. сила, мощь; 2. энергия, электроэнергия; 3. способность, возможность ... 13. мощность, производительность. Examples: high-power deep discharge – глубокий разряд повышенной мощности; absolute thermoelectric power – коэффициент термоэлектродвижущей силы; absorbing power – абсорбирующая способность; абсорбционная способность; поглощающая способность; сила поглощения. *Capacity* – вместимость, ёмкость, объём; *capacity building* – наращивание потенциала; *capacity extension* – наращивание ёмкости; *capacity factor* – коэффициент загрузки. *Capacity* in Collins Free Online Translator is rendered as ′мощность′. The ability to differentiate basic terms and correctly apply the knowledge of terminological system from the area of thermal power engineering assist students in choosing the appropriate option and strategy of solving translation problems.

3 Creative tasks and projects aimed at mastering translation competence in thermal power engineering students

The advanced level of mastering translation competence is realized within the context of the additional educational program ′Translator in the area of professional communication′ delivered at the Department of Foreign Languages, the Institute of Power Engineering at the National Research Tomsk Polytechnic University, Russia. The curriculum includes courses in the use of Russian for the purpose of academic and research communication, abundant practice in mastering English language skills, courses in English lexicology, stylistics and pragmatics, and the most part of the curriculum is devoted to the theory and practice of written translation from English into Russian. Students are motivated to do translation projects and get involved in translation competitions. One of them is a traditional competition held under the support of the Russian National Committee Cigre, Youth Division [12]. The translation competition involves three rounds: first, written translation from English into Russian of the realistic research paper presented at the conference devoted to the modern trends and problems of thermal power engineering, second, presentation of the key aspects of the translated paper in English using the format of PowerPoint presentation, and third, written translation of an article focusing on trends and tendencies in power engineering in general and precise translation of a set of specific terms. If the first task is individual, that is every competitor receives an individual research paper for translation, then in the final round students are offered one common text for translation within a fixed period of time under the jury supervision. At all stages of the competition participants are allowed to use dictionaries and reference sources of their choice. Presentation of the translated paper is of vital importance not only for the purpose of evaluating the quality of translation but also promotes and deepens competitors′ knowledge of the subject area and the way it is transferred in the English language.

At the level of the master′s educational cycle students are prepared to be involved in individual or group projects that provide abundant practice in improving translation competence and perfecting communication in the English language. Example 1 is a research task-based project aimed at developing a virtual laboratory of thermal power engineering. The laboratory should be presented online and should give an idea of the state-of-the-art trends and developments in the area, be part of the online resources of students′ department and optionally, can be recommended as a learning tool. Interactivity, feedback and visual appeal are welcomed. Example 2 is also a research task-based project aimed at designing a virtual translation environment supporting studies, which provides opportunities for producing student-generated learning materials and resources. Project team members are to decide which materials are of prior importance to ensure successful and user-friendly learning and how they can be used to help students benefit from the way they are presented. You may also want to consider possibilities of editing these materials by visitors (Web 2.0, wikis and others). The expose to realistic task-based projects can only be realized at an advanced level of efficient communication in English and close to expert application of translation competence.

4 Conclusion

The efficiency of communication in English and well-developed translation competence are important components of successful professional career of the 21st century engineering graduate. Basic practical tasks at the beginning of the learning process and more sophisticated creative assignments at a later stage provide necessary milestones in the process of becoming a confident translator in the area of thermal power engineering.

References

1. New pedagogies for deep learning: a global partnership – 2016. – Retrieved from: http://www.education.vic.gov.au/school/teachers/support/Pages/deeplearning.aspx
2. Fullan, M. Langworthy, M. A rich seam. How new pedagogies find deep learning. – 2014. – Retrieved from: http: //www.michaelfullan.ca/wp-content/uploads/2014/01/3897.Rich_Seam_web.pdf
3. Assessment and teaching of 21st century skills. – 2010. – Retrieved from: http://www.cisco.com/c/dam/en_us/about/citizenship/socio-economic/docs/ATC21S_Exec_Summary.pdf
4. PACTE research group. Departament de Traducció i d'Interpretació. – 2016. – Retrieved from: http://grupsderecerca.uab.cat/pacte/en
5. A. Beeby, L. Berenguer, D. Ensinger, O. Fox, A. Hurtado Albir, N. Martínez Melis, W. Neunzig, M. Orozco, M. Presas, R. Vega, *Acquiring translation competence: hypotheses and methodological problems of a research project*. In: *Investigating translation*, ed. By A. Beeby, D. Ensinger, M. Presas (Universitat Autònoma de Barcelona Amsterdam, 2000)
6. A.Beeby, M.Fernández, O.Fox, A.Hurtado Albir, I.Kozlova, W.Neunzig, M.Presas, P.Rodríguez-Inés, L.Romero, Investigating Translation Competence: Conceptual and Methodological Issues *Meta,* **50**, 609 (2005).
7. European standard. Translation services – Service requirements. – 2004. – Retrieved from: http://web.letras.up.pt/egalvao/prEN-15038.pdf
8. Collins Free Online Translator. – 2016. – Retrieved from: http://www.collinsdictionary.com/translator
9. Cambridge Dictionary. – 2016. – Retrieved from: http://dictionary.cambridge.org/
10. Lingvo Live. Online Dictionary from ABBYY. – 2016. – Retrieved from: https://www.lingvolive.com/
11. Electronic Dictionary Multitran. – 2016. – Retrieved from: http://www.multitran.ru/
12. Russian National Committee Cigre Youth Division. – 2016. – Retrieved from: http://www.cigre.ru/rnk/youth/

MATEC Web of Conferences **91**, 01056 (2017) DOI: 10.1051/matecconf/20179101056
Smart Grids 2016

FEATURES OF FUNCTIONING THE INTEGRATED BUILDING THERMAL MODEL

Maxim N. Morozov[1,*], *Alena* O. Zhdanova[1], and *Maksim* V. Piskunov[1]

[1] National Research Tomsk Polytechnic University, 634050 Tomsk, Russia

Abstract. A model of the building heating system, consisting of energy source, a distributed automatic control system, elements of individual heating unit and heating system is designed. Application Simulink of mathematical package Matlab is selected as a platform for the model. There are the specialized application Simscape libraries in aggregate with a wide range of Matlab mathematical tools allow to apply the "acausal" modeling concept. Implementation the "physical" representation of the object model gave improving the accuracy of the models. Principle of operation and features of the functioning of the thermal model is described. The investigations of building cooling dynamics were carried out.

1 Introduction

When creating complex models of a heating system, as a rule, the occurrence of difficulties is usually caused by following requirements (for example, to highlight works [1-6]):

1) simulation of individual components of building life support systems (for example, the control system, heating appliance, heat exchanger, etc.);

2) maintenance of model functioning in a wide temperature range.

Solution of the first problem is to find a compromise between the model description completeness and its adequacy and accuracy of studied system reproduction. The choice of internal and external influencing factors (which must be also adequately reproduced) variety is an important step in developing of the model. Excessive accuracy influences negatively on the capabilities of a computer, increasing simulation time. When solving the problem as a part of this work, there were used approaches, described in [7, 8].

Another no less important problem is to extend the temperature range of model operating. This is necessary condition for the investigation of district heating system reliability. The problem resides in narrow range of individual model component operating temperatures. For example, the temperature difference of the coolant reaches 90 °C during simulation. At temperatures of approximately 0 °C, the decrease of modeling accuracy is inevitable. This problem is solved by optimization of thermal component models.

The aim of the present work is to investigate functioning of the building thermal model for further optimization of computing resources and for enhance the functionality and model accuracy.

* Corresponding author: tpu_chm@tpu.ru

2 Research object

The object of research is a three-floored administration building (total area of 9870 m^2 and volume of 35.926 m^3) which is located in West Siberia (Tomsk city). Heat supply of the building is provided by the central heating unit at the temperature chart 95/70 °C. The individual heating unit has been designed under the "independent scheme" [9]. Water is used as a coolant in the heating system internal circuit [10].

The thermal model of the building has been developed the actual operation of the building. The application Simulink of the mathematical package Matlab [11] has been selected as a platform for a model. This choice is motivated by the presence of software wide capabilities: tools for analyzing systems dynamics over time, connection with Workspace and other Matlab applications (for example, Control System Toolbox), simulation of various disturbances (determined or random with different distribution laws) and situations (regular or non-regular, including emergency) which occur at the technical system operation. Block-oriented approach has been applied to simulation and "visual" programming of building heating system in combination with a wide range of Matlab mathematical tools. An extension package Simscape, which implements a "physical" simulation concept [12], has also been applied to develop the model.

3 Investigation of the building thermal model

The thermal model structure of the building district heating system is shown in Fig. 1.

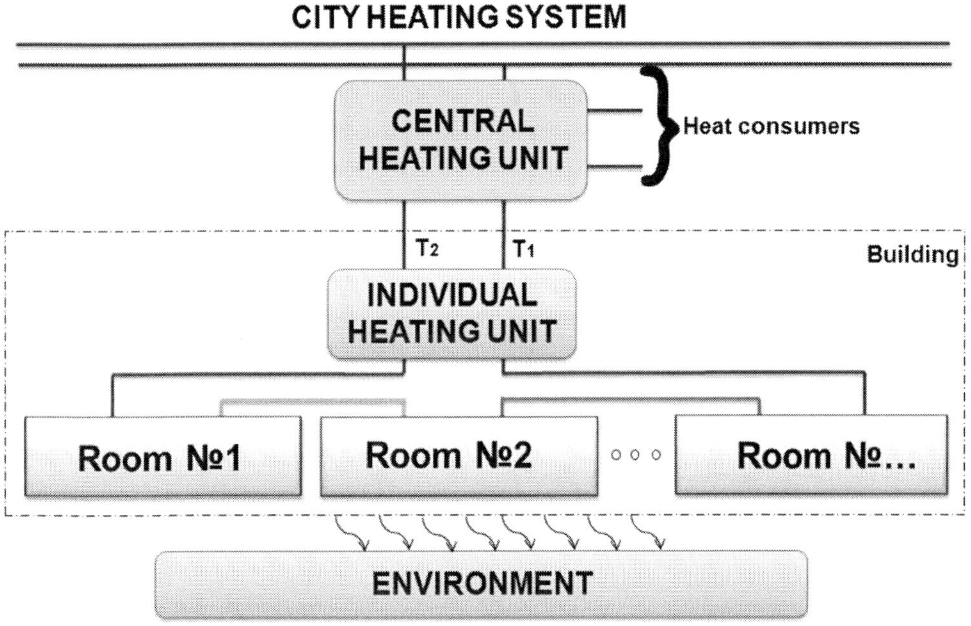

Fig. 1. Structure of the district heating system model: T_1 – direct coolant temperature, T_2 – return coolant temperature.

The source of heat supply is a centralized urban heating network. After passing the individual heating unit, a coolant is directed to the building heating system. Generally, the model of an auditorium is characterized by the presence of heat sources and consumers. Thermal energy delivery into the premise derives from heating appliances of the heating system, and its decline – by means of heat transfer processes through exterior envelopments to the environment. This approach allows scaling the typical premise model to form the total thermal mode in the building.

Fig. 2 shows an operation algorithm flowchart of the developed thermal model. A model calculation process is made in several stages. At the initial stage, model initialization takes place: connecting the library blocks, defining the signals dimensions, data types, block parameters estimation, as well as optimization of functional elements execution order and the random-access storage allocation to perform the calculation.

Further, initial data loading into the Workspace and Simulink application takes place, after that simulation cycle implementation begins. Each cycle begins with a data import and calculation of simulation time steps values. The fulfillment of uninterrupted heat supply conditions is verified, as well as of heating system operating mode.

After that, model current parameters (temperature, heat flow) are calculated, and a central regulator, relating to the individual heating unit, is activated. Actual heating system heat flow, directed to heat consumers (premise), is calculated. Inside of each premise a local control loop is functioning, headed by room controller, which is responsible for maintenance of the individual thermal mode.

At the stage of heat loss calculation, the thermal processes library Simscape is connected (sequentially, it is implemented the local calculators initialization, calculation of heat consumption and climate parameters of each premise). Further, the output heat flow parameters calculation takes place.

It is important to note that specific functional blocks are calculated in the order, defined in the initialization stage, at each step of simulation. Upon that, there are involved functions, computing the block state variables, state variables derivatives and outputs during the current cycle.

As stop conditions at cyclic simulation there were taken: the finish of preset simulation time, the emergence of "algebraic loops" and other errors during the simulation process, as well as computer available memory shortage.

Automatic processing of obtained results is performed after completing the iterative simulation process. A data export in various formats allows applying various exterior information processing tools.

As a rule, thermal processes in buildings are described by mathematical equations, reflecting the heat transfer stationary mode. Upon that, heat flows are constant and don't depend on time, and thermal processes parameters (for example, coefficients of external and internal envelopments heat transfer, heat irradiation from heating appliance, etc.) are chosen from the reference literature.

In practice a stationary mode is observed only for a very limited time, due to constantly changing external disturbing factors. It is important to note, that the developed model considers dynamics of the real object functioning mode, as well as thermal processes nonlinearity and non-stationarity.

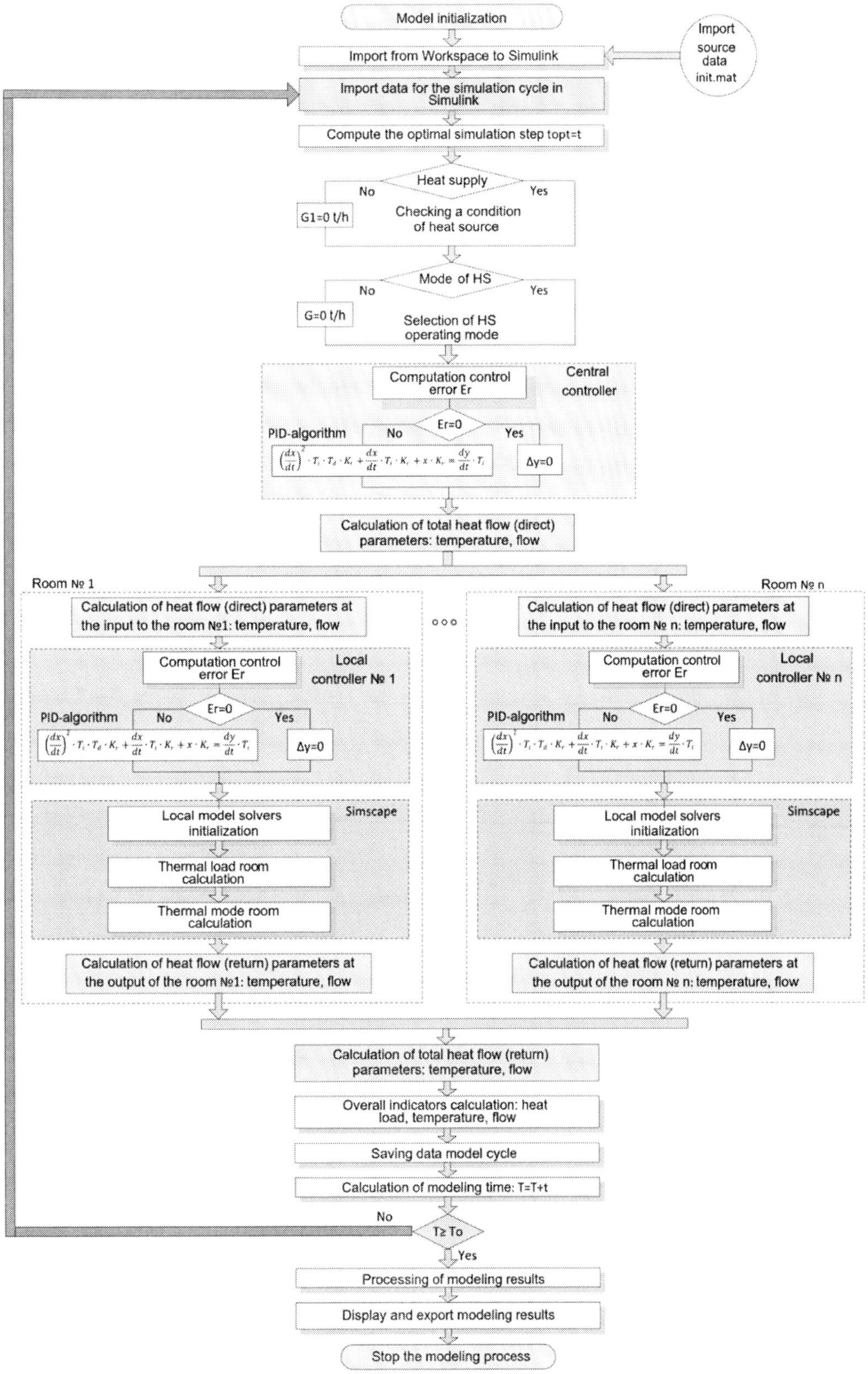

Fig. 2. An operation algorithm flowchart of the thermal model.

DOI: 10.1051/matecconf/20179101056

4 Investigation of building cooling dynamic

Carried out investigations of building cooling dynamics is the best proof of model performance. Two groups of premise have been selected for prognostic investigation – from the corner and central parts of the building. The following numeration has been admitted:

1) premise no. 1–3 are "corner" on the 1–3th floors respectively;
2) no. 4–6 are "central" premises on the 1–3th floors respectively.

Heating appliances within each group are connected in series (vertical single-tube configuration with overhead coolant distribution). Premises are characterized by the area of 52 m^2 and identical layout. "Corner" premises have one additional window. Simulation of heating system emergency operation is at heat supply restriction from the central heating unit. As an example, the results of this simulation of the heating system operating mode at heat supply restriction at 18 o'clock are given below.

Fig. 3 shows the cooling dynamics of the heating system coolant at T_{ext} = -40 °C in case of its continuous circulation. In this case, the minimum water cooling to "threshold" temperature time was recorded for the "corner" premise on the first floor and it was 46 h 45 min. The maximum cooling time of 56 h 13 min is observed near the "central" premise on the first floor. At the absence of coolant circulation the time of its cooling to the critical temperature is comparable to the time of air temperature reduction.

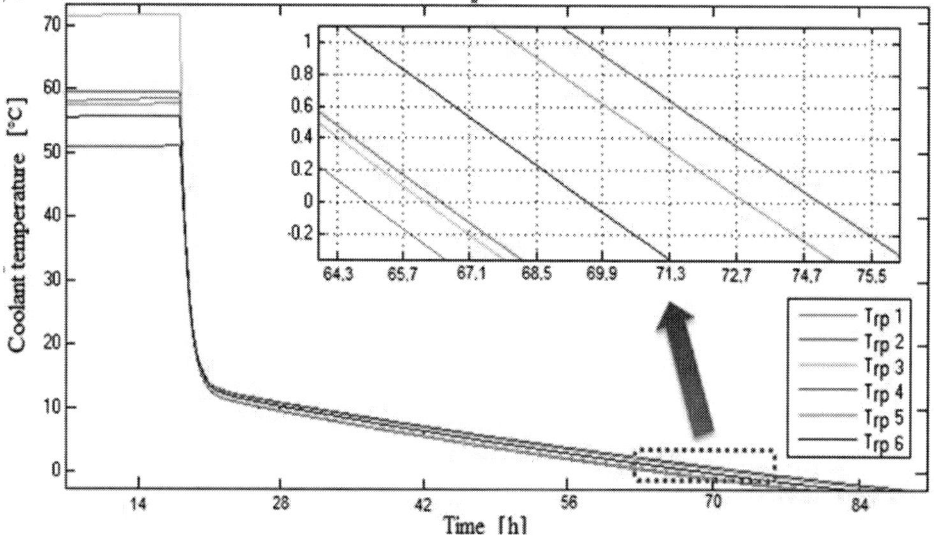

Fig. 3. Cooling dynamics of the heating system coolant at T_{ext} = -40 °C in case of its constant circulation: $T_{rp\,i}$ – heating system coolant temperature of the *i*-th premise, where *i* = 1, 2,…6.

5 Summary

A model of the district heating system of the building, which allows investigating the stability of typical buildings and constructions thermal mode (to analyze cooling processes of the heating system coolant, air temperature reduction in premises of various configurations, as well as exterior envelopments) during heat supply failure, has been developed. It is also possible to use the model for energy efficiency investigation of different buildings.

Obtained simulation results (a model, dependences, conclusions and ratios) can be used for preventive measures formulation to optimize the basic operations and procedures, aimed at restoring the thermal modes of buildings and constructions in case of heat supply failure.

The reported study is supported by the Russian Foundation for Basic Research RFBR (contract 16-38-00628).

References

1. M. Wallace, R. McBride, S. Aumi, P. Mhaskar, J. House, T. Salsbury, Chem. Eng. Sci. **69**, 45 (2012)
2. S. Privara, Z. Vana, E. Zacekova, Energ. Buildings **55**, 341 (2012)
3. M.A. Mishin, Polzunovskii vestnik **1**, 146 (2010)
4. V.V. Burstev, A.S. Basi, N.M. Baitinger, Polzunovskii vestnik **1**, 147 (2004)
5. A.V. Pugovkin, S.V. Kuprekov, D.V. Abushkin, I.A. Zarechnaya, N.I. Muslimova, Lectures of TUSUR University: Control, computing equipment **22**, 293 (2010)
6. M. Maasoumy, M. Razmara, M. Shahbakhti, A.S. Vincentelli, Energ. Buildings **77**, 377 (2014)
7. P. Lauenburg, J. Wollerstrand, *Proceedings of the 11th International Building Performance Simulation Association Conference*, 150 (Glasgow, Scotland, 2009)
8. T. Persson, Proceedings of the 45th International Conference of Scandinavian Simulation Society, 199 (Copenhagen, Denmark, 2004)
9. M. Zago, A. Casalegno, R. Marchesi, F. Rinaldi, Energies **4**, 2115 (2011)
10. O. Ibrahim, F. Fardoun, R. Younes, H. Louahlia-Gualous, Build. Environ. **72**, 259 (2014)
11. M. Kiyan, E. Bingöl, M. Melikoglu, A. Albostan, Energy Convers. Manage. **72**, 147 (2013)

IMPLEMENTATION OF AN INTEGRATED THERMAL BUILDING MODEL FOR INVESTIGATIONS OF HEAT FLOWS

Maxim N. Morozov[1,*], *Alena* O. Zhdanova[1], and *Maksim* V. Piskunov[1]

[1] National Research Tomsk Polytechnic University, 634050 Tomsk, Russia

Abstract. A model of the building heating system, consisting of energy source, a distributed automatic control system, elements of individual heating unit and heating system is designed. Application Simulink of mathematical package Matlab is selected as a platform for the model. There are the specialized application Simscape libraries in aggregate with a wide range of Matlab mathematical tools allow to apply the "acausal" modeling concept. Implementation the "physical" representation of the object model gave improving the accuracy of the models.

1 Introduction

When creating complex models of a heating system, as a rule, the occurrence of difficulties is usually caused by following requirements (for example, to highlight works [1-6]):

1) simulation of individual components of building life support systems (for example, the control system, heating appliance, heat exchanger, etc.);

2) maintenance of model functioning in a wide temperature range.

Solution of the first problem is to find a compromise between the model description completeness and its adequacy and accuracy of studied system reproduction. The choice of internal and external influencing factors (which must be also adequately reproduced) variety is an important step in developing of the model. Excessive accuracy influences negatively on the capabilities of a computer, increasing simulation time. When solving the problem as a part of this work, there were used approaches, described in [7, 8].

Another no less important problem is to extend the temperature range of model operating. This is necessary condition for the investigation of district heating system reliability. The problem resides in narrow range of individual model component operating temperatures. For example, the temperature difference of the coolant reaches 90 °C during simulation. At temperatures of approximately 0 °C, the decrease of modeling accuracy is inevitable. This problem is solved by optimization of thermal component models.

The aim of the present work is to develop a thermal model of the building.

2 Research object

* Corresponding author: tpu_chm@tpu.ru

The object of research is a three-floored administration building (total area of 9870 m^2 and volume of 35.926 m^3) which is located in West Siberia (Tomsk city). Heat supply of the building is provided by the central heating unit at the temperature chart 95/70 °C. The individual heating unit has been designed under the "independent scheme" [9]. Water is used as a coolant in the heating system internal circuit [10].

The thermal model of the building has been developed the actual operation of the building. The application Simulink of the mathematical package Matlab [11] has been selected as a platform for a model. Block-oriented approach has been applied to simulation and "visual" programming of building heating system in combination with a wide range of Matlab mathematical tools. An extension package Simscape, which implements a "physical" simulation concept [12], has also been applied to develop the model.

3 Thermal model of building

As a rule, thermal processes in buildings are described by mathematical equations, reflecting the heat transfer stationary mode. Upon that, heat flows are constant and don't depend on time, and thermal processes parameters (for example, coefficients of external and internal envelopments heat transfer, heat irradiation from heating appliance, etc.) are chosen from the reference literature.

In practice a stationary mode is observed only for a very limited time, due to constantly changing external disturbing factors. It is important to note, that the developed model considers dynamics of the real object functioning mode, as well as thermal processes nonlinearity and non-stationarity.

The law of energy conservation for a typical premise is at the basis of the thermal model of the building heating system there [7]:

$$dE_a / dt = Q_{hs} - Q_{ec} \tag{1}$$

$$E_a = V_a \cdot \rho_a \cdot c_a \cdot T_a) \tag{2}$$

where

E_a internal energy of air in the premise [J]

V_a volume of air in the premise [m^3]

ρ_a air density [kg/m^3]

c_a air heat capacity [J/(kg·K)]

T_a air temperature in the premise [°C]

Q_{hs} heat input to the premise from the heating system [W]

Q_{ec} total heat loss of the premise through exterior envelopments [W]

During simulation, the following assumptions were accepted:

- specific heat capacity and density are characterized by constant values;
- coolant loss are lacked (equality of input and output expenditures);
- heat exchange with adjacent premises is lacked.

Considering the accepted assumptions, the law of energy conservation will be as follows:

$$m_a \cdot c_a \cdot (dT_a / dt) = Q_{hs} - Q_{ec} \tag{3}$$

$$Q_{hs} = g_{ha} \cdot c_{ha} \cdot (T_1 - T_2) \tag{4}$$

where

m_a mass of air, contained in the premise [kg]

g_{ha} coolant mass flow rate [kg/h]
c_{ha} coolant heat capacity [J/(kg·K)]
T_1 initial coolant temperature [°C]
T_2 finite coolant temperature [°C]

The following laws were applied to simulate different mechanisms of heat transfer between model components:

- Newton-Rikhman Law, describing the convective heat exchange:

$$q_{к} = k \cdot A \cdot \mathrm{d}T \quad (5)$$

where
$q_{к}$ heat flow [W]
k heat irradiation coefficient [W/(m²·K)]
A surface area of heat irradiation [m²]
$\mathrm{d}T$ temperature difference [°C]

- Fourier's Law, describing the conductive heat exchange:

$$q_{т} = (\lambda / b) \cdot A \cdot \mathrm{d}T \quad (6)$$

where
$q_{т}$ is heat flow [W]
λ is thermal conductivity coefficient [W/(m·K)]
b is layer thickness of heat exchange surface [m]
A is surface area, perpendicular to heat flow [m²]
$\mathrm{d}T$ is temperature difference [°C]

Model implementation in the Matlab package is based on representing of each individual functional component in the form of s-functions and subsystems blocks. In the process of simulation, the redistribution of heat flow is observed between sources, thermal energy transfer medium and its consumers.

Fig. 1-3 shows basic blocks of the developed thermal model of the district heating system of the building. The model comprises the following main functional components [8]:

- a heating appliance (radiator);
- a mixer tap;
- a valve;
- a thermal protection block of the premise;
- a weather block;
- a block of the premise temperature mode.

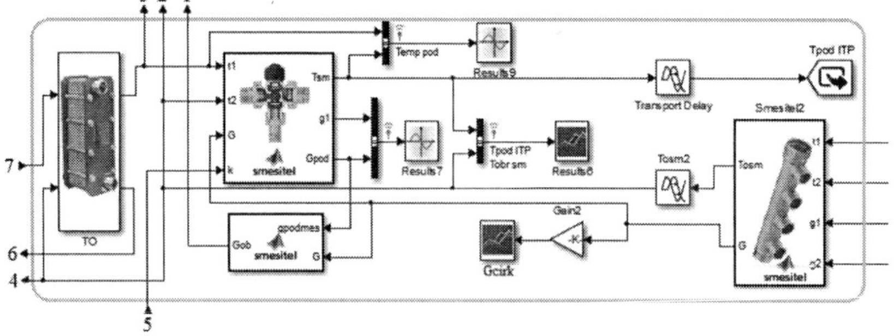

Fig. 1. The main thermal model blocks of the district heating system of the building: block of individual heating unit elements.

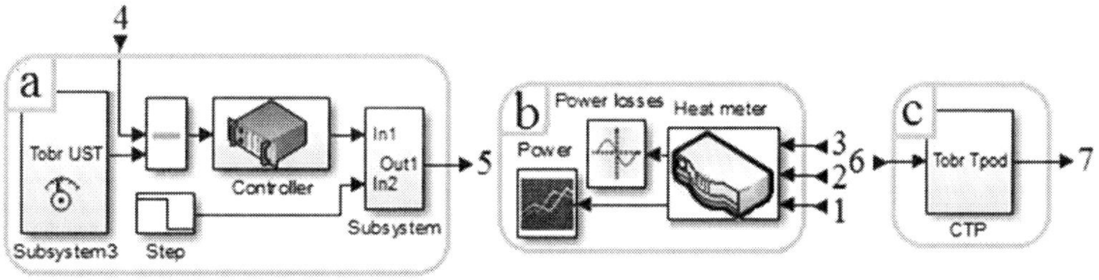

Fig. 2. The main thermal model blocks of the district heating system of the building: a – heat consumption controller, b – heat calculator, c –heat supply source.

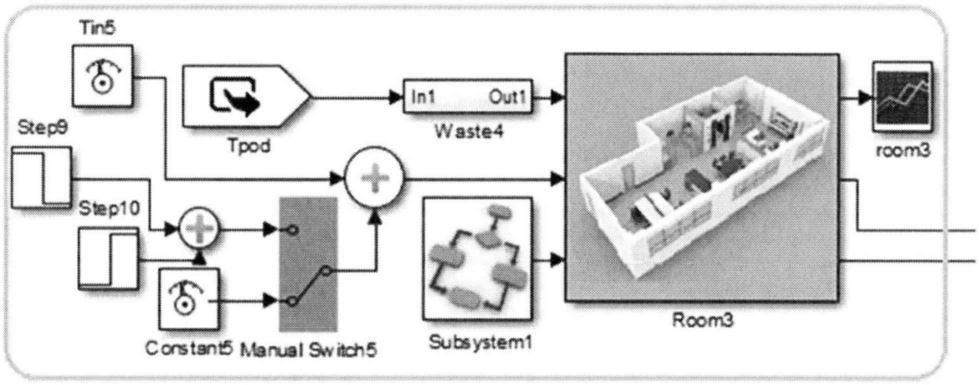

Fig. 3. The main thermal model blocks of the district heating system of the building: block of premise model.

A heating appliance

The process of heat transfer from the heating system coolant to the environment is carried out through the radiator case. Under that, different heat exchange mechanisms are involved:

1) convection and conductive heat exchange are realized at the "coolant – radiator wall" site;
2) heat transfer through the wall is carried out by thermal conductivity;
3) the "radiator wall – premise" site is characterized simultaneously by radiant heat exchange, convection and thermal conductivity.

Heat transfer rate of all of these sites reflects the total heat transfer coefficient K_{bat}.

In general, heat transfer simulation of a heating appliance is a complex problem that requires the use of mathematical physics methods to analyze the temperature distribution on the radiator surface. The most optimal approach at simulation is to use the average logarithmic temperature difference between the radiator surface and air inside the premise [7]:

$$T_{\log} = (T_1 - T_2) / \ln [(T_1 - T_{\text{in}}) / (T_2 - T_{\text{in}})] \quad (7)$$

where

T_1 coolant temperature at the radiator input [°C]
T_2 coolant temperature at the radiator output [°C]
T_{in} temperature inside the premise [°C]

The model of the heating appliance is based on the law of energy conservation:

$$m_{\text{ha}} \cdot c_{\text{ha}} \cdot (\text{d}T_2 / \text{dt}) = Q_{\text{hs}} - Q_{\text{bat}} \quad (8)$$

$$Q_{\text{bat}} = K_{\text{bat}} \cdot A_{\text{bat}} \cdot T_{\log} \quad (9)$$

 DOI: 10.1051/matecconf/20179101057

where

m_{ha} weight of the coolant inside the radiator case [kg]
c_{ha} coolant heat capacity [J/(kg·K)]
Q_{hs} heat input from the heating system coolant [W]
Q_{bat} heat irradiation of the heating appliance [W]
K_{bat} heat transfer total coefficient [W/(m²·K)]
A_{bat} heat irradiation surface area of radiator [m²]

Eq. (8) allows describing the change in coolant energy at radiator output (the left side of the equation), depending on balance between radiator heat input and heat irradiation (the right side).

Model implementation of the heating appliance using Simulink is shown in Fig. 4.

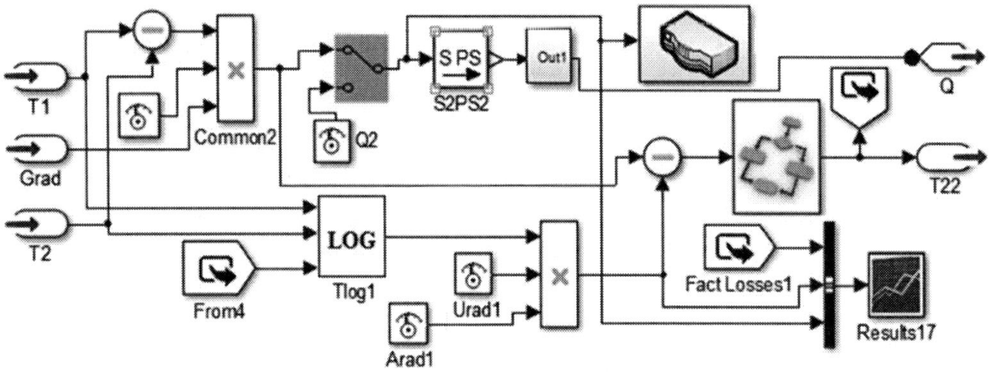

Fig. 4. A model of the heating appliance.

A thermal protection block of the premise

This block is a set of external envelopments (walls, windows, ceiling) models, as well as air inside the premise. The functioning of the block is described by the following equation:

$$Q_{ec} = [(K_w \cdot A_w) + (K_c \cdot A_c) + (K_{wd} \cdot A_{wd})] \cdot (T_{in} - T_{ext}) \tag{10}$$

where

Q_{ec} total premise heat loss through external envelopments [W]
T_{in} air temperature inside the premise [°C]
T_{ext} outside air temperature [°C]
K_i total heat transfer coefficient [W/(m²·K)]
A_i surface area of envelopments [m²]

Indexes w, c, wd of parameters K_i and A_i means respectively wall, ceiling and window.

Given equation permit determining the total heat loss of the premise and is used at the Eq. (3). In turn, the total coefficient of single-layer construction heat exchange is determined by geometric and thermophysical characteristics:

$$\frac{1}{K_i} = \frac{1}{k_{i\ in}} + \frac{b_i}{\lambda_i} + \frac{1}{k_{i\ ext}}$$

where

$k_{i\ in}$ heat irradiation coefficient on the inner surface [W/(m²·K)]
$k_{i\ ext}$ heat irradiation coefficient on the outer surface [W/(m²·K)]
b_i layer thickness of a heat exchange surface [m]
λ_i thermal conductivity coefficient of layer [W/(m·K)]

According to the law of energy conservation (by analogy to the Eq. (3)), the thermal inertia of envelopments is considered in the block, at that, characteristics such as weight, heat capacity and the initial temperature of constructions elements are used. Block implementation using Simulink is shown in Fig. 5.

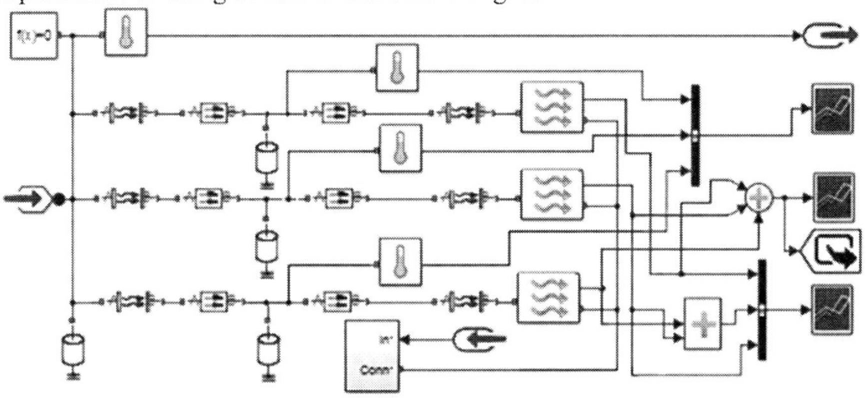

Fig. 5. A model of the block of the heating protection of the premise.

4 Summary

A model of the district heating system of the building, which allows investigating the stability of typical buildings and constructions thermal mode (to analyze cooling processes of the heating system coolant, air temperature reduction in premises of various configurations, as well as exterior envelopments) during heat supply failure, has been developed. It is also possible to use the model for energy efficiency investigation of different buildings.

The reported study is supported by the Russian Foundation for Basic Research RFBR (contract 16-38-00628).

References

1. M. Wallace, R. McBride, S. Aumi, P. Mhaskar, J. House, T. Salsbury, Chem. Eng. Sci. **69**, 45 (2012)
2. S. Privara, Z. Vana, E. Zacekova, Energ. Buildings **55**, 341 (2012)
3. M.A. Mishin, Polzunovskii vestnik **1**, 146 (2010)
4. V.V. Burstev, A.S. Basi, N.M. Baitinger, Polzunovskii vestnik **1**, 147 (2004)
5. A.V. Pugovkin, S.V. Kuprekov, D.V. Abushkin, I.A. Zarechnaya, N.I. Muslimova, Lectures of TUSUR University: Control, computing equipment **22**, 293 (2010)
6. M. Maasoumy, M. Razmara, M. Shahbakhti, A.S. Vincentelli, Energ. Buildings **77**, 377 (2014)
7. P. Lauenburg, J. Wollerstrand, *Proceedings of the 11th International Building Performance Simulation Association Conference*, 150 (Glasgow, Scotland, 2009)
8. T. Persson, Proceedings of the 45th International Conference of Scandinavian Simulation Society, 199 (Copenhagen, Denmark, 2004)
9. M. Zago, A. Casalegno, R. Marchesi, F. Rinaldi, Energies **4**, 2115 (2011)
10. O. Ibrahim, F. Fardoun, R. Younes, H. Louahlia-Gualous, Build. Environ. **72**, 259 (2014)
11. M. Kiyan, E. Bingöl, M. Melikoglu, A. Albostan, Energy Convers. Manage. **72**, 147 (2013)

AUTHOR INDEX

AUTHOR INDEX